ARC WELDING AUTOMATION

MANUFACTURING ENGINEERING AND MATERIALS PROCESSING

A Series of Reference Books and Textbooks

1. Computers in Manufacturing, *U. Rembold, M. Seth, and J. S. Weinstein*
2. Cold Rolling of Steel, *William L. Roberts*
3. Strengthening of Ceramics: Treatments, Tests, and Design Applications, *Harry P. Kirchner*
4. Metal Forming: The Application of Limit Analysis, *Betzalel Avitzur*
5. Improving Productivity by Classification, Coding, and Data Base Standardization: The Key to Maximizing CAD/CAM and Group Technology, *William F. Hyde*
6. Automatic Assembly, *Geoffrey Boothroyd, Corrado Poli, and Laurence E. Murch*
7. Manufacturing Engineering Processes, *Leo Alting*
8. Modern Ceramic Engineering: Properties, Processing, and Use in Design, *David W. Richerson*
9. Interface Technology for Computer-Controlled Manufacturing Processes, *Ulrich Rembold, Karl Armbruster, and Wolfgang Ülzmann*
10. Hot Rolling of Steel, *William L. Roberts*
11. Adhesives in Manufacturing, *edited by Gerald L. Schneberger*
12. Understanding the Manufacturing Process: Key to Successful CAD/ CAM Implementation, *Joseph Harrington, Jr.*
13. Industrial Materials Science and Engineering, *edited by Lawrence E. Murr*
14. Lubricants and Lubrication in Metalworking Operations, *Elliot S. Nachtman and Serope Kalpakjian*
15. Manufacturing Engineering: An Introduction to the Basic Functions, *John P. Tanner*
16. Computer-Integrated Manufacturing Technology and Systems, *Ulrich Rembold, Christian Blume, and Ruediger Dillman*
17. Connections in Electronic Assemblies, *Anthony J. Bilotta*

18. Automation for Press Feed Operations: Applications and Economics, *Edward Walker*
19. Nontraditional Manufacturing Processes, *Gary F. Benedict*
20. Programmable Controllers for Factory Automation, *David G. Johnson*
21. Printed Circuit Assembly Manufacturing, *Fred W. Kear*
22. Manufacturing High Technology Handbook, *edited by Donatas Tijunelis and Keith E. McKee*
23. Factory Information Systems: Design and Implementation for CIM Management and Control, *John Gaylord*
24. Flat Processing of Steel, *William L. Roberts*
25. Soldering for Electronic Assemblies, *Leo P. Lambert*
26. Flexible Manufacturing Systems in Practice: Applications, Design, and Simulation, *Joseph Talavage and Roger G. Hannam*
27. Flexible Manufacturing Systems: Benefits for the Low Inventory Factory, *John E. Lenz*
28. Fundamentals of Machining and Machine Tools: Second Edition, *Geoffrey Boothroyd and Winston A. Knight*
29. Computer-Automated Process Planning for World-Class Manufacturing, *James Nolen*
30. Steel-Rolling Technology: Theory and Practice, *Vladimir B. Ginzburg*
31. Computer Integrated Electronics Manufacturing and Testing, *Jack Arabian*
32. In-Process Measurement and Control, *Stephan D. Murphy*
33. Assembly Line Design: Methodology and Applications, *We-Min Chow*
34. Robot Technology and Applications, *edited by Ulrich Rembold*
35. Mechanical Deburring and Surface Finishing Technology, *Alfred F. Scheider*
36. Manufacturing Engineering: An Introduction to the Basic Functions, Second Edition, Revised and Expanded, *John P. Tanner*
37. Assembly Automation and Product Design, *Geoffrey Boothroyd*
38. Hybrid Assemblies and Multichip Modules, *Fred W. Kear*
39. High-Quality Steel Rolling: Theory and Practice, *Vladimir B. Ginzburg*
40. Manufacturing Engineering Processes: Second Edition, Revised and Expanded, *Leo Alting*
41. Metalworking Fluids, *edited by Jerry P. Byers*
42. Coordinate Measuring Machines and Systems, *edited by John A. Bosch*
43. Arc Welding Automation, *Howard B. Cary*
44. Facilities Planning and Materials Handling: Methods and Requirements, *Vijay S. Sheth*
45. Continuous Flow Manufacturing: Quality in Design and Processes, *Pierre C. Guerindon*

Additional Volumes in Preparation

ARC WELDING AUTOMATION

HOWARD B. CARY
Consultant
Troy, Ohio

Marcel Dekker, Inc. New York • Basel • Hong Kong

Library of Congress Cataloging-in-Publication Data

Cary, Howard B.
Arc welding automation / Howard B. Cary.
p. cm. — (Manufacturing engineering and materials processing; 43)
Includes bibliographical references and index.
ISBN 0-8247-9645-4 (alk. paper)
1. Electric welding—Automation. I. Title. II. Series.
TK4660.C36 1995
671.5'212'028—dc20 95-9483
CIP

The publisher offers discounts on this book when ordered in bulk quantities. For more information, write to Special Sales/Professional Marketing at the address below.

This book is printed on acid-free paper.

Marcel Dekker, Inc.
270 Madison Avenue, New York, New York 10016

Current printing (last digit):
10 9 8 7 6 5 4 3 2 1

PRINTED IN THE UNITED STATES OF AMERICA

Preface

Welding is the most effective and economical way to permanently join metals, costing less than castings and forgings or riveted and bolted joints. Almost every metal product manufactured today is welded, from kitchen appliances to jet engines. Welding experts agree that as much as 50% of all arc welding performed in the United States should be automated. This could reduce welding costs by 50 to 75%, helping to keep U.S. industry competitive in the global marketplace. Robotics experts estimate that only 1% of arc welding is robotized today, and only 5% is mechanized to any degree. One major reason that welding automation has lagged is the lack of engineers who have knowledge of computers and robotics as well as welding application and manufacturing experience. It is the aim of this book to familiarize practicing engineers with the possibilities of automated welding.

Arc Welding Automation details the basics of arc welding and other welding processes and provides sufficient information to help make the transition from manual and semiautomatic welding to full ("hard") automation. Many tables, illustrations, and photos are presented to complement the thorough discussion of automated welding topics, including equipment and materials selection, process requirements, and specific applications and procedures.

Automated welding encompasses mechanized, automatic, adaptive control, and robotic welding. These and related processes, such as thermal cutting, are fully described. Chapter 6 discusses specific welding procedures, offering welding schedules and explaining applicable variables and parameters, while Chapter 7 addresses techniques the engineer can use to ensure a high quality weldment. A more detailed discussion of welding equipment, both standardized and dedicated, for all levels of automation can be found in Chapters 8–12. Chapter 13 provides numerous illustrative case studies of automated welding systems. Economic justification; design, tooling, and safety issues; materials selection; and installation, maintenance, and repair of equipment are also considered in this comprehensive volume. Finally, Chapter 20 looks ahead to the future of automated arc welding. There is no question that the factory of the future will use automated and robotic arc welding, and the factory of the future is coming faster than we thought.

To make *Arc Welding Automation* technically accurate, the official terminology of the American Welding Society (AWS) is used. This book includes information from many AWS standards, codes, and manuals. AWS has graciously allowed the use of these data to help us all communicate welding information more accurately. Also, the Robotic Industries Association has given permission to abridge their Glossary of Terms relating to robotics (see Appendix 1). For permission to use data, illustrations, and other materials to make this text as complete as possible, I would like to thank Hobart Brothers, specifically William H. Hobart, Jr., and Glenn Nally, as well as Phil Monnin, Jim Leschansky, and Zane Michael of Motoman, Inc. The typing of the manuscript was done by Gayle Meyers, who endured many, many revisions. My thanks for a job well done. Thanks also to the many, many people who furnished information and pictures; the list is long and I hope I haven't missed anyone.

Howard B. Cary

Acknowledgments

A. O. Smith Automotive Products, Milwaukee, WI
Accra-Weld Controls, Rockford, MI
Alexander Binzel, Corp., Frederick, MD
American Development Corporation, N. Charleston, SC
American Welding Society, Miami, FL
Bancroft Corporation, Waukesha, WI
Banner Welder Inc., Germantown, WI
Berkley Davis, Danville, IL
Bettermann of America, Pittsburgh, PA
Bortech Corporation, Alstead, NH
Bug-O Systems Intl. (Weld Tooling Corp.), Pittsburgh, PA
Cayuga Machine & Fabrication Co., Buffalo, NY
Cecil C. Peck Co., Cleveland, OH
Cincinnati Milacron, Cincinnati, OH
Clinton Tool Co., Oden, MI
CRC–Crutcher Resources Inc., Houston, TX
Cybo Robots, Indianapolis, IN
Fanuc Robotics, Auburn Hills, MI

Fusion Incorporated, Willoughby, OH
General Electric, Cincinnati, OH
Genesis Systems Group, Davenport, IA
Guild International, Bedford, OH
Gullco International Ltd., Toronto, Ont., Canada
Hobart Brothers Company, Troy, OH
igm Robotic Systems Inc., Menomee Falls, WI
Ingersoll-Rand, Water Jet Division, Baxter Springs, KS
Jenzano Incorporated, Port Orange, FL
Jetline Engineering, Inc., Irvine, CA
K. N. Aronson, Inc., Arcade, NY
KOHOL Systems, Inc., Dayton, OH
Lincoln Electric Company, Cleveland, OH
Melton Machine & Control Co., Washington, MO
L-Tec, Florence, SC
Laramy Products Co., Lyndonville, VT
Leybold Inficon Inc., East Syracuse, NY
Liberty Tool Co., Troy, OH
Mavrix Automatic Welding Inc., Muskego, WI
McCreery Corporation, Wickliffe, OH
Merrick Engineering Inc., Nashville, TN
Miller Electric Mfg. Co., Appleton, WI
Milton Machine & Control Co., Washington, MO
Nelson Stud Welding, Div. TRW, Elyria, OH
Newcor Bay City, Bay City, MI
Newport News Shipbuilding, Newport News, VA
Panasonic, Franklin Park, IL
Pandjiris Inc., St. Louis, MO
Penton Publishing, Cleveland, OH
PHOENIX Products Company, Inc., Milwaukee, WI
Plastronic Inc., Troy, OH
Plymovent Co., Edison, NJ
PMI Food Equipment Group, Troy, OH
Preston Easton, Inc., Tulsa, OK
Process Equipment Co., Tipp City, OH
Ransom Co., Div. of Big Three Industries, Houston, TX
Redman Controls & Electronic, Berkshire, England
Reis Machines Incorporated, Elgin, IL
Robotic Accessories, Tipp City, OH
Robotic Industries Association, Ann Arbor, MI
Rofin-Sinar, Plymouth, MI
TAFA, Inc., Concord, NH
Taylor-Winfield, Brookfield, OH
The Welding Institute, Cambridge, England
Thermadyne Industries, Inc., St. Louis, MO
Torsteknik, Torsis, Sweden

Webb Corporation, Webb City, MO
Weld Tooling Corp., Pittsburgh, PA
Weldex, Inc., Warren, MI
Weldmation, Inc., Madison Heights, MI
Yaskawa Electric America, Inc., West Carrolton, OH

Contents

1

Why Automate Welding?

1.1 Welding Is Everywhere

Welding is the most economical and efficient way to permanently join metals. It is the only way of joining two or more pieces of metal to make them act as a single piece, as a monolithic structure. Welding produces leakproof joints. Welding is used to join all of the commercial metals and to join metals of different types and strengths. Welding is vital to our economy.

Welding is a very important manufacturing method. Almost everything made of metal that we use daily is welded, including the stainless steel coffee pot, bicycles and toys, and jet engines for airplanes. Our automobiles would be much more expensive if it were not for welding. Oceangoing ships, river barges, railroad cars, and the external liquid fuel tank for the space shuttle are welded. Bulldozers, excavators, farm tractors, combines—in fact, all types of agricultural equipment and construction equipment—are welded.

Welding is a construction method. The cross-country pipelines that bring natural gas to our homes for heating and cooking are welded. Pipelines carrying oil are welded. Offshore drilling and production platforms for obtaining oil and gas from the ocean floor are welded. Elevated water shortage tanks and underground oil storage tanks are welded. Long bridges are welded, and the world's tallest building is welded.

Welding is an element of design. Intelligently designed weldments will always be less expensive than products made by other manufacturing methods. Weldments are less expensive than castings and forgings or riveted or bolted assemblies for similar applications. The metal thickness of castings is often heavier than would otherwise be needed to allow for the flow of molten metal. Forgings have weight and accessibility limitations. The design of riveted or bolted assemblies must provide for loss of strength due to the removal of metal for bolt holes. Welding allows the design of composite products that make use of different properties for specific requirements where needed.

Welding is the most important method for maintenance and repair of plants and equipment. Surfaces can be overlaid with special metals to meet specific requirements. Worn surfaces can be rebuilt to their original size and shape, broken and damaged parts can be quickly repaired, and downtime can be minimized with the use of welding.

Welding ranks high among the metalworking processes, which include machining, forging, forming, and casting. Just about everything made of metal is welded sooner or later, and nonmetal products are made in welded equipment. Welding is everywhere because it is the most efficient way to join metals.

1.2 Welding Is Cost-Effective

Welded products are less expensive than similar nonwelded metal products. For example, in piping work a welded assembly is less expensive than threaded pipe construction because it uses thinner walled pipe throughout the entire system. Making threads cuts away about half the wall thickness of the pipe, as shown by Fig. 1.1. The threaded assembly therefore requires pipe with a thicker or heavier wall than the welded assembly throughout the entire system. The material cost for the total system will be approximately double that of welded pipe. In addition, welded joints are leakproof, and threaded joints may not be.

A similar situation applies to riveted and bolted assemblies. The required overlap of the parts makes the product heavier and may add up to 10% to the material cost. Holes required for the bolts or rivets reduce the strength of the parts being joined. The width or thickness of the part must be increased to compensate for the loss of metal in the holes. This can add another 15% to material cost. In tension members this extra material cost applies to the entire length of the member; thus more weight and expense is involved. In addition, the welded joint produces a monolithic structure, whereas mechanical joints may loosen in time

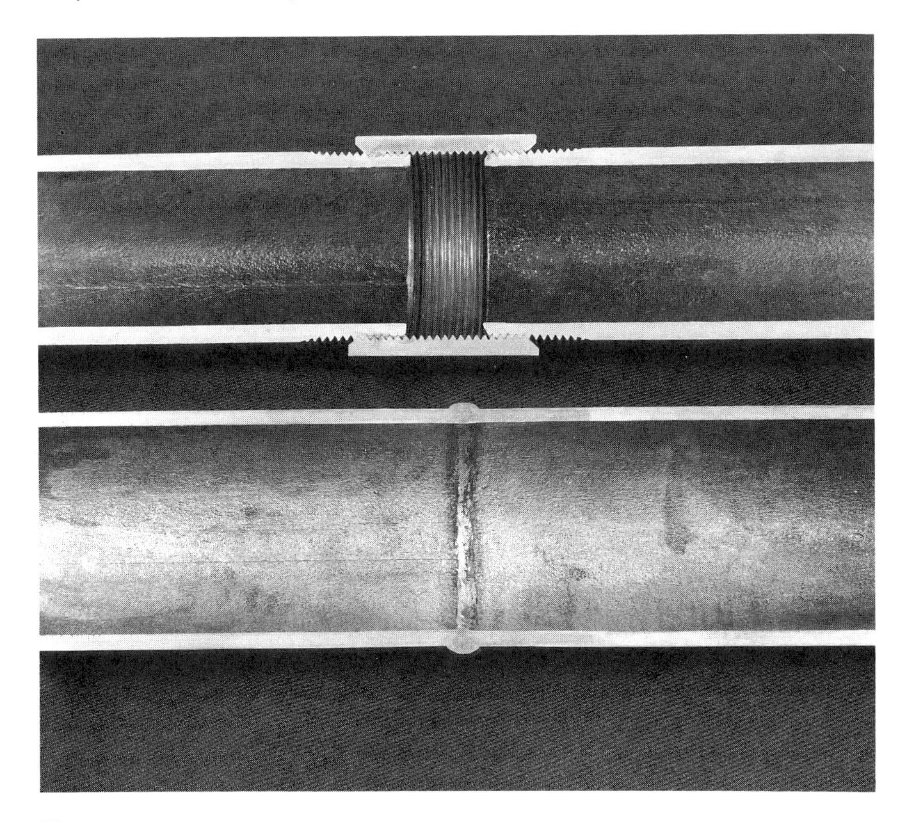

Figure 1.1 Pipe joint—welded joint (bottom) versus screw threads (top).

and require continual maintenance—another expense to be considered. This is especially true in heavy dynamically loaded structures. These two factors affect the cost of metal storage tanks. The overlap adds extra metal, and the material lost in holes must be compensated for. In addition, riveted or bolted joints may not be leakproof and can be a source of corrosion. Welded joints are always leakproof. Because of this, welding is used for industrial piping systems, dynamically loaded structural steel products, and storage tanks of all types.

Weldments provide greater design flexibility. The designer can place the metal at the exact location dictated by the function of the part and the stresses involved. The design can incorporate thick or thin sections of metal with specific characteristics. Machinery parts manufactured by welding are less expensive than castings. There are a large number of factors that make weldments a less expensive product. The designer has the freedom to make running changes on a weldment. Section thicknesses can be changed, metals of higher strength can be used, a stiffener can be added, clearances can be increased, and the design change

Figure 1.2 Brake shoe—casting and weldment.

time is minimal. The changed product can be immediately introduced into production without scrapping patterns, dies, cores, tools, etc. Weldments are lighter. They are slim, trim, strong, and tough. Thinner sections can provide the strength required, and the reduced weight cuts operating costs and shipping costs while increasing energy efficiency. Weldments have clean, trim lines that fit the modern-day concept of functional design. They use smooth rolled steel plates, for example, to achieve a streamlined appearance. Weldments are made with rolled steel, which requires a minimum of finishing. Rolled products are made to close dimensions. Good design can produce a weldment that requires a minimum of machining. Machining costs of weldment are much lower than those of a similar steel casting. Metal finishing is reduced, in some cases eliminated, and many weldments can be painted immediately after welding, which reduces manufacturing cycle time.

Since weldments are made from standardized plates and shapes, the same raw materials can be used for any weldment. Thermal cutting, the usual method of parts preparation, can be automated to provide any variety of pieces for the weldment. The raw material inventory and the manufacturing inventory are reduced, and manufacturing cycle time is shorter.

Castings are heavy, and almost every casting requires a pattern. Patterns and cores are expensive, and changing patterns is very costly. In addition, patterns deteriorate due to hard use in a foundry. Molten metal flow problems create voids in castings that have to be repaired. Castings require the use of risers where metal enters the cavity. These must be removed, which is costly and time-consuming. Castings tend to warp and shrink, and extra metal is required to allow for cleanup machining. The surfaces of steel sand castings are not smooth and require additional finishing. If die castings are used, the dies and patterns are extremely expensive, and the cost can be justified only for high-volume production.

An example of the difference between a casting and a weldment is the truck brake shoe shown in Fig. 1.2. Note the thinner uniform section of the weldment and the fact that it is much lighter in weight than the casting.

In spite of the fact that manual welding is labor-intensive, the production of weldments is not. This is because weldments are produced from hot rolled steel plates, structural shapes, bars, etc., that are made by mass production automated steel rolling.

1.3 Methods of Application of Welding

In the early days of arc welding, efforts were made to develop automatic methods for applying the weld. The first known application was by the automobile industry to manufacture rear axles. Figure 1.3 shows the early automatic welding equipment used for producing rear axles in 1920 [1]. Automatic welding did not gain popularity because covered electrodes were introduced at about the same time. Covered electrode "stick" welding, manually applied, was much more flexible in application and produced deposited weld metal of a higher quality.

The differences in methods of applying an arc welding process relate to whether they are controlled by the individual welder or by the welding machine. These affect the level of fatigue of the individual. When more functions are taken over by the machine, fatigue levels are reduced and productivity is increased. These functions are:

1. *Starts and maintains the arc*: includes striking the arc and maintaining the correct arc length.
2. *Feeds the electrode into the arc*: feeding the electrode or cold filler wire into the arc or weld pool.
3. *Controls the heat for proper penetration*: involves manipulating the electrode or torch for proper control of the molten weld metal.
4. *Moves the arc along the joint* (*travels*): provides relative motion at the proper velocity along the joint.
5. *Guides the arc along the joint*: tracks or follows the joint.

Figure 1.3 First automatic arc welding machine.

6. *Manipulates the torch to direct the arc*: manipulates the electrode or torch to direct the arc to the proper place for bead placement and joint fill.
7. *Corrects the arc to overcome deviations*: senses abnormalities and makes changes in welding parameters.

In arc welding, these functions must be coordinated and carried out simultaneously by the individual or by the machine or by a combination of human and machine to make a quality weld.

Both manual (MA) welding and semiautomatic welding are done by an individual known as a *welder.* In both instances the welding operation is under the continuous control of the welder; hence, it is a closed-loop system. The welder must start the arc, maintain a uniform arc, feed the electrode into the arc, obtain penetration and control the pool, move and guide the arc along the joint, manipulate the electrode or gun for correct placement of the weld metal, and overcome variations in order to ensure a quality weld. While doing this, the welder is watching a very bright light and is exposed to heat, smoke, and sparks. The fatigue factor is very high; therefore, the operator factor or duty cycle is relatively low. Additionally, in manual shielded metal arc welding the electrode is normally 14 in. long. When the electrode is consumed the arc stops and the welder must discard the short stub and insert a new electrode in the holder. This is a built-in stopping point. In addition, the welder must remove the slag covering left on the weld.

The semiautomatic (SA) welding method was developed in the early 1950s to increase welding productivity yet maintain flexibility. The welding gun is held by the welder, and the electrode wire is automatically fed through the gun into the arc. Semiautomatic welding has become the most popular arc welding method. It does not require the welder to maintain the arc, discard the stub, or change the electrodes. It still involves a relatively high fatigue level because the welder must watch the arc, hold the gun to control the placement of the molten weld metal, and follow the weld joint. Slag removal is largely eliminated, and the amount of smoke and fumes is reduced, but arc radiation and heat are still present. Semiautomatic welding has greatly improved the duty cycle of arc welding, and it is a closed-loop system.

Mechanized (ME) welding apparatus maintains the arc, feeds the electrode, moves along the weld joint, and provides for the maintenance of the arc, feeding of the electrode, and moving along the weld joint. It still requires the attention of the welding operator to make sure the arc is accurately following, fusing, and filling the joint for a quality weld. The operator is still involved, but the fatigue factor is much less and the duty cycle is improved. Because of the operator involvement, it is considered a closed-loop system.

Automatic (AU) welding apparatus is built to provide the required motion pattern and preset welding parameters. In most cases the parts are sufficiently accurate that changes are not required in the welding conditions or the motion pattern. Good quality welds result because the inherent tolerance of the welding process will accommodate minor joint variations. If the joint location or geometry is beyond normal variations, a defective weld may result. It is an open-loop system because the weld is not under constant observation by the operator.

In robotic (RO) welding, a robot with a controller provides the motion path of the arc and the welding parameters for making the weld. Robotic welding can be simple, with the same welding conditions prevailing throughout the entire weld sequence. The path, once established, will always be identically followed. This type of equipment will make quality weldments with well-prepared pieceparts over and over. This is an open-loop system, as the weld is not under constant observation by the operator. Robotic welding can be combined with adaptive control welding to provide a closed-loop system that closely simulates human control.

Adaptive control (AD) welding apparatus requires sensing devices. The data provided by these devices go to the computer controller. If joint conditions

Method of Application / Arc Welding Elements/Function	MA Manual (closed loop)	SA Semiautomatic (closed loop)	ME Mechanized (closed loop)	AU Automatic (open loop)	RO Robotic (open or closed loop)	AD Adaptive Control (closed loop)
Starts and maintains the arc	Person	Machine	Machine	Machine	Machine (with sensor)	Machine (robot)
Feeds the electrode into the arc	Person	Machine	Machine	Machine	Machine	Machine
Controls the heat for proper penetration	Person	Person	Machine	Machine	Machine (with sensor)	Machine (robot) (only with sensor)
Moves the arc along the joint (travels)	Person	Person	Machine	Machine	Machine (with sensor)	Machine (robot)
Guides the arc along the joint	Person	Person	Person	Machine via prearranged path	Machine (with sensor)	Machine (robot) (only with sensor)
Manipulates the torch to direct the arc	Person	Person	Person	Machine	Machine (with sensor)	Machine (robot)
Corrects the arc to overcome deviations	Person	Person	Person	Does not correct, hence potential weld imperfections	Machine (with sensor)	Machine (robot) (only with sensor)

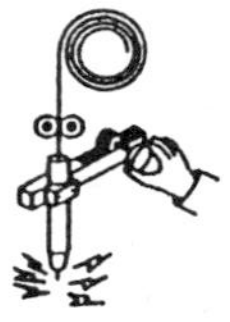

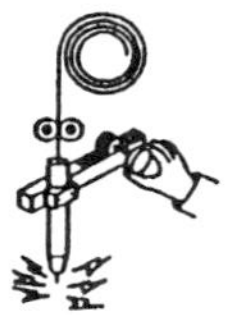

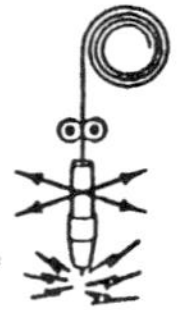

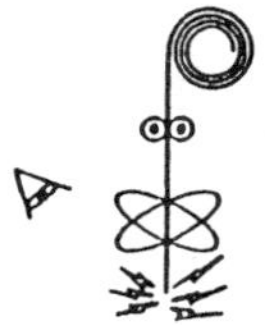

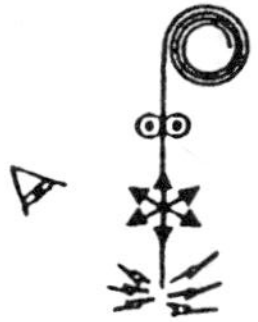

Figure 1.4 Method of applying welding.

change, the controller changes the welding parameters to provide a quality weld. Different sensing devices are required for different situations, and the degree of adaptive control may vary. This is a closed-loop system, as the weld is constantly monitored by sensors.

The methods of applying welding are illustrated in Fig. 1.4. In manual welding the individual has control over all of the functions, whereas in adaptive control welding the same functions are controlled by the machine. The skill required by the individual and operator fatigue are greatest when all functions are controlled by the person and diminish as the functions are taken over by the machine.

The American Welding Society (AWS) defines the various methods of applying welding processes as follows [2]:

Manual (MA) welding: welding with the torch, gun, or electrode holder held and manipulated by hand. (See Fig. 1.5.)

Semiautomatic (SA) welding: manual welding with equipment that automatically controls one or more of the welding conditions. (See Fig. 1.6.)

Mechanized (ME) welding: welding with equipment that requires manual adjustments of the equipment controls in response to visual observations of the welder, with the torch, gun, or electrode holder held by the mechanical device. (See Fig. 1.7.)

Figure 1.5 Manual welding.

Figure 1.6 Semiautomatic welding.

Automatic (AU) welding: welding with equipment that requires only occasional or no observation of the welding and no manual adjustments of the equipment controls. (See Fig. 1.8.)

Robotic (RO) welding: welding that is performed and controlled by robotic equipment. (See Fig. 1.9.)

Adaptive control (AD) welding: welding with a process control system that automatically determines changes in welding conditions and directs the equipment to take appropriate action. (See Fig. 1.10.)

An automatic or automated welding system consists of at least the following:

1. *Welding arc*: requires a welding power source and its control, an electrode wire feeder and its control, the welding gun assembly, and necessary interfacing hardware.
2. *Master controller*: controls all functions of the system; can be the robot controller or a separate controller.
3. *Arc motion device*: can be the robot manipulator, a dedicated welding machine, or a standardized welding machine; may involve several axes.
4. *Work motion device*: can be a standardized device such as a tilt-table positioner, a rotating turntable, or a dedicated fixture; may involve several axes.
5. *Work holding fixture*: must be customized or dedicated to accommodate

Figure 1.7 Mechanized welding.

the specific weldment to be produced; may be mounted on the work motion device.

6. *Welding program*: requires the development of the welding procedure and the software to operate the master controller to produce the weldment.
7. *Consumables*: the electrode wire or filler metal, the shielding medium (normally gas), and possibly a tungsten electrode.

There are differences in the degree of adaptive control of a welding system, depending on the number of sensors employed to monitor conditions. Sensors are needed to find the joint, provide root penetration, place the bead, follow the joint, ensure joint fill, and so on. Adaptive control requires sensing devices and computerized circuits that alter the motion and the value of a particular parameter to satisfy the new requirements. Many sensing devices will be required to provide

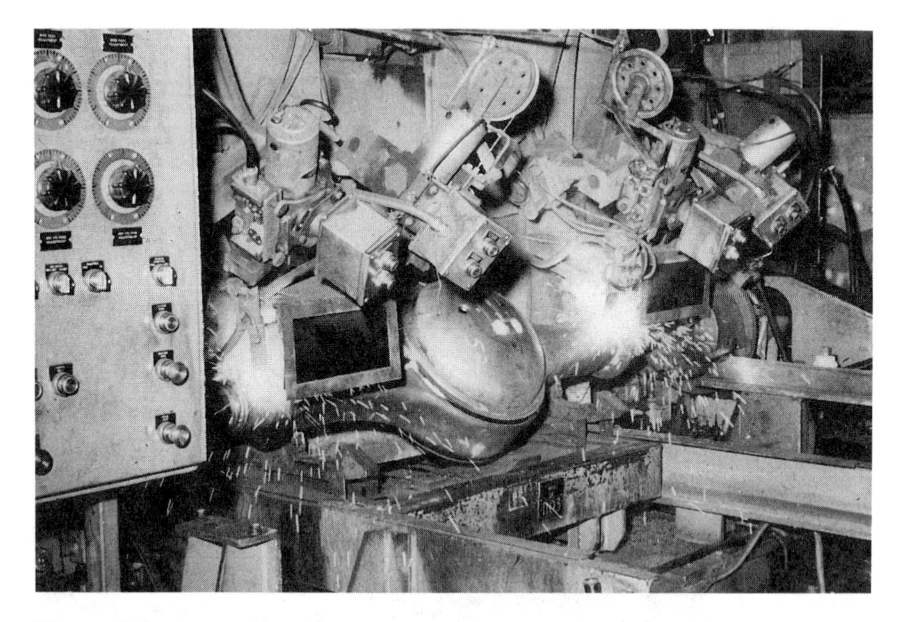

Figure 1.8 Automatic welding.

Figure 1.9 Robotic welding.

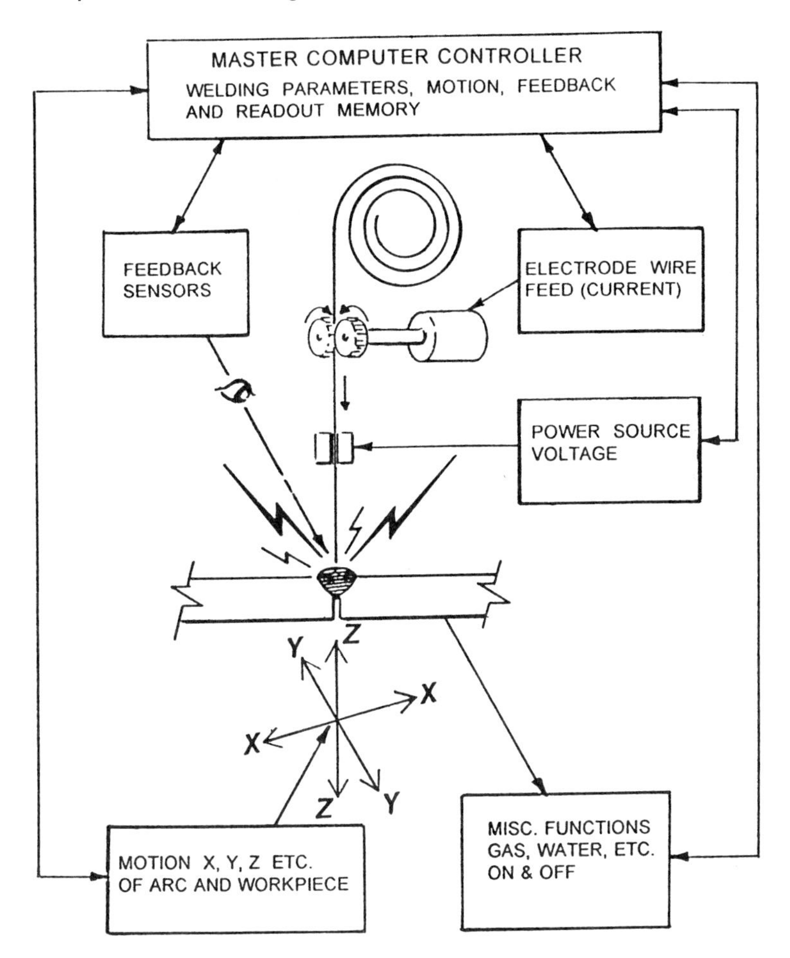

Figure 1.10 Adaptive control welding.

total adaptive control. Sensing devices and adaptive controls must be "real-time" for shop production use.

In adaptive control welding the equipment will provide a high quality weld even if there are deviations such as inaccurate piecepart preparation or inaccurate location of the joints. This will be done without human operation, supervision, or intervention. It is a closed-loop system.

There are differences in the degree of adaptive control that is included in the welding equipment. These relate to the number and type of sensors employed to monitor specific functions. It would be uneconomical to provide sensors for conditions that did not vary beyond acceptable tolerances. Sensing devices and adaptive controls are expensive.

Robotic welding equipment with appropriate sensors and feedback capabilities can follow the path even if it is not the original memorized path. Work motion devices can move the workpiece to coordinate motion under the direction of the robot controller. Welding conditions will change if the sensors detect changes in the joint configuration, for example. A robotic welding system with adaptive control is a closed-loop system and can be as complicated as necessary to control all parameters to make a perfect weld in every situation. It approaches the ability of a human welder to compensate for changes during the welding operation.

Adaptive control welding will become more widely used in the future. It will be required for large weldments, particularly those made of heavy materials requiring multipass welds. Adaptive control welding will become widespread because the sensing devices are becoming less expensive, more sensitive, and more rugged. In addition, the master controllers now have an open architecture, so numerous different sensors can be plugged into them.

1.4 Improving the Productivity of Welding

There is a continuing need to reduce the manufacturing costs of all products. This is also true of weldments. Efforts to reduce the cost of weldments involve many interrelated factors such as the design of the weldment, the design of the weld and joints, the selection of the welding process, the development of a welding procedure, shop practice, and the method of application of welding.

One factor stands out above all the rest. This is the cost of labor, which is continually rising. Labor is involved not only in making welds, but also in preparing the parts that make up the weldment and the cleanup, machining, and metal finishing required to complete the weldment.

The design of the weldment is another major cost factor. Intelligent designs produced by experienced designers will produce the most economical weldments. The following factors should be considered. The least expensive weld joint is no joint at all. Eliminating a weld joint by using a bend or a rolled structural section can reduce the amount of welding required. The incorporation of rolled structural sections and formed members into the weldment is good design practice. The use of heavy thermal-cut sections for bosses and the use of simple castings also help reduce the weldment cost. The use of weldable steels that do not require special attention reduces shop costs and total weldment costs.

Weld designs and joints relate to the junctions between parts of the weldment. Parts come together in the form of a butt, corner, lap, edge, or tee, and joining can be accomplished by means of a weld or a bend or a rolled structural section. The weld for every joint in the weldment must be the most economical possible. The economy of the weld directly relates to the amount of filler metal employed to produce the weld. The weld requiring the least amount of filler metal would be the least expensive; however, this must be weighed against the strength requirement of the weld. In most cases, a full penetration weld is required because it must develop the total strength of the members being joined.

The most popular type of weld is the fillet weld [3]. Fillets are sized by leg length, but strength relates to the throat dimension. Designers should know that as fillet weld size and strength double, the amount of weld metal required quadruples. This is graphically shown by Fig. 1.11. The smallest practical fillet should be employed. Wherever possible, double fillets should be used as they provide the same strength with half the weld metal. Because of this strength/size relationship, intermittent fillet welds should not be used.

Deep penetrating processes, when applied automatically, can ensure greater strengths because they develop a larger cross-sectional area. This relates to fillet welds and to certain types of groove welds. Figure. 1.12 shows a comparison of

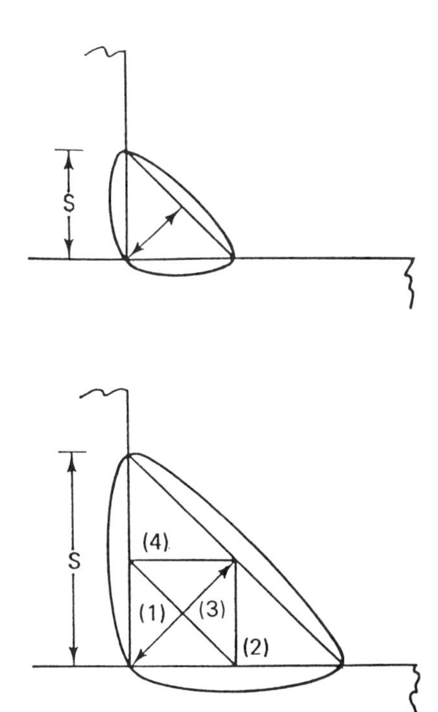

Figure 1.11 Fillet weld size.

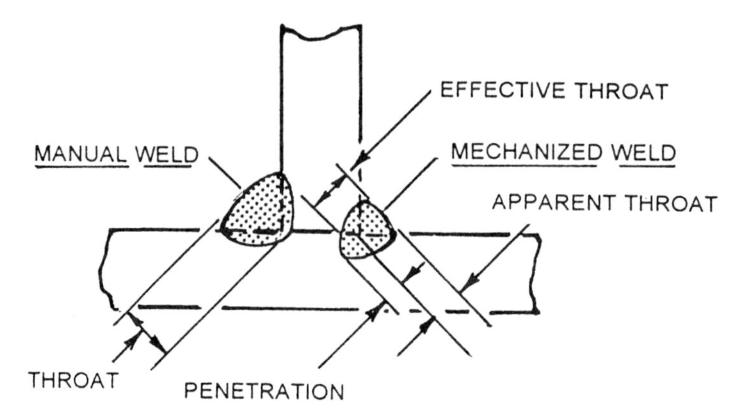

Figure 1.12 Penetration—manual versus mechanized application.

fillet weld penetration when the weld is applied manually and when it is applied by a mechanized method.

Groove welds are the second most popular type of weld. There are seven basic types of groove welds, and each can be single or double. The selection of the groove weld detail relates to the thickness of the metals being welded, accessibility of the joint, and the preparation required. The most economical weld is the one that uses a minimum amount of filler metal and requires the minimum amount of preparation. Weld details also relate to the process employed.

Each arc welding process can be used for specific applications. The weld must match the type of metal being welded, the thickness of the metal, and the welding position. A good knowledge of welding process capabilities is necessary to make an intelligent selection.

To improve welding productivity, select the process with the greatest weld metal deposition rate. The deposition rate is the amount of weld metal deposited on the weldment during a specific time period, usually measured in pounds per hour. Figure 1.13 illustrates the ranges of deposition rates for the various welding processes.

Someone with a limited amount of welding knowledge might choose to employ the submerged arc or electroslag process because of their high deposition rates. However, these processes are unsuitable for thin materials, for out-of-position welding, or for welding nonferrous materials such as aluminum. A good understanding of welding processes is necessary to make an intelligent selection.

The filler metal used for each process has a particular utilization factor—identified as the amount of weld metal actually deposited on the weldment compared to the amount of filler metal purchased. Welding electrodes and fluxes are relatively inexpensive, but they do affect the overall cost of the weldment. Utilization rates of the different types of fillet materials are shown by Fig. 1.14.

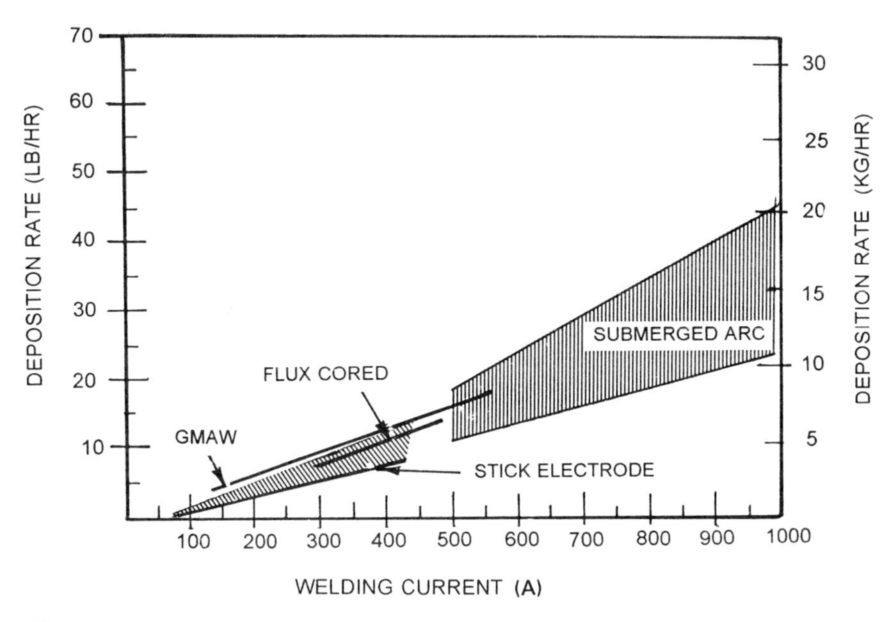

Figure 1.13 Process versus deposition rates. GMAW = gas metal arc welding.

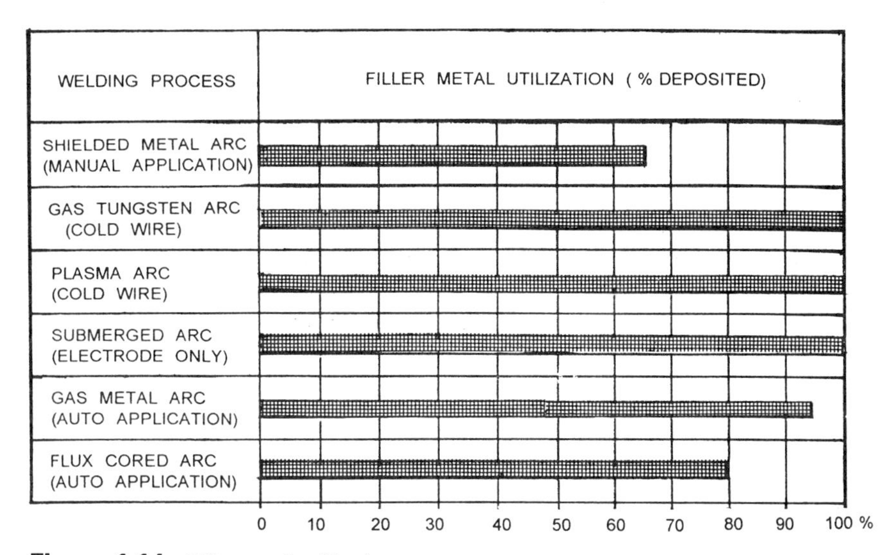

Figure 1.14 Filler metal utilization.

This figure shows that flux-cored electrodes usually deposit 85–90% of the filler metal purchased. The electrodes used for gas metal arc welding deposit up to 95% of the weld metal purchased, but submerged arc welding, which uses an equal weight of flux for the weight of weld metal deposited, provides a utilization of only 50% considering both the electrode and the flux.

A welding procedure is a written record of the factors involved in producing a weld. These include all of the welding parameters such as welding current, welding voltage, and travel speed. Other factors that must be recorded include the welding process, the method of application, the electrode type and size, and design, pass sequence, weld bead placement, and other details. Many of these items can be adjusted or changed to improve weld productivity.

The most important factor affecting weld productivity is the welding current. As welding current is increased, the deposition rate increases (Fig. 1.13). At the same time, greater welding currents increase weld penetration, which has a large effect on productivity. Welding voltage also affects penetration and weld bead contour as well as travel speed. Welding position is also a factor. For example, in the flat or down-hand position, higher welding currents can be used and larger beads may result. For a multipass weld, this can reduce the number of beads required to make the weld and thus lower its cost. Mechanical systems can travel at higher speeds, can use much higher currents, and can be more uniform in travel. These all help increase productivity. Mechanical movement systems use larger diameter electrodes and higher current, greatly increasing the deposition rate and productivity. With higher current and greater penetration, weld designs can use less weld metal yet provide the full penetration required. With higher speed, the weld is done more quickly, which improves productivity. The welding procedure should be tested and modified to increase current, travel speed, and penetration to improve weld productivity.

Shop practice in a welding department or factory has a great effect on welding productivity. This relates to such things as accuracy of parts preparation, clean joint surfaces, uniformity of the welding operation, the accuracy of setting up or of fixtures, and the post-weld cleaning or metal finishing operations.

Piecepart preparation is more important to welding productivity than any other factor. If parts are incorrectly prepared, they may be smaller than designed, which will result in excessive root openings or gaps between members. This requires a larger amount of weld metal, which is the most expensive metal of the weldment. On the other hand, if the pieceparts are larger than designed, the joint will be jeopardized, proper penetration may not result, or the piece may be so large that it must be trimmed prior to use. These result in extra expense or in failure of the weldment.

The accuracy of all operations prior to welding affects weldment quality and is very important for automatic welding applications. The accuracy of fixtures or setup operations contributes to excessive or tight gaps between members. Accuracy of pieceparts helps to provide sufficient metal for machining cleanup and other purposes. The uniformity of welding, particularly the uniformity of travel speed, has a great effect on the uniformity of weld size. Oversize fillets contribute to the deposition of excess weld metal, which greatly increases weld costs. Welds with long horizontal legs and short vertical legs require additional weld metal to attain design strength. This is due to poor welding technique. Oversize fillets and nonuniform fillets contribute a large percentage of extra cost of weldments.

Finally, the finishing required for weldment, including slag and spatter removal, increase the cost of the weldment. These can be reduced with proper procedures and techniques to reduce metal finishing costs. Other factors such as fixturing, warpage, and machining allowances must also be considered. Good shop practice greatly reduces weldment costs.

The method of application, discussed in Section 1.3, should be part of the welding procedure. The method of application affects the duty cycle or operator factor of the welding operation. The operator factor is the ratio of "arc on" time to total time. Weld metal is deposited in the joint only while the arc is ignited. Welders become fatigued, and there are many incidental tasks that must be performed that reduce the operator factor. In addition, slag must be removed from each pass prior to depositing the next one. Time is also lost for personal reasons. The operator factors for the different methods of application are shown in Fig. 1.15. These data are based on broad studies in different factories and represent averages. To improve productivity of welding, the operator factor must be improved.

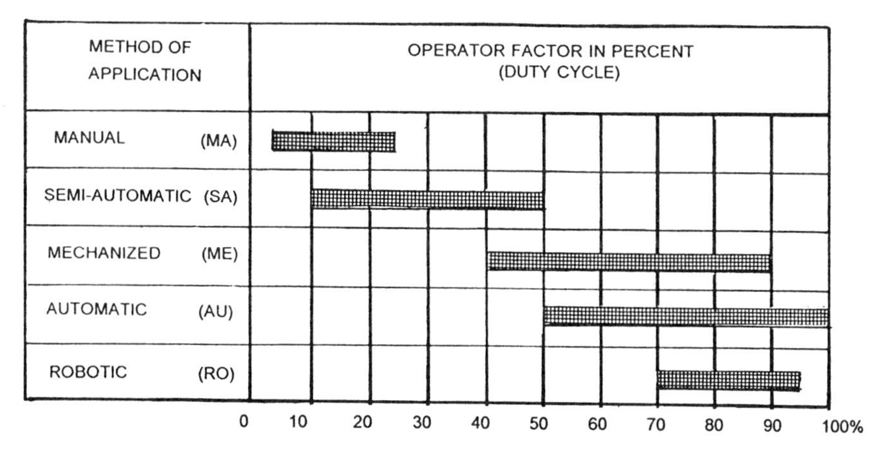

Figure 1.15 Operator factor.

1.5 The Advantages of Welding Automation

The previous section described ways to increase the productivity of welding. Many of these relate to the labor of actually making arc welds and to the frailties of the human welder. One of our objectives is to remove the human welder from the immediate arc area. The fumes, heat, muscle fatigue, bright arc, etc., are undesirable, and removing or reducing them improves the welder's job environment. Removing the person from the immediate welding operation will increase the arc "on time" or duty cycle by eliminating these fatigue factors.

The other reasons for increased productivity relate to the method of applying arc welds. Mechanical equipment can carry heavier tools than a person; torches and cables can be heavier, allowing the use of higher currents. Higher welding current, which produces more heat, makes possible a higher deposition rate and deeper penetration welds. It may also reduce the number of passes. Mechanical motion provides higher and more uniform travel speeds. Faster travel gets the job finished sooner, and steady travel speed gives consistent weld size.

All of these factors make arc welding less labor-intensive. They all help to speed up the welding operation and reduce the cost of welding.

In addition to the advantages mentioned above, automatic welding overcomes many of the problems associated with individually controlled manual or semiautomatic welding. However, automatic arc welding also has some problems.

Automatic welding equipment is of two types: (1) standardized equipment for specific types of welds or joints and (2) dedicated equipment for specific welded products. Both types result in higher quality welds and improved productivity than manual application. Standardized automatic equipment is more expensive than semiautomatic equipment and must be used continuously to be economically acceptable. For example, an orbital tube-to-sheet welding machine, shown in Fig. 1.16, will pay for itself very quickly if there are sufficient tube-to-sheet welds to be made. The same is true of other automatic machines used for making a large number of similar welds. Standardized automatic welding equipment must be kept busy making the kinds of welds it is designed to produce.

Dedicated automatic welding equipment also has problems. Dedicated machines are designed to weld a specific product such as a small tank, a torque converter housing, or an automobile axle assembly. Some dedicated machines will accommodate different sizes of the same basic product. A typical dedicated automatic welding machine with two heads for making 400-gallon truck fuel tanks is shown in Fig. 1.17.

Dedicated automatic welding equipment is economically feasible when a large number of identical parts are manufactured. It must be in operation most of the time to pay its way.

Most dedicated automatic welding equipment has limited flexibility. When the product is changed, the equipment must be replaced or redesigned. Rebuilding

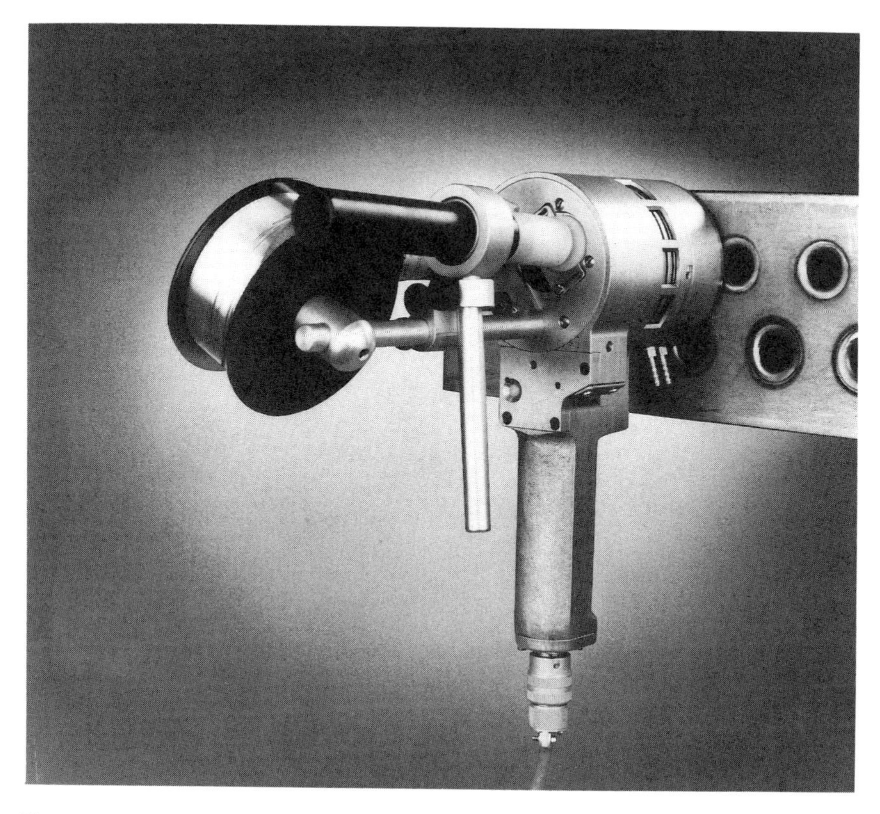

Figure 1.16 Orbital tube-to-sheet welding machine.

is expensive and is a problem in the automotive industry, where models change almost every year.

Another disadvantage of dedicated automatic welding equipment is its normal inability to compensate for variations in pieceparts. The production operations prior to the automatic welding operations may produce inconsistent parts. Springback can be different on different heats of steel, bend lines can be slightly off, thermal-cut parts can vary due to heat distortion or deterioration of torch tips, blanking dies wear, etc. In addition, parts may warp during welding, and holding fixtures may be weak and allow the joint to vary from its expected location. All of these factors may change the location or fit-up of the joint. If these variations exceed the tolerance of the welding process, the resulting weldment may have unacceptable weld quality.

The demands for higher quality and improved productivity and the desire to remove the human welder from the immediate arc area can be met only with the new method of application—*adaptive control welding*. This is a method of

Figure 1.17 Truck fuel tank welding machine.

application that is more advanced than automatic welding. It provides complete control of the welding operation, including accommodation for poorly fitted joints and the ability to modify taught travel patterns. The difference between automatic welding and adaptive control welding is the use of powerful controllers with sensing devices. These devices, along with the arc or torch manipulation and work motion systems, will produce quality welds even when the fit-up and accuracy of the weldment are poor. Adaptive control welding is a closed-loop system and is the closest to manual welding for overcoming fit-up problems.

Unfortunately, most dedicated automatic welding equipment does not provide for torch movement or work motion movement to deviate from the established system. In other words, adaptive control welding is not normally possible with dedicated automatic welding equipment. For this reason, manufacturing variations must be held to a minimum so that the normal arc welding tolerance will accommodate fit-up inaccuracies.

Robotic arc welding systems will become more popular and will take over jobs previously handled by dedicated automatic welding fixtures. Robotic welding equipment can be reprogrammed for different weldments. Programs can be stored in memory, and by the use of different holding fixtures many different weldments can be produced by the same robot. In this way, small production lots can be run efficiently. Quick changeover time from one weldment to another will keep the equipment busy the majority of the time. This provides rapid payback and meets economic justification requirements.

The arc welding robot can be either an automatic or adaptive control welding system. It can be reprogrammed to weld many different items in the course of a day. A robotic arc welding system is basically an automatic welding system; however, if sensors and an adaptive computer controller are included, it becomes an adaptive control welding system.

References

1. H. L. Uniand, Automatic electric arc welding machine, *The Welding Engineer,* July 1920.
2. *Standard Welding Terms and Definitions,* AWS A3.0-89, The American Welding Society, Miami, FL.
3. H. B. Cary, *Modern Welding Technology,* 3rd ed., Prentice-Hall, Englewood Cliffs, NJ, 1994.

2

Welding Technology

2.1 Welding Is a Technology

Welding is considered to be the most complex of all manufacturing technologies. It is much more than an occupation requiring a high level of manipulative skill. Welding is a technology based on the physical laws of nature. It is an engineering specialty that embraces many scientific disciplines. In order to transform welding from a manual skill level occupation to an automated production process, it is necessary to understand the scientific principles involved.

Welding conforms to all of the laws of physics. (Physics deals with energy and motion and is subdivided into such subjects as electricity, mechanics, sound, light, friction, magnetism, and heat.) The electric circuit for arc welding follows the same rules as any other electric circuit. In every electric circuit there are three factors. *Current,* in units of amperes (A), is a measure of the amount of electricity that flows through a conductor in 1 sec. *Electrical pressure,* measured in volts (V), is the specific force that causes current to flow. *Resistance,* measured in ohms (often symbolized as Ω), is the restriction to the flow of current.

Some of the other electrical factors that apply to welding circuits are impedance, frequency, and capacitance. The welding power source takes current from utility lines and converts it to the electric power proper for welding. The arc is a resistance. The cables are the conductors, and the contactors are switches.

The laws of mechanics form the basis of the design of weldments. They also relate to the strength of materials vital to weldment design.

Sound energy is transmitted through most materials, traveling at different rates through metals, gases, and liquids. It is used not only to make welds but also for looking inside welds to determine defects. One welding process and some nondestructive examination techniques use ultrasound.

Welding also involves the science of light or electromagnetic radiation. The laser welding process uses coherent light. A popular heating method uses infrared radiation. The radiographic examination system uses X-rays, which are in the electromagnetic spectrum but of a different wavelength than visible light.

Welding involves friction. Friction is an important element of one welding process; it also relates to wear, which is largely repaired by welding.

Magnetism produced by current flow has much to do with welding. Magnetism is used to generate electricity and also to transform it. Magnetism can create problems for welding in that unbalanced magnetic fields cause the arc to wander from its intended path. It is also used as an inspection technique.

The sciences of chemistry and metallurgy are also involved in welding. The choice of the gases used for fueling flames and for shielding arcs and the composition of fluxing elements and fluxes in electrodes depend on a knowledge of chemistry. Chemical knowledge of atoms and their structure is largely responsible for the success of welds. Metallurgy is the study of the makeup of metals and their relationship with welding.

It is necessary to become familiar with the basic terminology used by the welding industry. The American Welding Society (AWS) provides the majority of definitions. AWS official definitions [1] are used throughout this book.

Welding is a materials-joining process that produces a weld. A *weld* is a localized coalescence of metals or nonmetals produced either by heating the materials to the welding temperature with or without the application of pressure, or by the application of pressure alone with or without the use of filler metal. Coalescence is the growing together into one body of the materials being welded.

A *weldment* is an assembly whose component parts are joined by welding. A weldment can be made of many or few metal parts. It may contain metals of different compositions, and the pieces may be in the form of rolled shapes, sheets, plates, pipes, forgings, or castings.

Filler metal is the metal to be added in making a welded, brazed, or soldered joint; it becomes the weld. In some processes the filler metal is carried across the arc and deposited in the weld. In others, the filler metal is not carried across the arc but is melted by the heat of the arc and upon solidification becomes the weld metal.

Base metal is defined as the material to be welded, soldered, or cut. For some processes, the base metal is called the *substrate*.

Welding can be done in different *positions,* and a *welding procedure* provides the detailed methods and practices, including joint design details, materials, and methods of application of welding, that describe how a particular weld or weldment is made.

2.2 Welds and Joints

In describing welds, it is necessary to distinguish between the joint and the weld. Each must be described to completely describe the weld joint. A *joint* is the junction of members or the edges of members that are to be joined or have been joined. This relates to any weldment.

There are five basic types of joints for bringing two members together for welding. These joint types are the same as those used by other trades. The five basic joints (Fig. 2.1) are distinguished by the relative positions of the parts being joined.

1. *Butt joint (B)*: between parts aligned in approximately the same plane.
2. *Corner joint (C)*: between parts located at approximately right angles to each other and at the edge of both parts.
3. *T joint (T)*: between two parts approximately at right angles in the form of a T.
4. *Lap joint (L)*: between two overlapping parts in parallel planes.
5. *Edge joint (E)*: between the edges of two or more parallel or nearly parallel parts.

There are many variations to these descriptions. Butt joints may not always be in the same plane, corner joints may not be at right angles, and T joints may not be at right angles. However, these are the official descriptive words to use.

There are also joints where more than two parts come together, for example the cruciform or cross joint.

A *weld* is a localized coalescence of metals or nonmetals produced either by heating the materials to the welding temperature with or without the application of pressure or by the application of pressure alone with or without the use of filler metal. A weld is used to join the members of a weldment at the joint.

There are many different types of welds, and they are best described by their cross-sectional shape. The most popular type of weld used is the *fillet weld,* shown in cross section in Fig. 2.2. The size of a fillet weld is measured by the

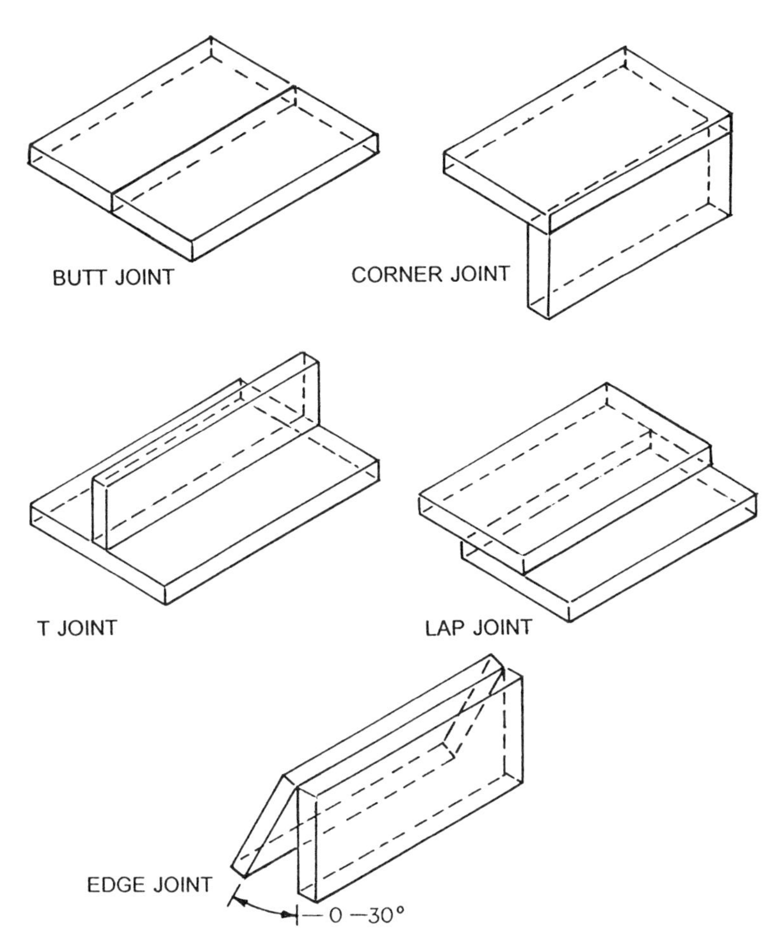

Figure 2.1 Five basic joints.

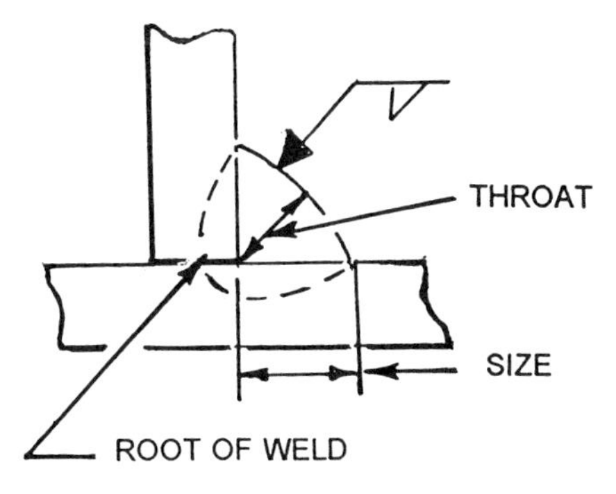

Figure 2.2 Fillet weld.

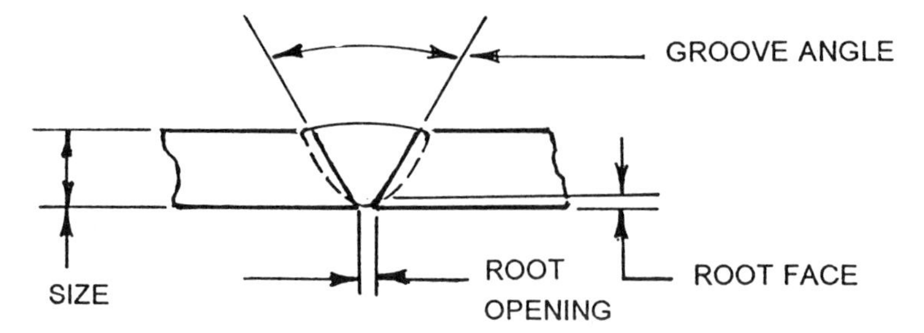

Figure 2.3 Groove weld.

largest inscribed right triangle that can be made within the cross section of the weld. The size of the weld is specified by its leg length even though its strength is based on the throat dimension of the fillet.

The second most popular type of weld is the *groove weld.* An example is shown in Fig. 2.3. There are seven basic types of groove welds, as shown in Fig. 2.4, and these are described in terms of the cross section of the groove. Groove welds have many different types of designs. The design relates to the bevel angles or the groove angle, the root face, the root opening, and whether they are single or double.

Other types of welds include the flange weld, the plug weld, the spot weld, the seam weld, bead welds, surfacing welds, and the backing weld. Joints are combined with welds to make *weld joints,* examples of which are shown in Fig. 2.5. The design of welds and weld joints will be more thoroughly covered in Chapter 15.

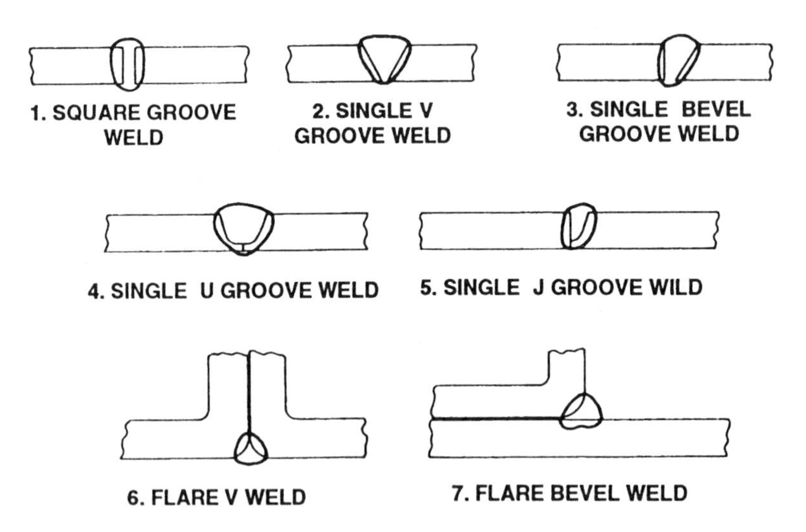

Figure 2.4 Seven basic groove welds.

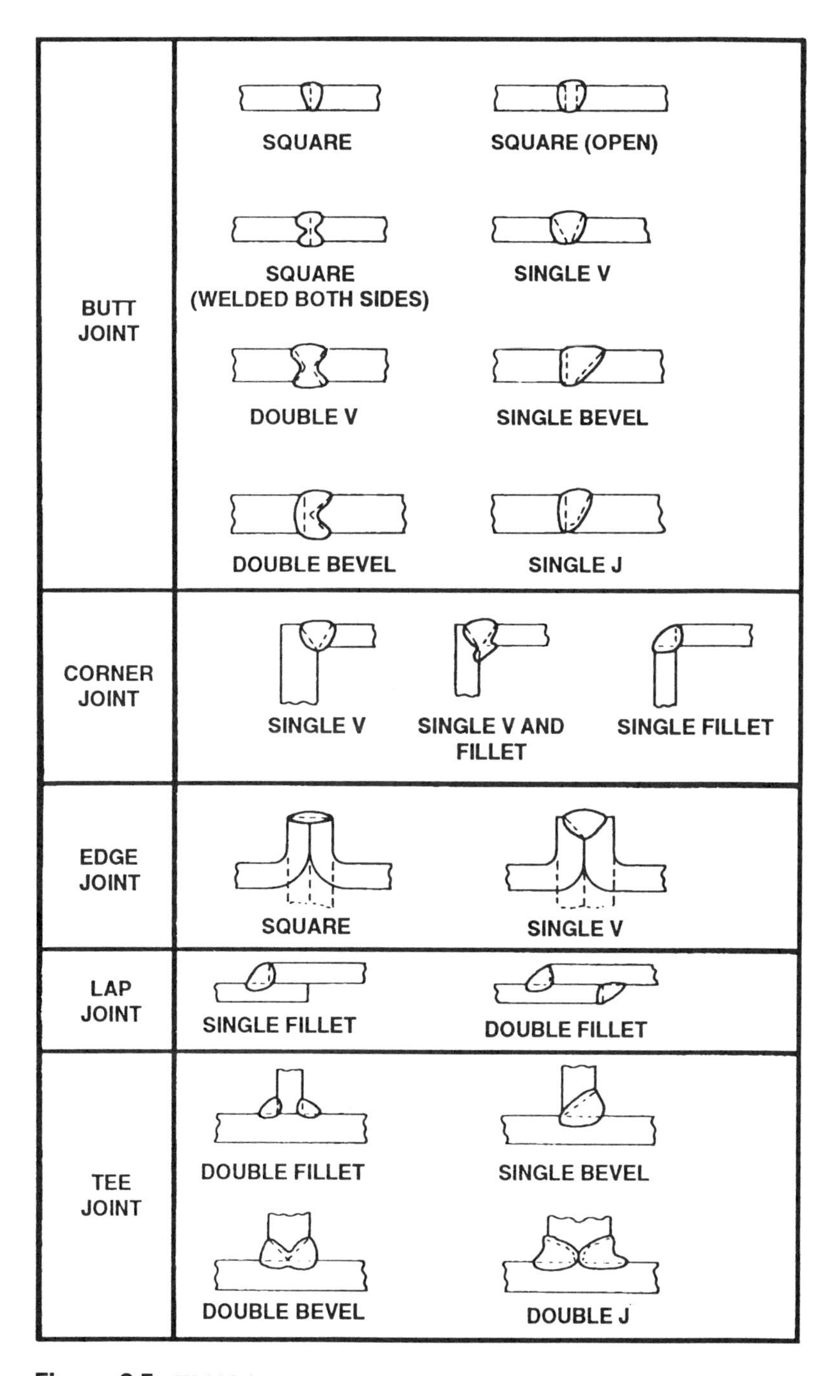

Figure 2.5 Weld joints.

2.3 Welding Positions

In describing welding, it is necessary to indicate welding positions. There are five basic welding positions as defined by the American Welding Society.

Flat welding position: the welding position used to weld from the upper side of the joint at a point where the weld axis is approximately horizontal and the weld face lies in an approximately horizontal plane.

Horizontal welding position (fillet weld): the welding position in which the weld is on the upper side of an approximately horizontal surface and against an approximately vertical surface.

Horizontal welding position (groove weld): the welding position in which the weld face lies in an approximately vertical plane and the weld axis at the point of welding is approximately horizontal.

Overhead welding position: the welding position in which welding is performed from the underside of the joint.

Vertical welding position: the welding position in which the weld axis at the point of welding is approximately vertical and the weld face lies in an approximately vertical plane.

A further subdivision related to vertical welding describes the technique in which "vertical up" or "vertical down" indicates the travel direction during welding.

A groove weld in a plate (Fig. 2.6) illustrates these definitions. Note that the different positions have been given letters and numbers. The G indicates groove weld, and the 1, 2, 3, or 4 indicated the particular position. The same numbers are used with the letter F indicating a fillet weld, shown in Fig. 2.7. In both fillet and groove welds, the axis of the weld can vary within specific limits,

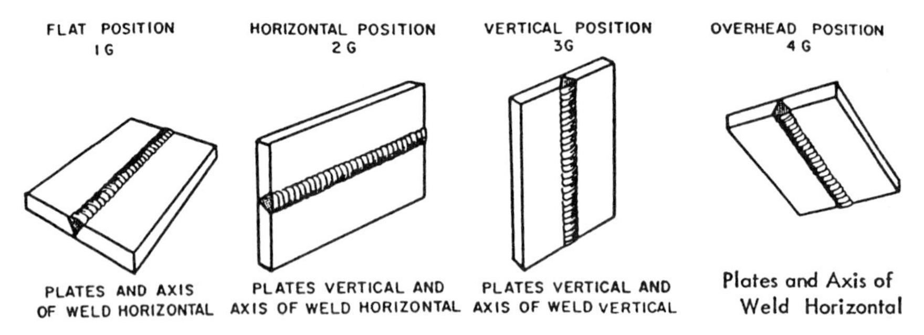

Figure 2.6 Welding position—singlegroove weld.

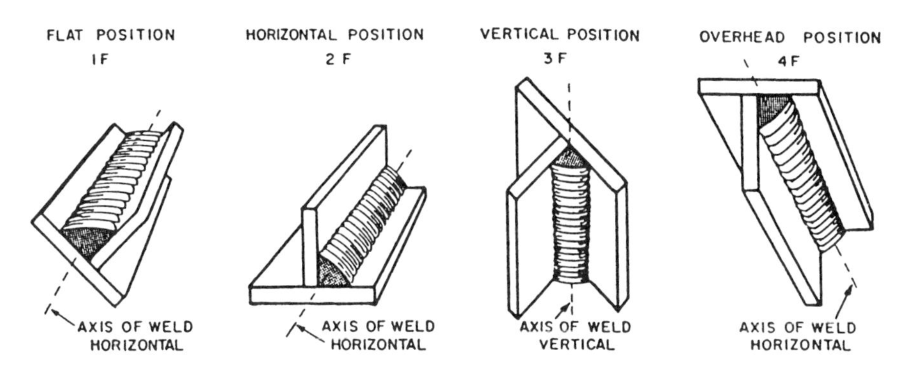

Figure 2.7 Welding position—fillet weld.

and the planes of the members being joined can also vary within limits. This is quite complex. For more information on this, see the AWS definitions book.

For pipe welding, the rules are slightly different (see Fig. 2.8). There are additional considerations for the horizontal welding position, here the axis of the pipe is vertical whereas the axis of the weld is horizontal. For roll welding, commonly known as flat or 1G welding, the weld is done at one particular position, normally the top of the joint, and the pipe is rotated while making the weld. This is known as the 1G position. However, for the same pipe position, that is, with the pipe axis horizontal, when the pipe is not rolled, it is known as the 5G position. This is extremely important with respect to orbital welding situations and double-ending welding systems. In position 6G the axis of the pipe is at an angle of approximately 45°, and the pipe is not rolled. This is used primarily for qualifications work.

Welding position is extremely important from several points of view. Much less manipulative skill is required in the flat position than in the vertical or overhead position. The flat position allows higher deposition rates, higher currents, and greater productivity. Most automated welding is performed in the flat position whenever possible. Welding positioners and manipulators are used to place the weld in the flat position to achieve higher productivity rates. In some applications, particularly for thinner metals and single-pass fillet welds, position is less important.

For pipe welding, a second system is used to indicate location of the weld while it is being made. Position is related to the face of a clock, with the twelve o'clock position at the top of the pipe, six o'clock at the bottom, and so forth.

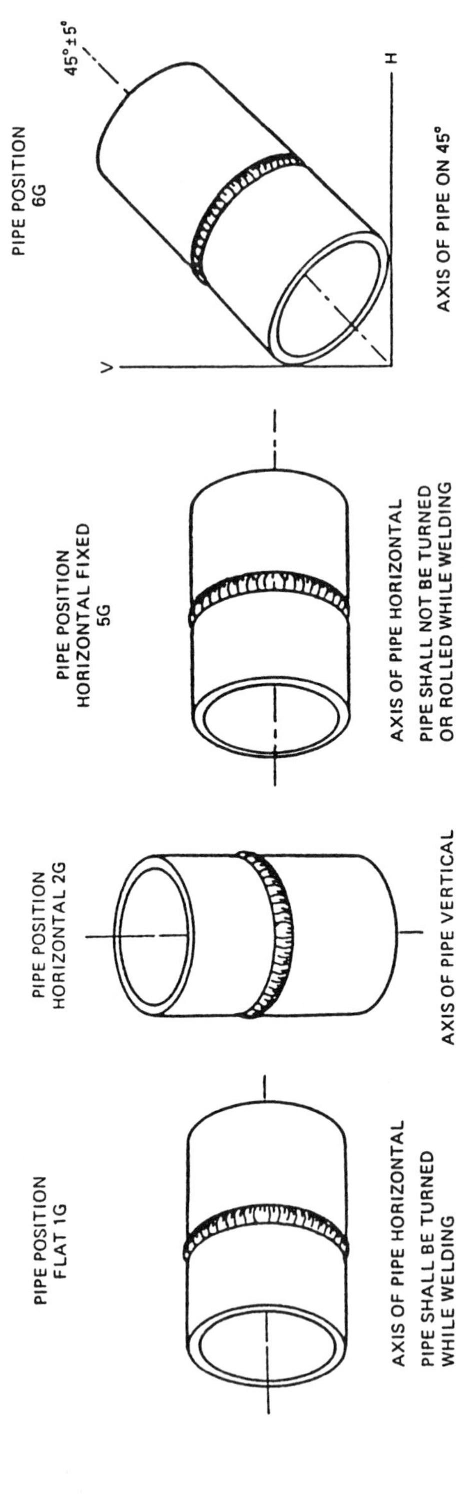

Figure 2.8 Welding position—pipe welds.

2.4 Welding Processes

There are over 100 different welding processes and process variations. They are defined by the American Welding Society and grouped as shown by Fig. 2.9, which is the AWS Master Chart of Welding and Allied Processes. AWS defines a welding process as "a distinctive progressive action or series of actions involved

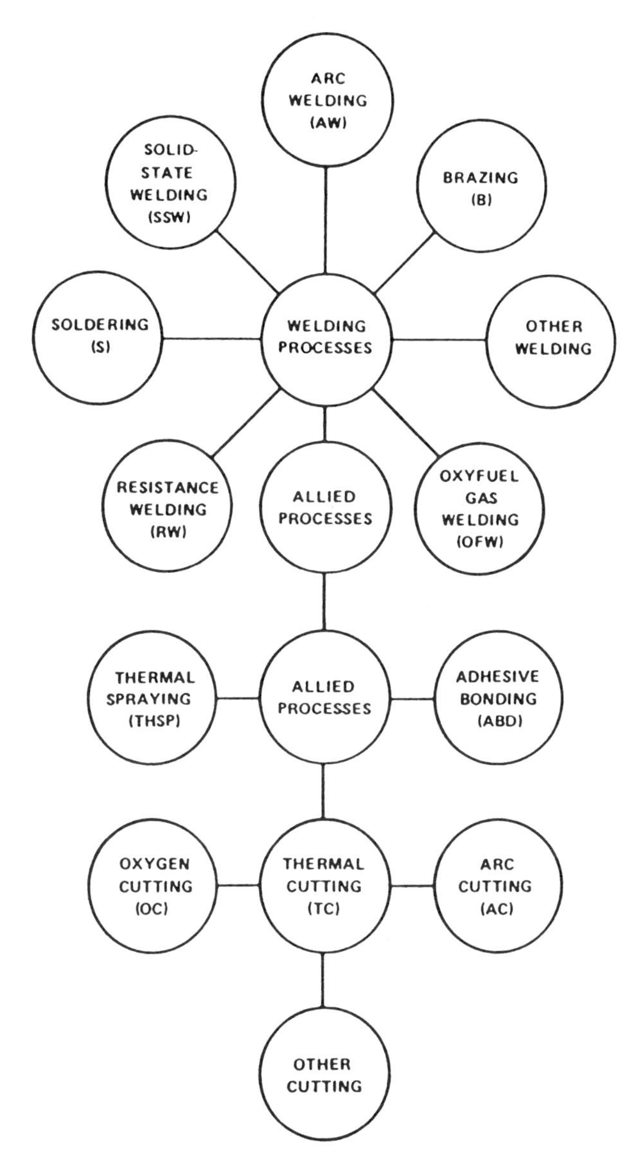

Figure 2.9 AWS Master Chart of Welding and Allied Processes.

in the course of producing a basic type of result." The AWS definitions describe the significant element of operation instead of significant metallurgical characteristics. A welding process is "a joining process which produces coalescence of materials by heating them to the welding temperature, with or without the application of pressure or by the application of pressure alone and with or without the use of filler metal." Processes are grouped according to the "mode of energy transfer" as the primary consideration. A secondary factor is the influence of capillary attraction in effecting distribution of filler metal in the joint. Capillary attraction distinguishes the welding processes grouped under brazing and soldering from the other processes. Allied processes include non-welding processes related to the occupation of welding.

In the master chart grouping, AWS has deliberately omitted designation of pressure or no pressure, because this factor is an element of operation of the specific welding process. The designation "fusion welding" is not recognized as a grouping because fusion is involved in many of the processes. Other facts such as types of current used and whether the electrode is consumable or nonconsumable or the current continuous or intermittent, are not included in process groupings. Neither is the method of application, although it is sometimes included in the process names in other countries.

Many of the processes have variations. The basic groupings and the more popular welding processes are shown by Table 2.1.

This book concentrates on arc welding processes. Each arc welding process is explained and related to the method of application. The remaining welding processes and thermal cutting and spraying processes are treated briefly to indicate that they can be applied by automatic methods.

All arc welding processes involve motion, either in making the weld or in moving from weld to weld. The change from manual to semiautomatic to mechanized welding and finally to automatic or automated welding is the transition from human control and motion to automatic control and motion. In arc welding there is relative motion between the arc and the material being welded. Not only does the arc travel relative to the work, but there is also oscillation and changes in the angle of the gun or torch with respect to the axis of the weld. Directing motion in welding also involves directing or pointing the electrode or torch to place the weld metal precisely where it is required and to control the weld profile. Under the control of a human welder all of these motions may be performed simultaneously. In mechanized welding, the controls are preadjusted and established before the weld is initiated. For automated welding these motions must be mechanized so they can be changed during the welding operation. This makes automated welding equipment very complex. The number of motions to be controlled depends on the degree of automation of the welding operation, and the complexity of the welding operation dictates the complexity of the automatic equipment.

Table 2.1 Popular Welding Processes by Basic Groups and Letter Designations

Group	Welding process	Letter designation
Arc welding	Carbon arc	CAW
	Electrogas	EGW
	Flux-cored arc	FCAW
	Gas metal arc	GMAW
	Gas tungsten arc	GTAW
	Plasma arc	PAW
	Shielded metal arc	SMAW
	Stud arc	SW
	Submerged arc	SAW
Brazing	Diffusion brazing	DFB
	Dip brazing	DB
	Furnace brazing	FB
	Induction brazing	IB
	Infrared brazing	IRB
	Resistance brazing	RB
	Torch brazing	TB
Oxyfuel gas welding	Oxyacetylene welding	OAW
	Oxyhydrogen welding	OHW
	Pressure gas welding	PGW
Resistance welding	Flash welding	FW
	Projection welding	RPW
	Resistance-seam welding	RSEW
	Resistance-spot welding	RSW
	Upset welding	UW
Solid-state welding	Cold welding	CW
	Diffusion welding	DFW
	Explosion welding	EXW
	Forge welding	FOW
	Friction welding	FRW
	Hot pressure welding	HPW
	Roll welding	ROW
	Ultrasonic welding	USW
Soldering	Dip soldering	DS
	Furnace soldering	FS
	Induction soldering	IS
	Infrared soldering	IRS
	Iron soldering	INS
	Resistance soldering	RS
	Torch soldering	TS
	Wave soldering	WS
Other welding processes	Electron beam	EBW
	Electroslag	ESW
	Flow	FLOW
	Induction	IW
	Laser beam	LBW
	Percussion	PEW
	Thermit	TW

2.5 The Welding Arc

The arc is the major component of an electric arc welding circuit. The relationship between welding current and arc voltage is shown by the volt–ampere characteristic curve of the welding arc (see Fig. 2.10). The curve is nonlinear and can be affected by many factors.

The power in the arc is the product of the welding current and the arc voltage. This power produces the heat necessary to melt metal to make a weld. The welding power source provides the current flow across the circuit. There are two types of welding arcs—one with a nonconsumable electrode and one with a consumable electrode. The nonconsumable electrode arc creates heat for melting the base metal and the filler metal, but the electrode itself does not melt. With a consumable electrode arc, the heat produced melts both the surface of the base metal and the consumable electrode metal, which then crosses the arc and becomes the deposited weld metal. This is described later.

The principle of the nonconsumable arc is shown in Fig. 2.11. Its most popular use is in the gas tungsten arc welding process (TIG). The arc occurs because electrons are emitted, or evaporated, from the surface of the cathode (negative pole) and flow across the gap, which is a region of hot electrically charged gas, to the anode (positive pole), where they are absorbed.

Arc voltage varies from less than 10 V to over 40 V in most welding applications. The arc length is proportional to the voltage across the arc; however, other factors affect the voltage, including the arc atmosphere, the metal of the

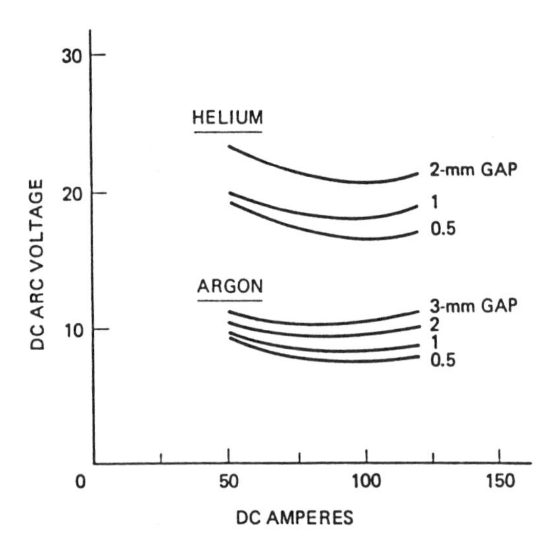

Figure 2.10 Volt–ampere characteristic curve—welding arc.

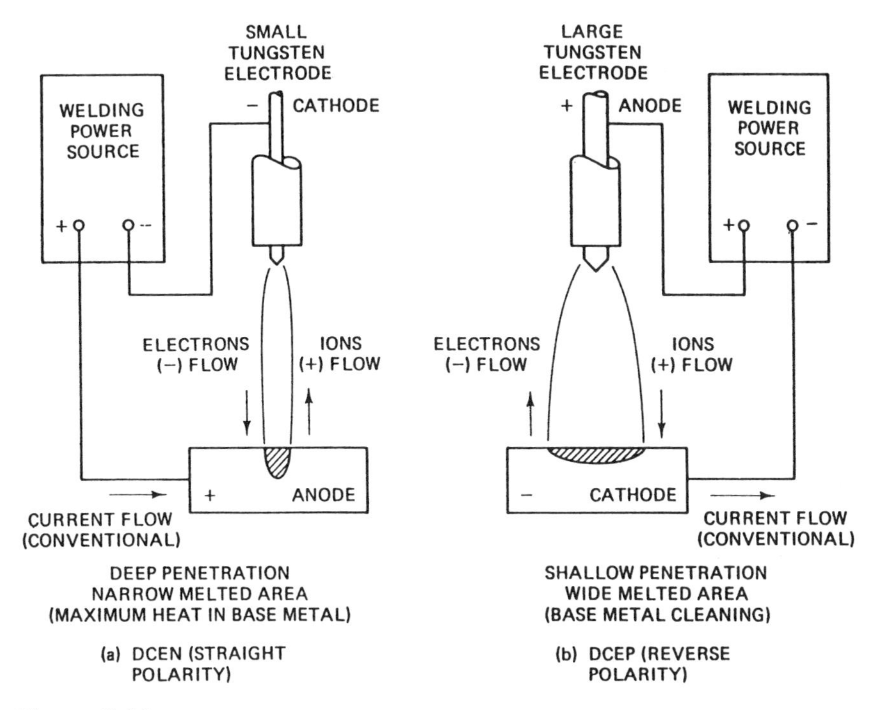

Figure 2.11 Nonconsumable welding arc circuit diagram.

poles of the arc, the size of the electrode, the polarity of the electrode, and the current density in the electrode. As arc length increases, voltage increases; however, the arc length must be maintained between certain limits for practical welding usability.

The arc column is normally round in cross section and is made of two concentric zones: an inner core or plasma and an outer flame. The plasma carries most of the current. The plasma of a high-current arc can reach a temperature of 5000–50,000°C. There are three distinct zones in the arc length, as shown by Fig. 2.12. The major drops in voltage across the arc are adjacent to the cathode and the anode.

Arcs operate under steady-state conditions. There are limits to the length of a useful welding arc, and there are limits to the size of the electrode. Both direct current and alternating current can be used for welding. The sinusoidal alternating current wave shape was originally used for welding. Recently, pulsing direct current and square-wave alternating current have become useful for welding. These give additional latitude to the adjustment of current levels in the welding arc and provide a method of controlling the power in the arc.

The ability of the electrode to emit electrons affects arc starting and its use with alternating current. Tungsten is an excellent emitter of electrons, while aluminum, particularly with an oxide coating, is a very poor emitter of electrons. This creates problems when welding aluminum with alternating current.

Heat produced by the arc can be accurately controlled by the use of a pulsed current system. This allows the average current to be changed yet still allows high current for penetration and low current for controlling the molten weld pool.

The consumable electrode welding arc (Fig. 2.12) forms between the tip end of the electrode and the workpiece, a point-to-plane configuration. The physics of consumable electrode welding is much more complex than that of the nonconsumable arc. Processes that use the consumable arc electrode are shielded metal arc welding (stick), gas metal arc welding (MIG), flux-cored arc welding, and submerged arc welding. These arcs, in general, use much higher currents than the gas tungsten arc. The consumable electrode welding arc is a steady-state condition maintained at the gap between the tip of the melting electrode and the molten pool of the workpiece. The electrode is continuously fed into the arc at

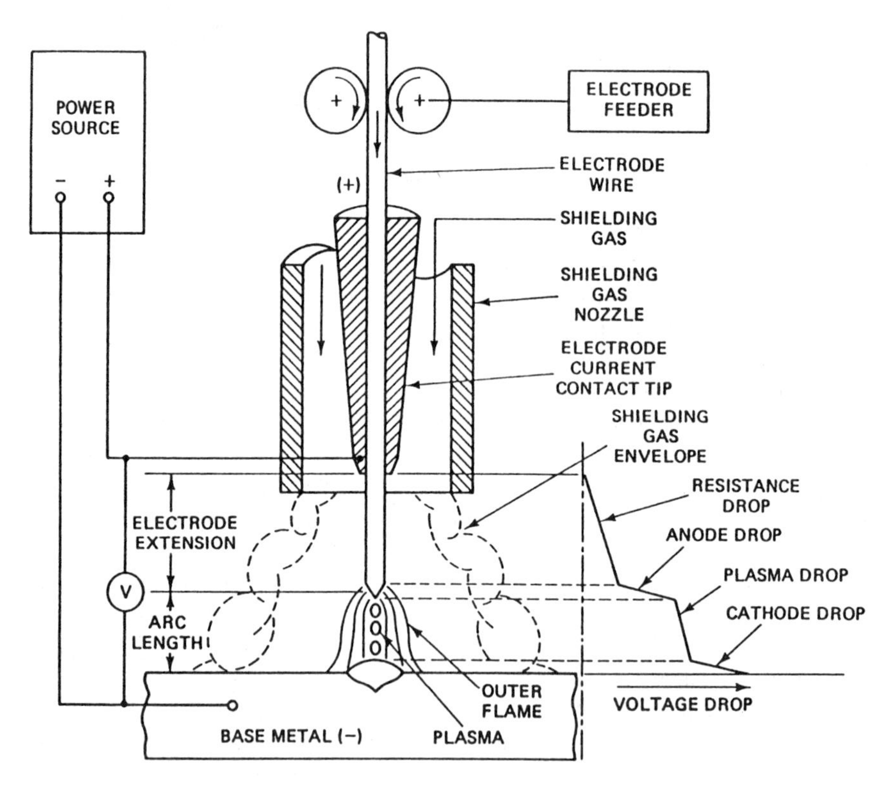

Figure 2.12 Consumable electrode welding arc circuit diagram.

the same rate at which it is melting, and thus a uniform arc length is maintained. The molten metal transfers across the arc. There are different modes of metal transfer, which affect the type of weld. The gas metal arc and flux-cored arc welding processes normally use direct current. In most applications the arc electrode is the positive terminal (anode).

References

1. *Standard Welding Terms and Definitions,* AWS 3.0-89, The American Welding Society, Miami, FL.

3

Arc Welding Processes Suitable for Automation

3.1 Arc Welding Processes

A special grouping of the arc welding processes is shown in Fig. 3.1. The top row of processes are those in which weld metal crosses the arc. These include shielded metal arc welding, gas metal arc welding (also known as MIG), flux-cored arc welding, and submerged arc welding (also known as sub-arc). The second row includes arc welding processes in which the weld metal does not cross the arc. The arc is used as a heat source to melt the surface of the base metal and the filler metal. These processes are gas tungsten arc welding (also known as TIG), carbon arc welding, and plasma arc welding. The two specialized arc processes in the bottom row are stud arc welding and electrogas welding. Electroslag welding, shown bottom right, is not an arc welding process but is included because it uses the same equipment.

Arc welding processes can be further categorized according to the *method of application,* which refers to the level or degree of mechanization and control. Several processes can be applied only manually, while others are semiautomatic,

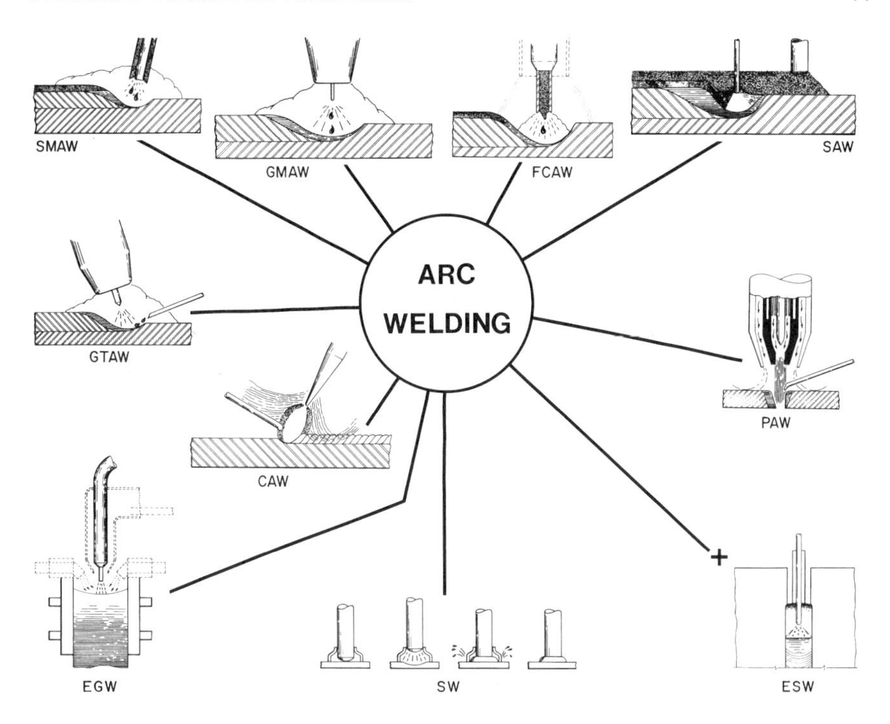

Figure 3.1 Grouping the arc welding processes.

mechanized, automatic, or robotic or use adaptive control. Figure 3.2 shows this relationship and the popular method of applying each process. In this chapter each arc welding process is briefly described and the degree of mechanization or automation possible is indicated.

The method of application impacts the cost of making a weld. It was pointed out earlier that welding occurs only while the arc is ignited. Hence, it is important to keep the arc operating with the highest possible duty cycle. It is also important to deposit metal as quickly as possible during the "arc on" period of time.

Each of the arc welding processes has a different range of deposition rates, with the rate varying according to the welding current employed and the electrode size and type. With a higher welding current, the electrode wire will melt off more rapidly and there will be more metal deposited per unit of time. There are limits to the maximum current that can be applied to each size and type of electrode. Usually larger electrodes can be used at higher current for greater deposition rates.

With a limited knowledge of welding, one might consider using the submerged arc process or the electroslag process for maximum productivity due to its high deposition rate. However, each arc welding process has specific

Arc Welding Elements/Function \ Method of Application	MA Manual (closed loop)	SA Semiautomatic (closed loop)	ME Mechanized (closed loop)	AU Automatic (open loop)	AD Adaptive Control (closed loop)	RO Robotic (open or closed loop)
Gas Metal Arc (GMAW)	Not Possible	Most Popular	Popular	Popular	Very Popular	Very Popular
Flux Cored Arc (FCAW)	Not Possible	Most Popular	Popular	Popular	Very Popular	Very Popular
Submerged Arc (SAW)	Not Possible	Little Used	Most Popular	Popular	Possible	Possible
Gas Tungsten Arc (GTAW)	Most Popular	Possible-Rare	Used	Popular	Used	Used
Plasma Arc (PAW)	Popular	Possible-Rare	Used	Used	Used	Used
Stud Welding (SW)	Not Possible	Used	Popular	Used	Not Popular	Used

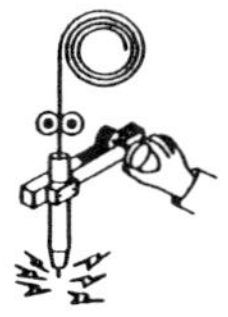

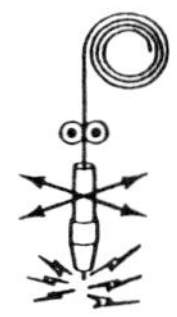

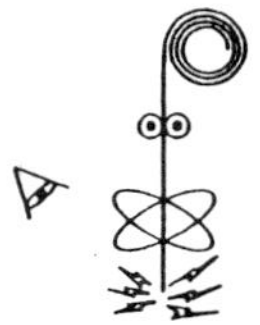

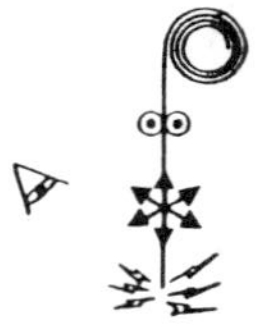

Figure 3.2 Possible methods of applying arc welding processes.

limitations as to position, metal thickness, etc. A particular process cannot be universally applied to all applications, and the submerged arc welding process cannot be used in the vertical or overhead position, nor can the electroslag process be used on thin materials. The gas tungsten arc, gas metal arc, and flux-cored arc welding processes can be used for any welding position; however, all processes cannot be used to weld on all metals. Hence, to obtain maximum productivity, an understanding of the arc welding processes is necessary in selecting the process for the particular job.

Another factor that relates to the cost of deposit weld metal is the utilization of the filler metal. When 100 lb of a particular filler metal is purchased, for example, the actual weight of metal deposited varies according to the process. The *utilization factor* is the ratio of the weight of filler metal deposited in the weld to the weight of filler metal purchased and is expressed as a percentage. For the covered electrode, the utilization factor is approximately 65%. The flux-cored electrode, with the inside fluxing, has a much higher utilization factor, approximately 85%; the core of material represents approximately 13% of the weight of the electrode, and the spatter may represent another 2–5%. For the solid gas metal arc electrode, utilization approaches 95%. For submerged arc welding and electroslag welding, utilization approaches 100% on the basis of the electrode alone; when flux is considered, the submerged arc process utilization falls to about 50%, and that of electroslag falls to approximately 90%. In the cold-wire processes, such as gas tungsten arc and plasma arc welding, utilization approaches 100%.

All of the arc welding processes can be mechanized, but it is doubtful that they all will be. Shielded metal arc welding (stick welding) is very popular but is a manual process. It can be mechanized by using gravity welding apparatus with inclined or spring-loaded electrode holders and extra long electrodes, but this application is not very common. Both electroslag and electrogas welding were developed as mechanized systems for vertical position welding. Both are used for special applications, and their use will never become widespread.

The carbon arc welding process was mechanized many years ago but has been replaced with shielded metal arc, gas metal arc, and flux-cored arc welding. The air carbon arc cutting and gouging processes have also been mechanized and have attained limited popularity as automatic systems.

3.2 Gas Metal Arc Welding

Gas metal arc welding (GMAW) uses an arc between a continuous filler metal (consumable) electrode and the weld pool. Shielding is provided by an externally supplied shielding gas. This process is also known as MIG welding, MAG welding, CO_2 welding, microwire welding, short arc welding, dip transfer welding, and wire welding.

Semiautomatic application of the gas metal arc welding process is shown in Fig. 3.3, and Fig. 3.4 is a diagram of the process. GMAW uses the heat of an arc between a continuously fed consumable electrode and the work to be welded. The heat melts the surface of the base metal and the end of the electrode. Metal that melts off the electrode is transferred across the arc to the work, where it solidifies and becomes the deposited weld metal. Externally supplied shielding gas surrounds the arc area and protects it from contamination from the atmosphere. The electrode is fed into the arc automatically from a coil. The arc length is maintained automatically, and travel can be accomplished manually or by machine.

Figure 3.3 GMAW—semiautomatic application.

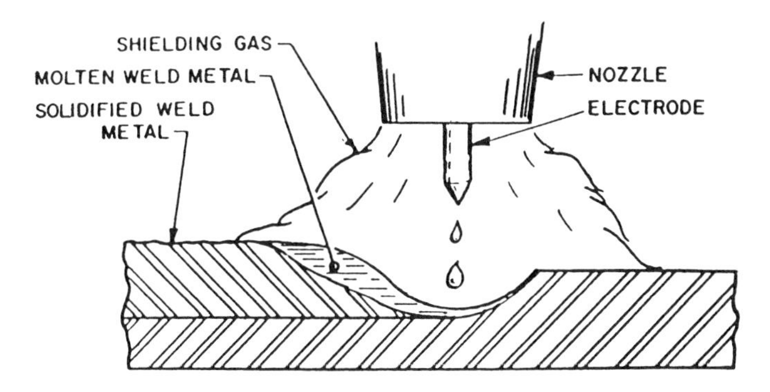

Figure 3.4 GMAW—process diagram.

The metal being welded dictates the composition of the electrode and the shielding gas. The shielding gas and the electrode wire diameter affect the type of metal transfer. The type of metal transfer is one way of classifying variations of the process.

Methods of application for gas metal arc welding include semiautomatic, mechanized, automatic, robotic, and adaptive control welding. Semiautomatic welding is the most popular. GMAW cannot be applied manually. It is the most popular method for automated systems because the electrode wire is continuous. As a continuous-wire process it has a high operator factor and high deposition efficiency, and its deposition rate is very similar to that of the shielded metal arc welding process. The arc is visible to the welder, and the skill level required for semiautomatic GMAW is less than that required for manual shielded metal arc welding.

GMAW is used by all industrial manufacturing operations. It is also used for field construction, including pipelines, and for maintenance and repair work.

Gas metal arc welding is an all-position welding process. However, its applicability depends on the diameter of the electrode wire and the particular variation.

Gas metal arc welding is used primarily to weld steels but can be used to weld most metals. Certain variations are restricted to specific metals; for example, carbon dioxide shielding is restricted to steels. Electrodes are matched to the base metals, and the shielding gas is selected according to the metal being welded. Metal thicknesses ranging from 0.010 in. (0.039 mm) upward can be welded. Thickness range and productivity depend on the variation being used. The inert gas version (MIG) is used for nonferrous metals. The spray arc and pulsed arc variations depend on gas type and on the power source design.

Gas metal arc welding can be used for the same joint designs as shielded metal arc or gas tungsten arc welding. Joints are detailed in Chapter 15. Because of the small diameter of the electrode wire, it is possible to get root penetration without wide root openings. The use of CO_2 shielding gas also gives deep penetration. The smaller diameter electrode wire also allows the use of a smaller groove angle, which improves the economics of the process.

The equipment used for GMAW is diagrammed in Fig. 3.5. Direct current is normally used, and the constant-voltage or flat-characteristic type of power source is recommended. The wire feeder should operate at a constant but adjustable speed. This provides a simplified control circuit, eliminates electrode burnback and stubbing, and is a self-regulating system. Welding voltage ranges from as low as 15 V to more than 40 V, and welding current can be as low as 20 A or as high as 500 A or more.

The welding torch for most automated welding applications is a straight machine type of torch. Robotic applications may use a gooseneck type. When gas mixtures containing CO_2 gas are used for shielding, the torches may be air-cooled.

For high-current gas MIG applications using inert gases, water-cooled torches are employed. The torch itself should be sized to the welding current. The power source for automated welding should always be 100% duty cycle. Transformer rectifier or solid-state power sources are preferred.

There are two materials consumed in making GMAW welds: the electrode wire and the shielding gas. Both must be carefully selected with regard to the base metal to be welded and the process variation. Electrode wires are identified by specifications established by the American Welding Society. Information concerning these materials is given in Chapter 16.

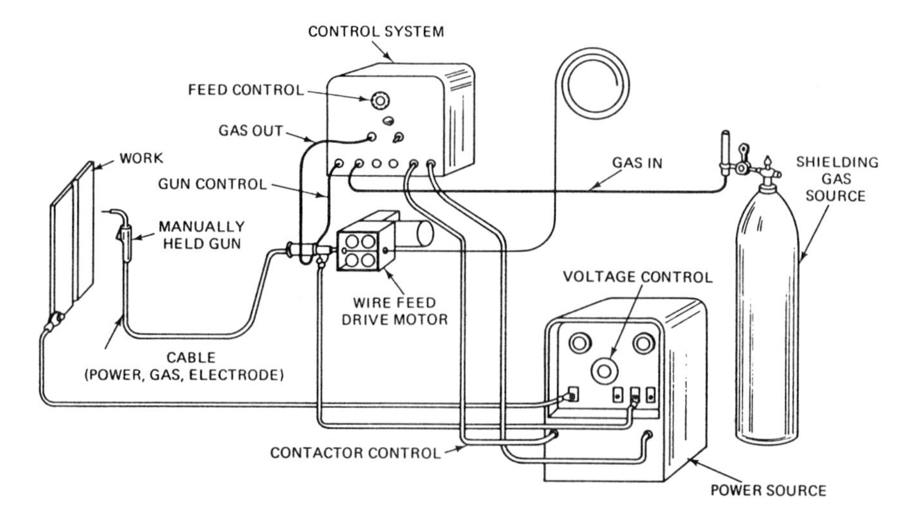

Figure 3.5 GMAW—equipment diagram.

There are few limitations to the GMAW process. Probably the most serious relates to problems of wind and drafts affecting the efficiency of the shielding gas envelope around the arc area. Adequate shielding of the welding operation will normally correct this situation. Another potential limitation is welding over oily and dirty surfaces. As with any arc welding process, the surface to be welded should be clean and dry.

3.3 Flux-Cored Arc Welding

Flux-cored arc welding (FCAW) is an electric arc welding process that uses an arc between a continuously fed flux-filled electrode and the weld pool. Shielding is obtained through decomposition of the flux within the tubular wire. Additional shielding may or may not be obtained from an externally supplied gas. The weld metal crosses the arc. The process is also known as Fabco, Dual-Shield, Innershield, and Fabshield Welding. Semiautomatic application of flux-cored arc welding is shown in Fig. 3.6.

The FCAW process is capable of welding carbon and alloy steels, cast and wrought iron, and some stainless steels. The process is also capable of producing hard-surfacing deposits. It is most commonly used to weld medium to thick steels because of the high deposition rates obtained with the larger electrode diameters. Metal melted off the electrode is transferred across the arc, where it solidifies and becomes the deposited weld metal.

The position in which welding can be carried out depends on the size of the electrode. Larger diameter wires are used in the flat and horizontal positions. Smaller diameter wires are used in the vertical and overhead positions. A layer of slag is often produced on the weld bead and must be removed after welding. Metal-cored electrode wires produce little or no slag.

The flux-cored arc welding process is usually applied semiautomatically. The equipment maintains the arc and feeds filler metal to the joint while the welder provides travel and guidance along the joint. The welder starts and stops the arc by using the trigger on the welding gun. Flux-cored arc welding is illustrated in Fig. 3.7.

The process can also be applied by mechanized, automatic, or robotic methods. In mechanized welding, the equipment controls all of the welding operations except for joint guidance. In robot or automatic adaptive control welding, the equipment also controls joint guidance. The welding operator monitors the system for these methods of application. The skill level needed for semiautomatic flux-cored arc welding is similar to that required for gas metal arc welding. The equipment for semiautomatic flux-cored arc welding (Fig. 3.8) consists of a power source, wire feeder, welding gun, and welding cables. A shielding gas system is added to the externally shielded variation. The power

Figure 3.6 FCAW—semiautomatic application.

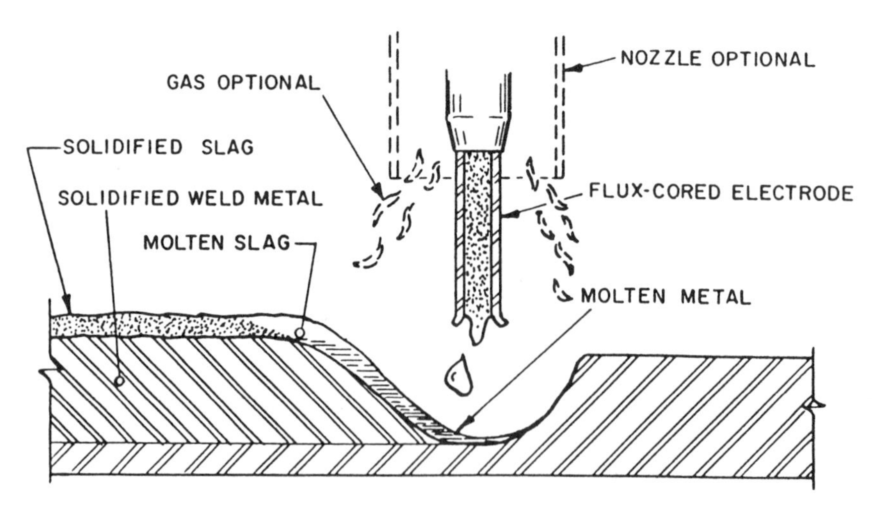

Figure 3.7 FCAW—process diagram.

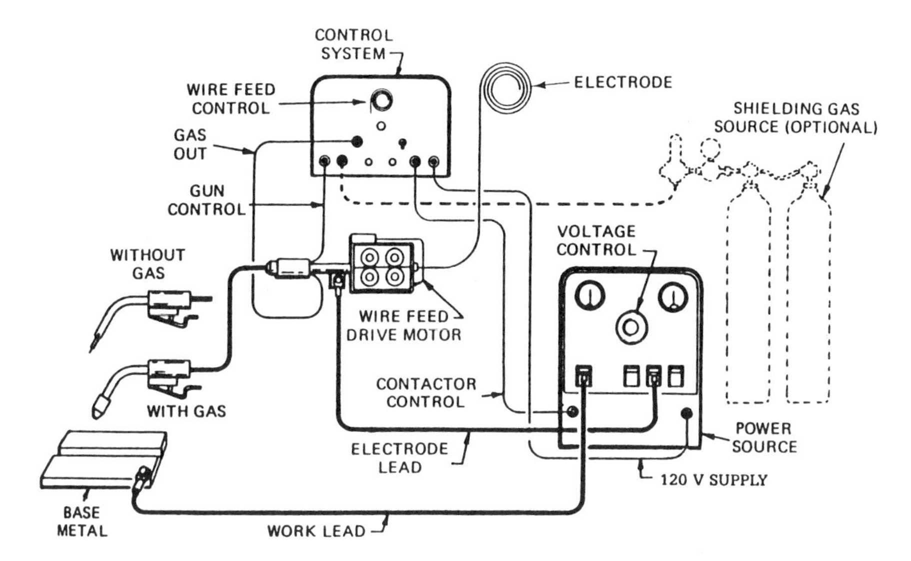

Figure 3.8 FCAW—equipment diagram.

source and wire feeder are basically the same as those used for gas metal arc welding. The materials required are the consumable flux-cored electrode wire and the shielding gas. This process is similar to gas metal arc welding, but the electrode wire is different.

3.4 Submerged Arc Welding

Submerged arc welding (SAW) is an arc welding process that fuses together the parts to be welded by heating them with one or more electric arcs between one or more bare electrodes and the work. The arc is shielded by a blanket of granular flux on the work. The filler metal is obtained by melting the solid electrode wire and sometimes by alloying elements in the flux. The weld metal crosses the arc. This process is also known as *welding under powder.* It is illustrated in Fig. 3.9.

The submerged arc welding process is capable of welding carbon steels, low-alloy steels, quenched and tempered steels, nickel and nickel alloys, copper and copper alloys, and some stainless steels. The process provides high deposition rates, which make it excellent for medium-thick and thick sections. It produces deep penetration, which means that less edge preparation is required to obtain penetration. Steels up to ½ in. (12.7 mm) thick can be welded without edge preparation.

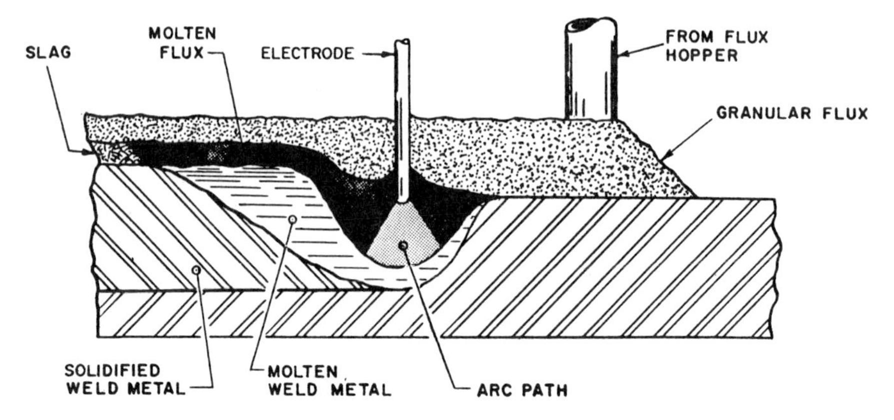

Figure 3.9 SAW—process diagram.

The process is limited to the flat and horizontal fillet positions because of the flux used to shield the weld puddle. However, with special flux dams, it can be used in the horizontal groove weld position.

Since the arc is hidden, only safety glasses are required by the welding operator. The process produces a smooth weld bead with no spatter. A layer of slag is produced on the weld bead and must be removed.

The submerged arc welding process is normally applied by the mechanized or automatic method. The welding operator must monitor the process during welding. Semiautomatic welding is possible but not widely used. Manual welding is not possible, and adaptive control and robotic welding are rarely used. Mechanized submerged arc welding is shown in Fig. 3.10.

Manual welding skills are not required. However, a technical understanding of the equipment and welding procedures is necessary to use this process successfully.

The equipment components required for automatic submerged arc welding are shown in Fig. 3.11. These are the welding machine (power source), the wire feeding mechanism and control, the welding torch, the flux hopper and flux feeding mechanism, and the travel mechanism and control system. A flux recovery system is usually included in an automatic installation.

The welding machine or power source for submerged arc welding can be either an AC or DC power source. For DC submerged arc welding, a constant-voltage (CV) or constant-current (CC) power source can be used. The CV type is more common for small-diameter electrode wires, usually 1⁄8 in. (3.2 mm) or less in diameter. The CC type is more commonly used for electrode wires of larger diameter, usually 5⁄32 in. (4.0 mm) or more. The wire feeder must be matched to the type of power source used. When alternating current is employed, the machine is a constant-current type.

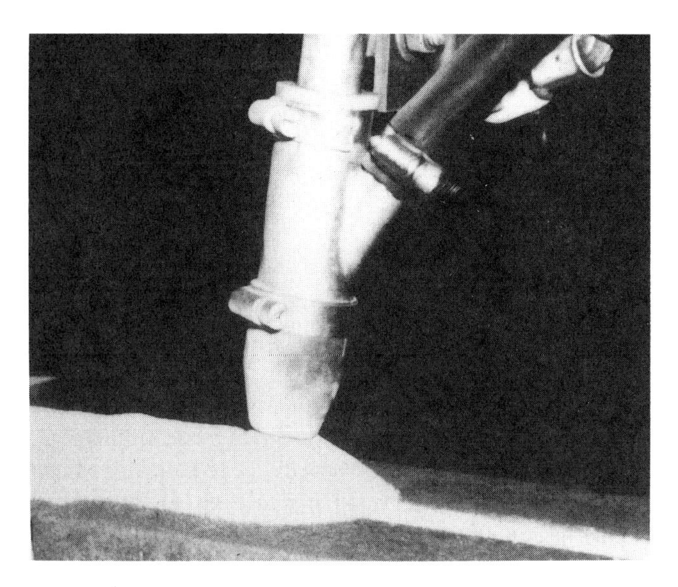

Figure 3.10 SAW—mechanized application.

The materials required are the consumable electrode wire and the granulated flux. These must be matched to the metal being welded.

The submerged arc process is a consumable electrode welding process that uses a granular mineral flux to cover the arc and liquid metal. The flux protects the arc from environmental conditions, visually obscures the arc, and performs

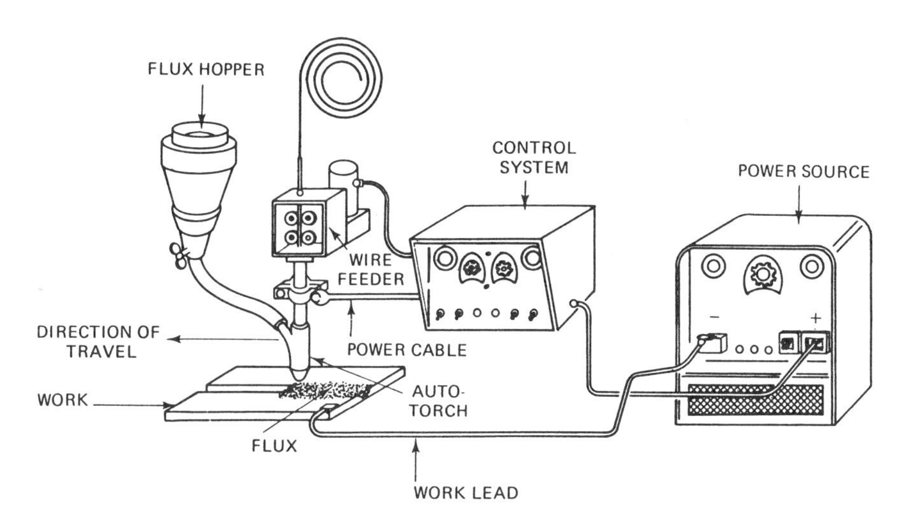

Figure 3.11 SAW—equipment diagram.

certain metallurgical functions. The flux is usually fed to the arc zone by gravity from a hopper. The rate of flow must be coordinated with the arc travel speed to ensure a proper, uniform flux burden over the weld pool.

The fluxes are formulated to react with the liquid metal, providing such reactions as deoxidation, alloying, and bead shape control. The flux also contains constituents to help in initiating and stabilizing the arc.

The process control unit performs such functions as initiating primary power, starting the arc, regulating filler metal feed, starting and controlling arc travel, and controlling the flux feeding.

Postweld cleaning is performed by removing the fused slag and vacuuming off the unfused flux.

The submerged arc process finds wide use in heavy fabrication industries, such as shipbuilding, and the manufacture of pressure vessels, line pipe, and earth-moving equipment. Its attractiveness lies in its economics, as the process is commonly used at high currents and can use up to three electrodes. Under these conditions, weld metal deposition rates of 30–100 lb/hr can be achieved.

The submerged arc welding process will never become popular for automated or robotic systems because of its use of granular flux. The flux is difficult to feed and produces a heavy slag coating that may be difficult to remove. The unused flux becomes a housekeeping problem on the floor and in fixtures. In addition, the flux is very abrasive and contributes to rapid wear of tooling.

3.5 Gas Tungsten Arc Welding

Gas tungsten arc welding (GTAW), also known as TIG, is an arc welding process that fuses together the parts to be welded by heating them with an arc between a nonconsumable tungsten electrode and the work. Filler metal may or may not be used with the process. Shielding is provided by an inert gas or inert gas mixture. The weld metal does not cross the arc. The process is also known as Heliarc, Heliweld, and WIG welding. It is illustrated in Fig. 3.12.

The GTAW process can be used to weld steel, stainless steel, aluminum, magnesium, copper, nickel, titanium, and other metals and can be used on a wide range of metal thicknesses. However, due to the relatively low production rates, thinner materials are most often welded. It is also popular for welding pipe and tubing.

This process can be used in all positions to produce high-quality welds. Since a shielding gas is used, the weld is clearly visible, no spatter is produced, postweld cleaning is reduced, and slag is not trapped in the weld.

The gas tungsten arc welding process is widely applied using the manual method, shown in Fig. 3.13. The welder controls the torch with one hand and

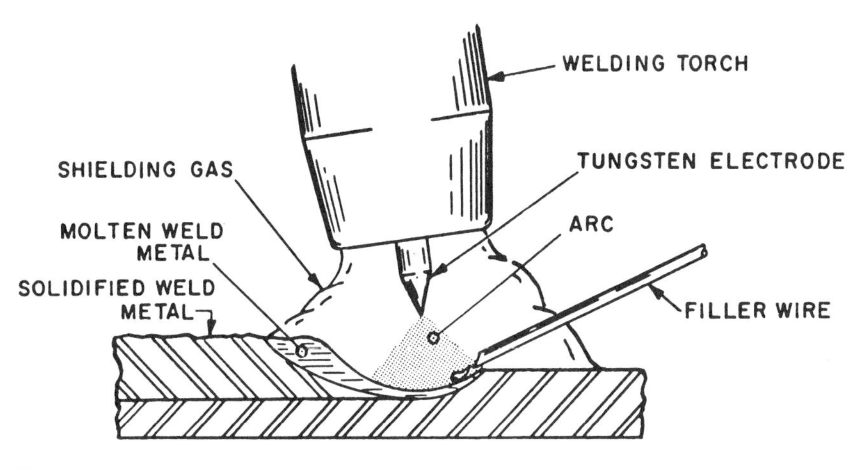

Figure 3.12 GTAW—process diagram.

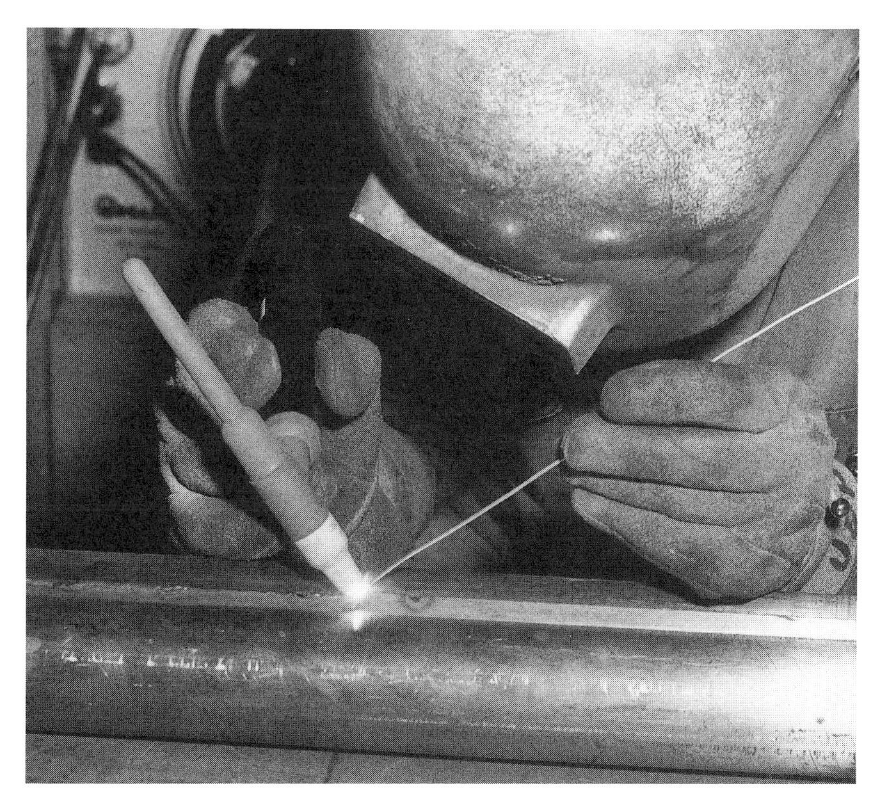

Figure 3.13 GTAW—manual welding.

feeds filler metal with the other. A relatively high degree of manual skill is required. The semiautomatic method is rarely used. The mechanized, automatic, and robotic methods are increasing in popularity. The filler metal is fed into the weld pool by a wire feeder with special controls. GTAW is ideally suited to automated welding.

The major components required for gas tungsten arc welding are shown in Fig. 3.14. These are the welding machine or power source, the gas tungsten arc welding torch, including the tungsten electrode, the shielding gas and controls, and the filler wire and feeder. There are several optional accessories available, including a foot rheostat that permits the welder to control current while welding. There are also arc timers and programmers, high-frequency units for arc starting, water-circulating systems, and other specialized devices.

A specially designed welding machine or power source is used for gas tungsten arc welding. The power sources have steep drooping characteristic curves. Both alternating current and direct current are used. A transformer or transformer/rectifier type of power source is normally employed. Special power sources with specific characteristics are widely used. The power source usually contains a high-frequency generator that is used to aid arc starting when welding with direct current and is used continuously when welding with alternating current. Program controllers are also included. Pulsing power sources are increasingly being used.

The materials required for gas tungsten arc welding are the consumable filler wire or rod and the shielding gas. The tungsten electrode is considered nonconsumable in use but must be replaced often.

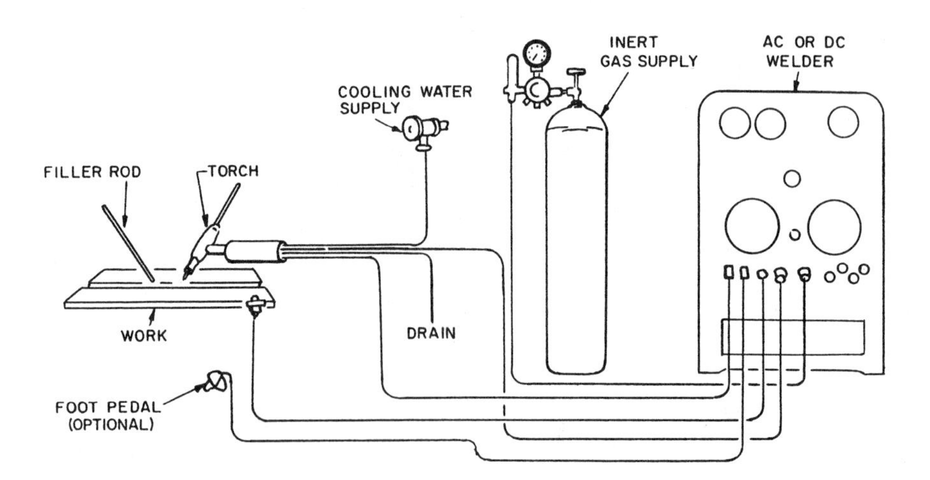

Figure 3.14 GTAW—equipment diagram.

3.6 Plasma Arc Welding

Plasma arc welding (PAW) is an electric arc welding process that fuses together the parts to be welded by heating them with a constricted arc between a nonconsumable tungsten electrode and the work (transferred arc). Shielding gas is obtained from the hot ionized gas issuing from the orifice. Auxiliary inert shielding gas or a mixture of inert gases is normally used. The weld metal does not cross the arc. The process is illustrated by Fig. 3.15.

Like gas tungsten arc welding, the plasma arc welding process can be used to weld most commercial metals, and it can be used for a wide variety of metal thicknesses. On thin material, from foil to ⅛ in. (3.2 mm), the process is attractive because of the low heat input. The process provides relatively constant heat input because arc length variations are not very critical. On material thicknesses greater than ⅛ in. (3.2 mm), a keyhole technique is often used to produce full-penetration single-pass welds. In the keyhole technique, the plasma completely penetrates the workpiece. The molten weld metal flows to the rear of the keyhole and solidifies as the torch moves on. The welds produced are characterized by deep, narrow penetration and a small weld face.

The PAW process is not easily deflected and tolerates greater variations in joint alignment than the gas tungsten arc welding process. Also, no tungsten spitting occurs with the process. Figure 3.16 shows a welder manually applying a plasma arc weld.

Another similarity to gas tungsten arc welding is that the plasma arc welding process is widely applied manually. The mechanized, robotic, or automatic adaptive control methods can also be used to increase welding speeds. This

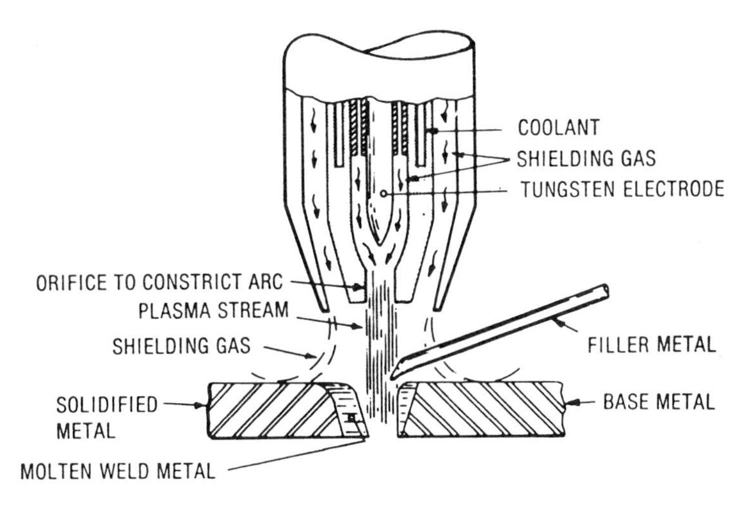

Figure 3.15 PAW—process diagram.

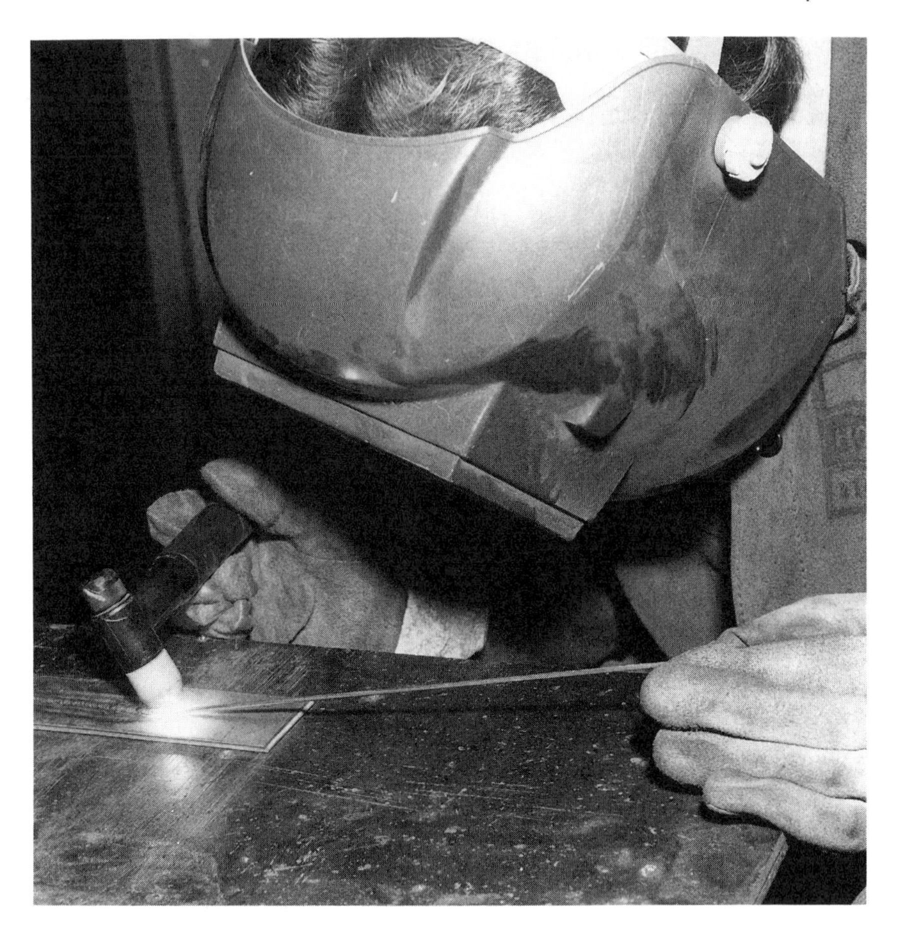

Figure 3.16 PAW—manual welding.

process requires a degree of welding skill similar to that required by gas tungsten arc welding. However, the equipment is more complex and requires a high degree of knowledge to set up and use.

The major equipment required for plasma arc welding (Fig. 3.17) includes a welding machine or power source, a special plasma arc control system, the motion system and control, the plasma welding torch (water cooled), the source of plasma and shielding gas, and filler material when required.

The power source for plasma arc welding is the constant-current type with a drooping output characteristic. A GTAW power source may be used for plasma welding, because it includes a contactor, remote current control, and provisions for shielding gas and cooling water. Pulsing power sources are increasingly being

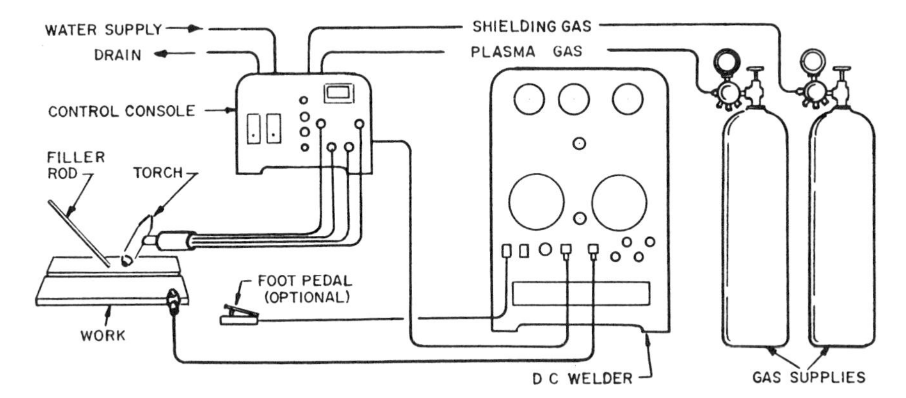

Figure 3.17 PAW—equipment diagram.

used. Automated operations normally employ programmed current control including upslope and downslope, and pulsing is normally used.

The materials required for plasma arc welding are the filler metal or rod and the shielding gas. The tungsten electrode, while considered nonconsumable, must occasionally be replaced, because it deteriorates in use.

3.7 Stud Welding

Arc stud welding (SW) is a special-purpose arc welding process that is used to attach specially designed studs to the work. Partial shielding is provided by a ceramic ferrule surrounding the stud. Arc stud welding is a machine welding process that uses a specialized gun to hold the stud and produce the weld. The process is shown by Fig. 3.18.

The stud gun holds the stud in contact with the workpiece (step 1) until the welding operator depresses the trigger, causing the welding current to flow from

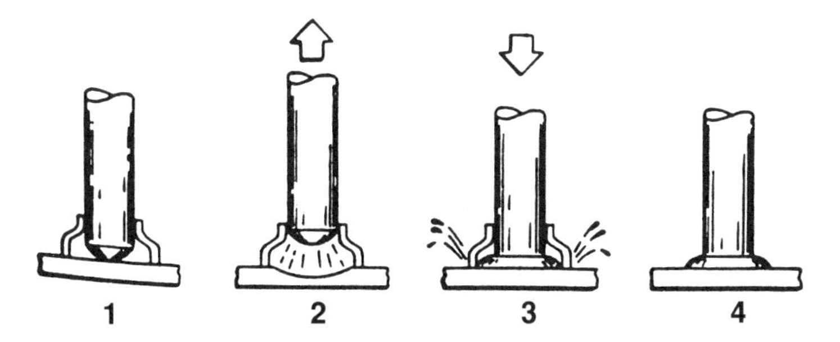

Figure 3.18 Stud welding—process diagram.

the power source through the stud (which acts as an electrode) to the work surface. The welding current activates a solenoid within the gun, which draws the stud away from the work surface (step 2) and establishes an arc. The intense arc heat melts both the work surface and the end of the stud at the same time. Arc duration is controlled by a timing device built into the control unit. When welding current is shut off, the gun solenoid releases its pull on the stud and a spring-loaded action pushes the stud down into the molten pool on the workpiece (step 3). The molten stud end solidifies with the molten pool on the work surface, and the stud weld is completed (step 4).

Stud welding applied semiautomatically (Fig. 3.19) is widely used for attaching studs and other similar devices to plates or structural members. Studs are normally thought of as threaded round fasteners. However, rectangular shapes, hooks, pins, brackets, and other configurations can also be stud welded. Steel studs range in diameter from ⅛ in. (3.2 mm) to 1 in. (25.4 mm), vary in length, and can be threaded or plain. The arcing end may contain a small amount of welding flux or some other means of shielding the arc area. The flux protects the weld and the arc from atmospheric contamination and contains scavengers that purify the melted metal. Stud welding is normally used for steels and stainless steels, but variations of the process can be used on nonferrous metals.

The equipment for stud welding is diagrammed in Fig. 3.20. It includes the welding machine or power source, the stud gun, the control unit, studs, and disposable ferrules.

Figure 3.19 Making a stud weld.

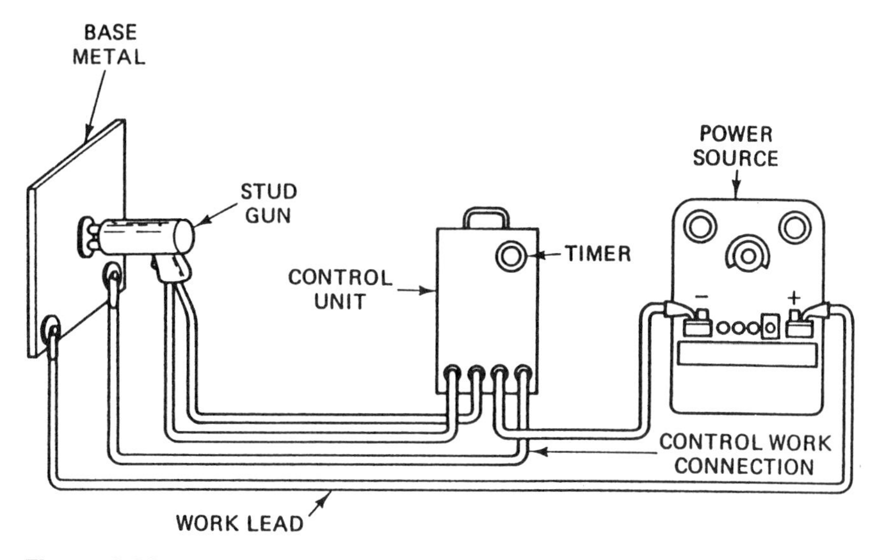

Figure 3.20 Stud welding—equipment diagram.

The welding machine or power source supplies direct current. A constant-current or drooping characteristic source is normally used. The welding current is dictated by the size or diameter of the stud. The electrode, or the stud, is the negative pole (straight polarity). Amperage required ranges from 200 to 500 A.

The gun holds the stud and has a switch that starts the control sequence. It also includes the solenoid, which provides the withdrawal or lift action to establish the arc. A spring mechanism within the gun applies the pressure required to plunge or push the stud into the molten pool on the workpiece. The gun should be adjusted to accommodate the size of the stud being used and to provide the correct arc length during the arc period. The control unit consists of a welding current contactor, a timing device, and the necessary interconnections. The welding current passes through the stud gun to provide power for the solenoid.

A ferrule is used with each stud. Ferrules are made of a ceramic material and are broken off and discarded after each weld is made. The ferrule concentrates the heat during welding and confines the molten metal to the weld area. It helps prevent oxidation of the molten metal during the arcing cycle, but it must be made to fit the studs being used. There is no specification for studs or ferrules.They are provided by the manufacturer of the stud gun.

There are two variations that are easy to automate. These are the capacitor discharge stud welding method and the drawn arc capacitor discharge stud welding method. The studs are slightly different. The capacitor discharge stud has a small tip at the end. In operation this tip is brought into contact with the base metal; then pressure is applied by the gun. The stored energy is discharged through the small projection, which rapidly disintegrates. This creates an arc that heats the surface of the base metal and the arcing end of the stud. The weld is completed in a very short period of time. It is done so quickly that heat buildup is minimal. It can be used for welding studs on thin sheet metal.

In the drawn arc capacitor discharge welding variation, the arc is initiated in the same manner as in arc stud welding. The stud is lifted from the base metal, the arc is established, and in a short time the stud is plunged into the molten pool and the weld is completed. Shielding gas is sometimes used with this variation, especially on aluminum. Special guns and controls are required.

For automation, the studs are sorted, positioned, and fed into a tube from an automatic feeder to the point of application. This feeder is shown in Fig. 3.21 welding studs on a rotary table. The automobile industry has used this method for attaching trim strips to the sides of automobile bodies. Figure 3.22 shows a robot

Figure 3.21 Automated stud welding.

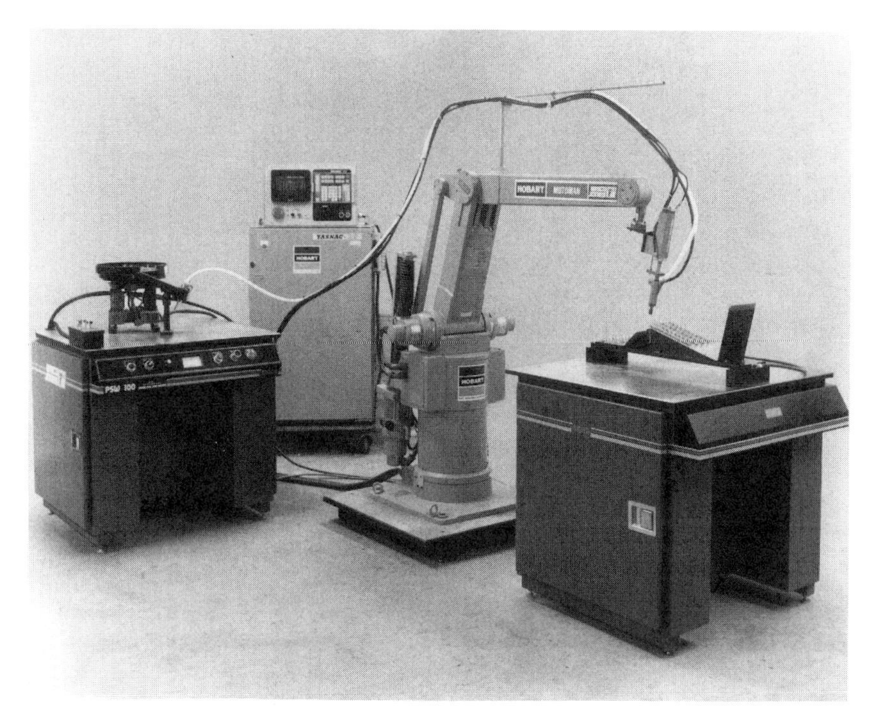

Figure 3.22 Robotic stud welding.

performing a stud welding operation. The robot controller controls the stud feeding apparatus as well as the stud welding locations.

3.8 Other Arc Welding Processes

Shielded metal arc welding is an electric arc welding process in which the heat for welding is generated by an electric arc between a covered metal electrode and the work. The filler metal is deposited from the electrode, and the electrode covering provides the shielding. The weld metal crosses the arc. The process is also known as *stick welding* or *stick electrode welding*. It is shown in Fig. 3.23.

Shielded metal arc welding is mentioned here because it is one of the oldest, most popular, simplest, and most versatile of the arc welding processes. It can be used to weld both ferrous and nonferrous metals and for metal thicknesses greater than about 18 gauge (1.2 mm) in all positions. The shielded metal arc welding process is manually applied. It requires a relatively high degree of welder skill. The arc is under the control of the welder and is visible.

This process leaves slag on the surface of the weld bead that must be

Figure 3.23 Shielded metal arc welding—stick.

removed. The equipment is extremely simple and rugged. Shielded metal arc welding is not suitable for automated welding because the arc is extinguished when a "stick" of electrode is consumed.

Carbon arc welding is the oldest arc welding process. It is an electric arc welding process that fuses metal parts together by heating them with an arc between a carbon electrode and the work. No shielding gas or flux is used. A variation of the process uses two carbon electrodes, and the arc forms between them. The carbon arc welding process has limited industrial use. It is not suitable for automated welding.

Electrogas welding is a mechanized arc welding system for welding medium-thick steel in the vertical position. It will not become very popular due to its limited application.

Electroslag welding is not an arc welding process but uses equipment similar to that used in the continuous electrode wire process. It is used for welding thick sections of steel in the vertical position. It is very specialized and will not become popular.

3.9 Process Selection

The criteria for selecting an arc welding process for any of the mechanized or automatic arc welding methods of application are not as complex as they may seem. The selected process must satisfy the following requirements.

1. It must use a continuous electrode or filler metal.
2. It must have the ability to weld the metals involved.
3. Its use must increase productivity over that of the manual or semiautomatic method currently in use.

The processes most widely used for mechanized welding include gas metal arc welding, flux-cored arc welding, submerged arc welding, gas tungsten arc welding, plasma arc welding, and stud arc welding. With the exception of stud welding, all of these processes allow the use of an electrode wire continuously fed from a reel or a coil. If studs are available in magazine form, stud welding or any other method that allows mechanized feeding of studs can be employed.

The metals to be welded must be weldable by a process that can be automated. The gas metal arc can be used for welding many of the common metals. Flux-cored arc welding is used primarily on the steels. Submerged arc welding is used primarily on steels, but is also used on nickel-based alloys. Gas tungsten and plasma arc welding can be used for most weldable metals. Table 3.1 indicates the use of the various welding processes for the most commonly welded metals.

As for the position capabilities of the process, all of the processes mentioned above, with the exception of submerged arc welding, can be used in all positions. The use of stud welding depends on the variation and the size of the studs to be welded. Welding position information is summarized in Fig. 3.24.

Finally, consider the productivity of the process. This is summarized by Fig. 3.25, which shows the deposition rates for the various processes. Note that in gas metal arc welding, the deposition rate depends on such welding parameters as the electrode wire size and the type of metal transfer. In submerged arc

Table 3.1 Summary of Metals Welded by Various Processes[a]

	Welding process				
Base metal	GMAW	FCAW	SAW	GTAW	PAW
Aluminum and aluminum alloys	A	No	No	A	A
Copper-base alloys					
Brasses	C	No	No	C	C
Bronzes	A	No	No	A	B
Copper alloy	A	No	No	A	A
Copper-nickel alloy	A	No	No	A	A
Irons					
Cast, malleable, nodular	B	B	No	B	B
Wrought	A	A	A	B	B
Lead	No	No	No	B	B
Magnesium alloys	A	No	No	A	B
Nickel-base alloys					
Inconel	A	No	No	A	A
Nickel alloy	A	No	C	A	A
Nickel-silver alloy	C	No	No	C	C
Monel	A	No	C	A	A
Precious metals	No	No	No	A	A
Steels					
Low-carbon steel	A	A	A	A	A
Low-alloy steel	A	A	A	A	A
High- and medium-carbon steel	A	A	B	A	A
Alloy steel	A	A	B	A	A
Stainless steel	A	B	A	A	A
Tool steels	C	No	No	A	A
Titanium	A	No	No	A	A
Tungsten	No	No	No	B	A
Zinc	No	No	No	C	C

[a]Metal or process rating: A = Recommended or easily weldable; B = acceptable but not best selection or weldable with precautions; C = possible—usable but not popular or restricted use or difficult to weld; No = not recommended or not weldable.

welding, the chart shows the rate for the usual single electrode and also for two electrodes, which are used in some specialized SMAW applications.

In general, the final decision lies between gas metal arc, flux-cored arc, gas tungsten arc, and plasma arc welding. The continuous electrode wire processes have the greatest productivity. However, the cold-wire processes can be used to weld almost all metals. By reviewing the factors mentioned above, it will soon become apparent which process should be used.

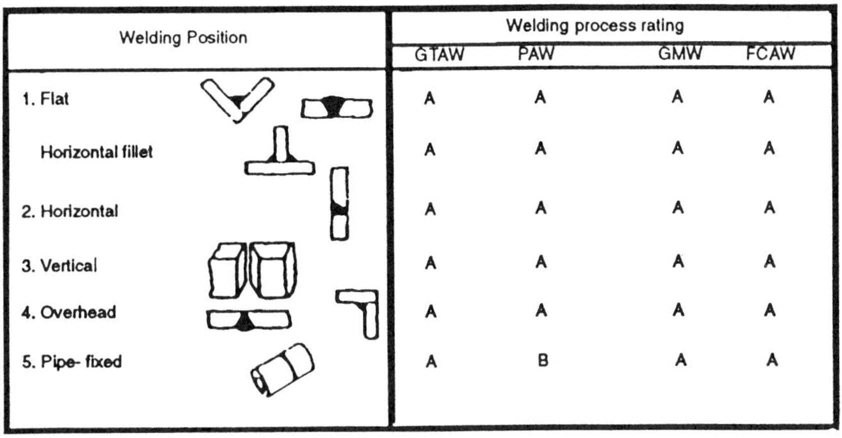

Welding Position	Welding process rating			
	GTAW	PAW	GMW	FCAW
1. Flat	A	A	A	A
Horizontal fillet	A	A	A	A
2. Horizontal	A	A	A	A
3. Vertical	A	A	A	A
4. Overhead	A	A	A	A
5. Pipe- fixed	A	B	A	A

Figure 3.24 Position capabilities.

This chapter has not made mention of electron beam welding or laser welding. Laser welding is becoming more popular due to the development of more efficient machines, better understanding of the process, and broader applications. There are certain applications where laser welding could be the method of choice. There are also some applications for electron beam welding. The interested reader should review the above information and refer also to Chapter 4.

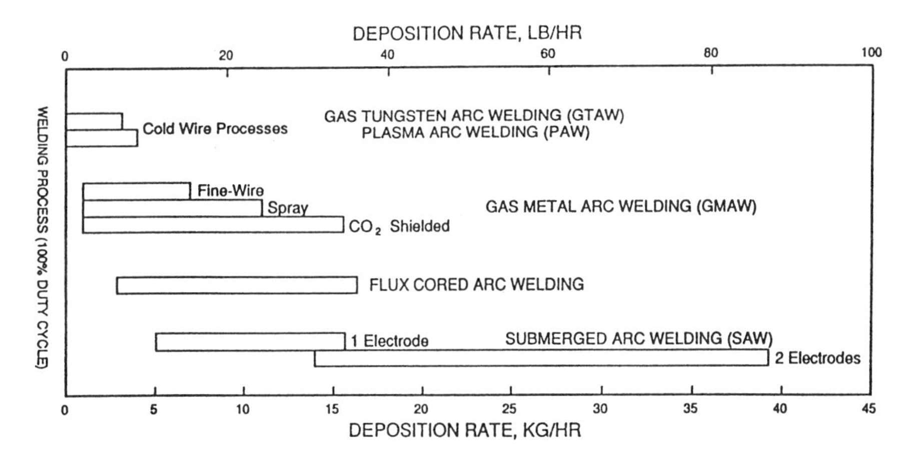

Figure 3.25 Deposition rate summary.

4

Automation of Other Welding Processes

4.1 Brazing and Soldering

The AWS Master Chart of Welding and Allied Processes (Fig. 2.9) shows brazing and soldering as families of processes. Brazing is a group of welding processes that produce coalescence of materials by heating them to the brazing temperature in the presence of a filler metal having a liquidus above 840°F (450°C) and below the solidus of the base metal.* The filler metal is distributed between faying (closely fitted) surfaces of the joint by capillary action. The definition for soldering is similar, except that the filler metal has a liquidus not exceeding 840°F (450°C) and below the solidus of the base metal. The difference between brazing and soldering lies in the melting/freezing temperatures of the filler and base metals. In view of this, these two process groups will be discussed together.

There are a large number of brazing and soldering processes. The primary

---

*The *liquidus* of a metal is the lowest temperature at which the metal is completely liquid, the temperature at which freezing starts. The *solidus* of a metal is the highest temperature at which the metal is completely solid, the temperature at which melting starts.

difference among them is the method used to apply heat to the assembly. Obviously, a higher temperature is required for brazing. Whether a process can be automated, either partially or entirely, depends on the heating method, the manner of moving the parts and providing filler metal and flux, and the design and fixturing of the parts.

Soldering and brazing can be used to join most metals. They are both widely used for joining nonferrous metals. The selection of filler metal, flux, furnace atmospheres, and joint design is beyond the scope of this book. The reader should consult the *AWS Soldering Manual* [1] or the *AWS Brazing Manual* [2] for detailed information.

There are two methods for dip brazing. One uses a molten chemical bath, and the other a molten filler metal bath. For dip soldering there is only the molten metal bath method. The molten material is heated by any suitable means, and the parts must be properly cleaned and should be self-jigging. With the molten metal bath, fluxing is usually required before the dip operation. The parts are preassembled, then dipped into the molten material, allowed to remain for a specific time, removed, and then cleaned. The dip time is accurately controlled. Dipping of parts into molten material and into cleaning baths, etc., is done with mechanical equipment operated by the controller. Loading and unloading the fixture and preplacing filler metal when required may be done automatically or manually. Dip brazing or soldering is normally used for small production batches.

Both brazing and soldering can be accomplished in a furnace. The assembly should be self-jigging, preassembled with preplaced filler material, and fluxed prior to heating. Furnaces can be heated by any suitable method, and the temperature must be accurately controlled. Production can be either batch or continuous. For continuous production, the parts are moved through the furnace mechanically. The furnace may employ special atmospheres or a vacuum. The rate of travel through the furnace depends on the mass and composition of the parts being brazed or soldered. The controller for processing must accurately control travel speed and furnace temperature. Loading and unloading are normally manual; however, automatic systems can be employed.

In induction brazing or soldering, the heat is obtained from the resistance to an electric current induced in the parts being processed. A high-frequency alternating current field is produced, and the parts are moved through this field. Special work coils are used to couple the high-frequency current to the parts being processed. Heating is rapid, and time is allowed for the filler metal to flow into the joint. A mechanized system is used to move the parts or the work coils to apply the induction field to the part. The temperature depends on the strength of the induced field and the travel speed of the parts or work coils. Sensors and complex controllers are used to automate processing. Manual loading and unloading are usually employed; however, in an automated continuous system these functions are automatic.

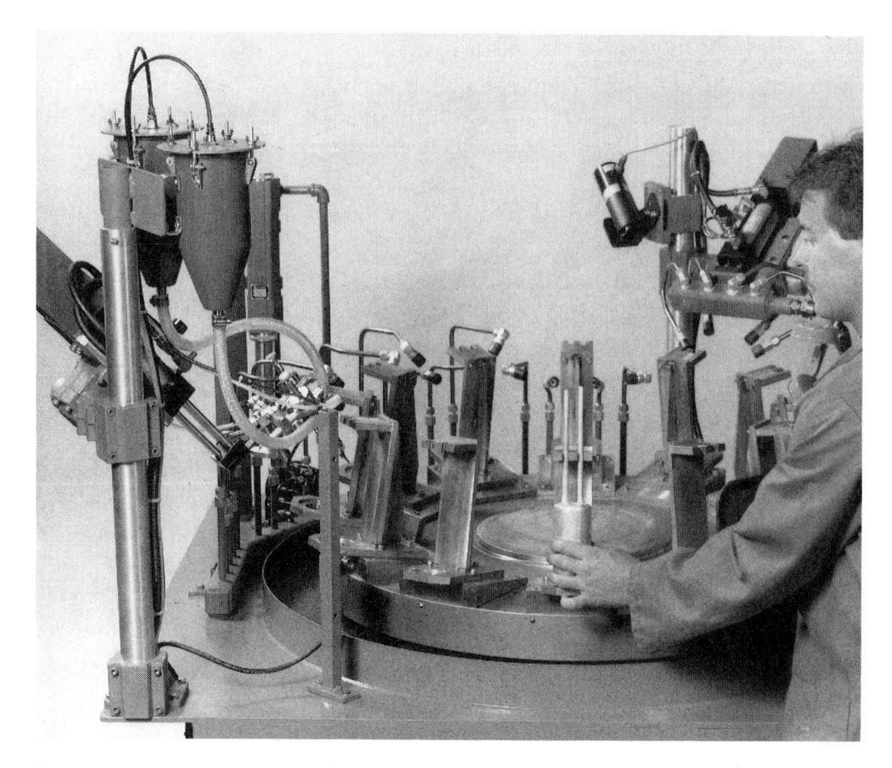

Figure 4.1 Automatic torch brazing operation.

In infrared brazing or soldering, the heat for the process is obtained from an infrared heat source, sometimes called "black heat." There is a minimum amount of visible light, as heating is done by the invisible "black" radiation. The infrared lamps use concentrating reflectors. The parts are positioned such that the radiant energy will impinge on them or on the joint. The parts must be self-jigging, cleaned, and fluxed and have preplaced filler materials. Infrared brazing is slower than induction brazing. It is usually used for continuous production and is controlled automatically.

In resistance brazing or soldering, the heat is obtained from the resistance to the flow of electric current through the parts. The parts become part of the electric circuit. Alternating current is normally used with high amperage and low voltage. The parts are held between the two electrodes while the electric current is applied. In the case of soldering, carbon blocks are employed as electrodes. The heat is generated at the joint interface, and its intensity depends on the resistance to current flow at the joint. The parts are preassembled, cleaned, and fluxed, and preplaced filler metal is required. A controller controls the electric current and duration of current flow. Resistance brazing/soldering is normally considered a mechanized operation and can be used for batch or continuous production. The degree of automation and control is related to production requirements, the method of loading and unloading, and similar factors.

Torch brazing or soldering uses heat provided by the flame of a fuel gas combustion torch. Automatic machines that move the parts through the heating flames are widely used. Figure 4.1 shows a machine that has a carousel for moving the parts. The degree of automation of the torch brazing/soldering operation depends on the complexity of the controller. Loading, unloading, etc., can be manual or automatic.

Wave soldering is similar to dip soldering. It is an automatic application in which the work is passed through a wave of molten solder. The wave of molten solder and the part being soldered may both move simultaneously. This system is widely used for the production of printed circuit boards. The assembled board is placed over a tank holding molten solder, and a wave of solder touches the lower side of the circuit board, soldering the metal of the circuit board to the leads of the electronic components. Fluxing is accomplished in a similar wave-type application. The controller regulates the heat of the solder bath and the motion of the system.

The other methods of brazing and soldering do not lend themselves to automatic or automated operation.

4.2 Resistance Welding

Resistance welding is one of the processes shown on the AWS Master Chart of Welding and Allied Processes. It is a group of welding processes that produce coalescence of the faying surfaces with the heat obtained from resistance of the workpieces to the flow of welding current in a circuit of which the workpieces are a part and by the application of pressure. This family of processes includes resistance spot welding (by far the most popular), projection welding, resistance seam welding, flash welding, and upset welding. For more information, refer to the *Resistance Welding Manual* [3].

Resistance spot welding is most widely used, commonly on low-carbon mild steel of sheet metal thickness. Most metals can be resistance spot welded. The basic spot welding process is shown schematically in Fig. 4.2a, with a close-up of the process in Fig. 4.2b.

Spot welding is one of the oldest welding processes; it was invented in 1877 and has been widely used ever since. The resistance welding processes vary considerably from one another. Spot welding, the most important, is given major emphasis here.

Spot welding was the first welding process to be mechanized. The first

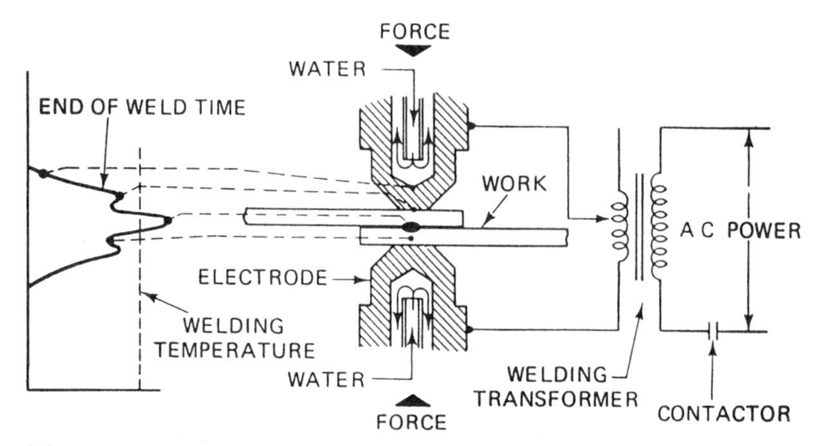
FORCE
WATER
END OF WELD TIME
WORK
A C POWER
ELECTRODE
WELDING
TEMPERATURE
WATER
FORCE
WELDING
TRANSFORMER
CONTACTOR

(a)

(b)

controllers and control systems were developed many years ago in the automotive industry for controlling spot welding operations.

Four factors are involved in making a resistance weld: (1) the amount of current that passes through the work, (2) the pressure that the electrodes transmit to the work, (3) the length of time the current flows through the work, and (4) the area of the electrode tip in contact with the work. Heat is generated by the passage of electric current through the resistance circuit. The maximum amount of heat is generated at the point of maximum resistance, which is at the surface between the parts being joined. High current at low voltage generates sufficient heat at this resistance point that the metal reaches a plastic state. The force applied before, during, and after the current flow forges the heated parts together so that coalescence will occur. Pressure is required throughout the welding cycle to ensure a continuous electric circuit. The time period for current flow is related to the heat input required to overcome heat losses and raise the temperature of the metal to welding temperature. Resistance welds are made very quickly, and the current, voltage, and pressure values are all closely regulated by the controller. The area or diameter of the weld is related to the geometry of the spot welding tip. The types and sizes of tips vary depending on the welding machine and the work to be welded.

Resistance welding is widely used in mass production. Welding operators normally load and unload the welding machine and initiate the welding operation. Resistance welding produces high-quality welds in high volume at high speeds.

Most of the common metals can be welded by most of the resistance welding processes. However, difficulties may be encountered with certain metals in heavy thicknesses. Weldability is controlled by three factors: resistivity, thermal conductivity, and melting temperature. All ferrous metals can be readily resistance welded. A spot welding system needs at least the following components:

1. Welding transformer for supplying power
2. A means of applying pressure
3. Electrode tips for conducting welding current to the work
4. A controller contactor
5. Material-moving devices

It is beyond the scope of this book to describe all of the different types of resistance welding machines in use. The more common ones will be briefly described.

Spot welding machines are available in two categories: single-point and multiple-point. Single-point spot welding machines include the rocker arm type, the press type, and portable gun welders. Multiple-point spot welding machines

---

Figure 4.2 The spot welding process. (a) Schematic diagram; (b) spot welding sheet metal.

are used for welding mass-produced products and are normally designed for a specific product or a family of similar products. A typical machine of this type is shown in Fig. 4.3. Machines of many, many different designs are made to produce specific items. They are widely used in the automobile industry. In some cases, press welders have electrode tips that can be moved to accommodate design or model changes. Each time the press cycles it makes a series of

Figure 4.3 Multiple-point spot welding machine.

welds for producing a specific spot welded part. Machines of this nature use controllers that program welding current, time, and pressure in a sequence to reduce the power line demand. Machines of this type have been in use for many, many years to make automobile components such as doors. They are very expensive.

Automated welding can also be performed with stationary rocker arm and press welders. In this activity the workpiece is placed on a table, which moves it to the correct position for a spot weld. The coordinated motion can be programmed, and human supervision is not required. Many spot welds can be made quickly with this system. An application of this type is shown in Fig. 4.4. The power sources and controller's contactor for spot welders and press welders are usually included in the welding machine. Table motion and workpiece movement by the spot welder are controlled by a computer.

The high cost of dedicated multipoint press welders can be avoided by using a robot and a coordinated positioner. Automation is accomplished in spot welding by attaching a gun welder to a robot. Gun welders are of two basic types. Some include the power transformer with the gun assembly, but high-power gun

Figure 4.4 Table moving part to spot welders.

welders may use a separate power source connected with concentric "kickless" cables. Gun welders are mounted on robots to provide flexible automated welding on a coordinated positioner. A good example is shown in Fig. 4.5, where two spot welding guns are shown welding on the same weldment held by fixtures on a positioner. Figure 4.6 is a close-up of making the weld. In this case, the expense is the cost of the fixtures holding the part. The robots and the positioner can be reprogrammed for model changes or for different parts.

Another way to automate spot welding is by means of a robot (Fig. 4.7). In this situation the tack welded assembly is picked up by a gripper attached to the robot. The robot is programmed to move it to the press spot welder. It is moved within the spot welder, which makes four rows of spot welds to internal stiffeners. The robot then moves the welded assembly to an outflow finished material rack, where it is put down. This operation could be completely automated if the tack welded assemblies were placed on individual shelves in the incoming material rack. The robot could be programmed to pick up a tack welded assembly from each shelf, make the spot welds, and place the welded assembly on an individual shelf in an outflow finished material rack. Having designated locations in the

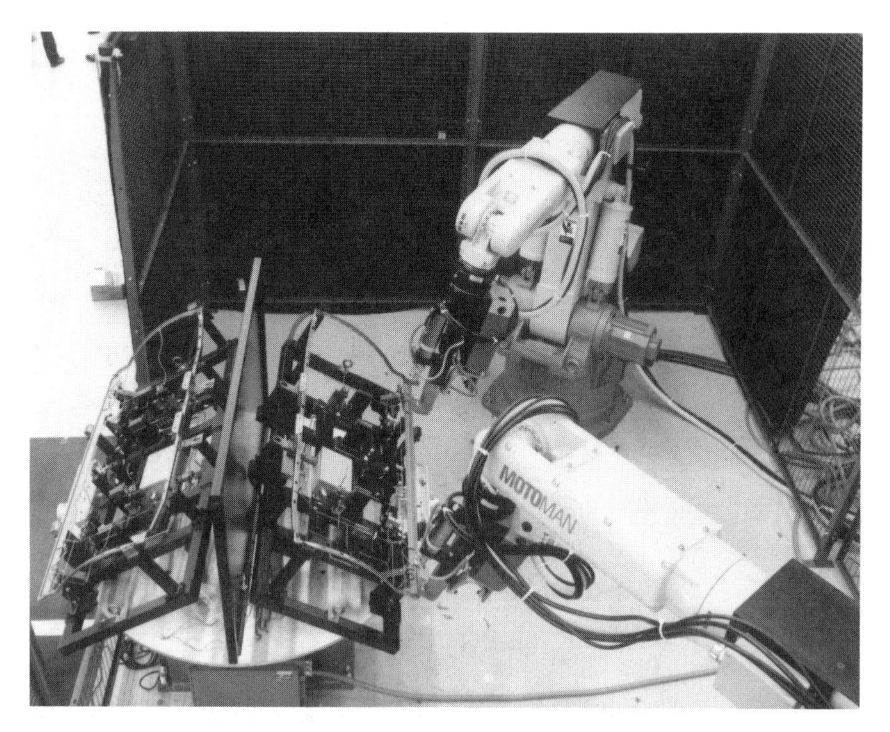

Figure 4.5 Spot weld gun on robot welding on positioner.

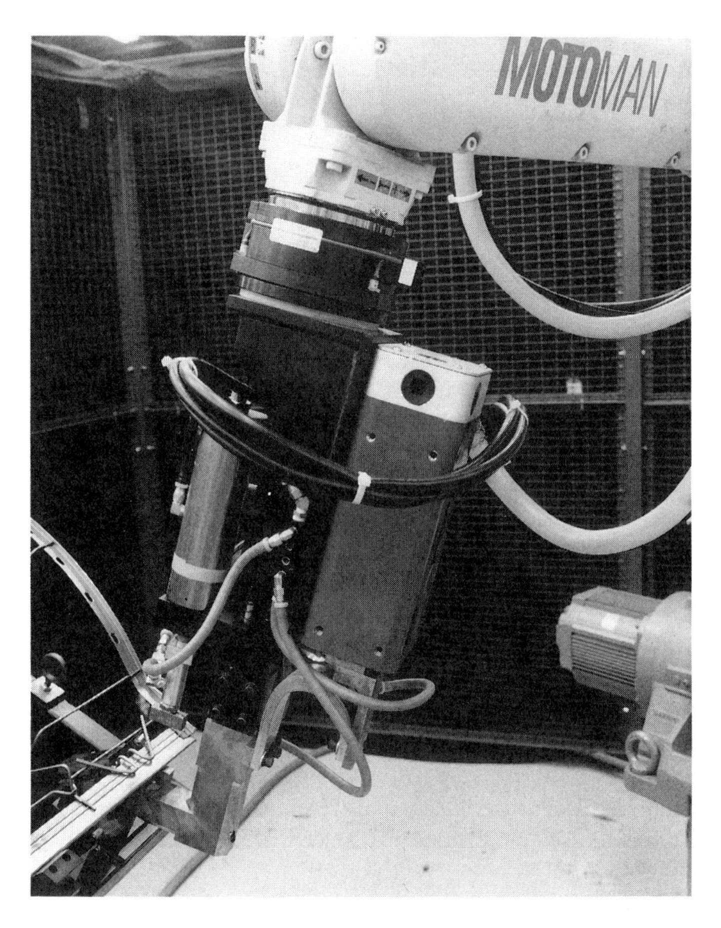

Figure 4.6 Close-up of gun making weld.

incoming and outgoing material racks allows unattended operation until the incoming material rack is emptied and the outgoing rack is filled. This spot welding robot cell can be reprogrammed to accommodate other assemblies that fit within the capacity of the racks, the robot work envelope, and the weight limit of the robot.

Various types of robot guns can be mounted on a robot. Figure 4.8 shows a scissor gun with its transformer mounted on a robot. Figure 4.9 shows a straight-action gun on a robot.

Different spot welding guns are needed for different jobs. Tool-changing systems are used for changing spot welding guns. This arrangement is shown by Fig. 4.10. The gun welder shown on the robot can be exchanged for another gun

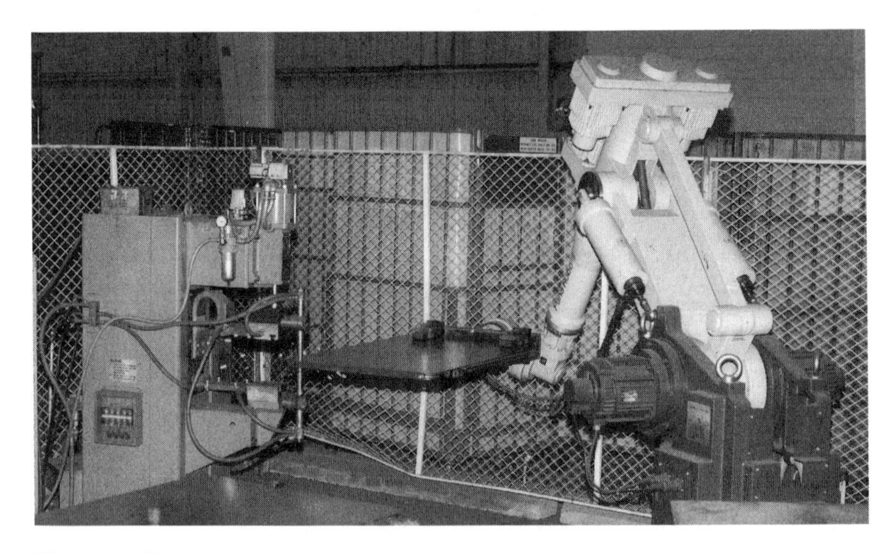

Figure 4.7 Robot holding the weldment.

welder hanging on the rack at the right. The robot carries a quick-change connector for hydraulic lines, electric power, and electric power control leads. Matching connectors are on the welding guns. The robot sets a gun on the rack and automatically disconnects from it; it then picks up another gun assembly and automatically connects to it.

Projection welding is very similar to spot welding. The difference is that one of the pieces has embossed projections that concentrate the heat for welding.

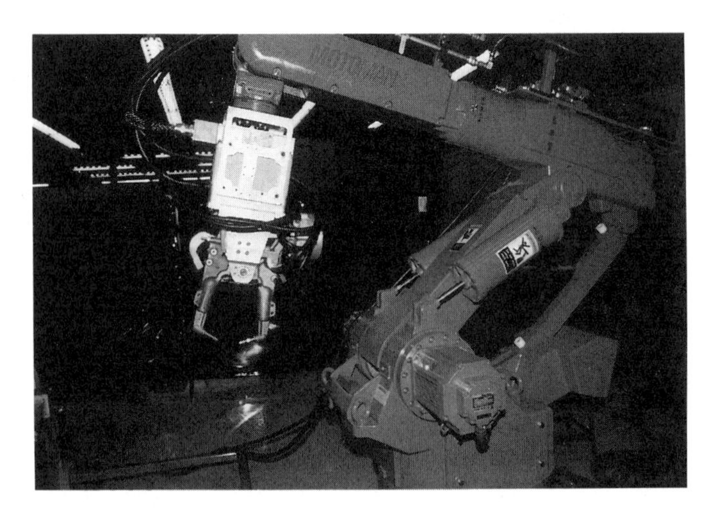

Figure 4.8 Scissor gun on robot.

Figure 4.9 Straight-action gun on a robot.

Figure 4.10 Gun-changing system.

The same basic equipment with special electrodes is used for projection welding. Motion can be accomplished by robots or dedicated motion devices. The controllers are essentially the same.

Resistance seam welding is similar to spot welding except that the parts being welded are moved through the machine by rotating round electrodes. Cooling these electrodes is more complex than for spot welding; hence external liquid cooling is usually employed. The controller on a seam welding machine controls the rotational speed of the round electrodes as well as pressure and welding current. Controllers can be adjusted to provide overlapping spot welds for a waterproof joint or separated spot welds when a leakproof joint is not desired. The rotating electrodes can be turned 90° to provide for other types of applications. Work can be mounted on a coordinated motion table to allow seam welding of irregularly shaped parts. The travel speed, current, pressure, programming, and motion are all under fully automatic control.

Flash welding is done in massive machines with automatic controls. Movement of the parts, application of pressure, and welding current are all programmed. The machine can be loaded and unloaded with automatic equipment for fully automated production.

Upset welding is similar to flash welding except that the flash does not occur; a weld is obtained by the exertion of forging pressure as material is brought to welding temperature. Equipment for upset welding is automatic and is controlled in much the same manner as flash welding equipment.

High-frequency resistance welding is a welding process normally used for the manufacture of pipe, tubing, and small structural beams. The welding current is induced in the material being welded via induction coils. The high-frequency current normally flows along the surface of the metal to be welded and is very localized. The material is automatically fed through a mill, which normally includes the forming rolls as well as the welding station. Welding current is introduced at the abutting surfaces to bring them to welding temperature, then passes through pressure rolls, where pressure is applied to produce a weld. The welding and forming mill is entirely automatic and operates at a high travel speed. The welding program is specified for the product and material type and thickness. The welding current, travel speed, pressure, etc., are all programmed by the controller.

4.3 Electron Beam Welding

Electron beam (EB) welding is a welding process that produces coalescence with a concentrated beam composed primarily of high-velocity electrons impinging on the joint. The process is used without shielding gas and without the application of pressure. It is a fusion welding process, with the base metal, possibly together with filler metal, melted to produce the weld. Heat is generated in the workpiece as it is bombarded by a high-velocity electron beam. The kinetic energy (energy of motion) of the electrons is transformed into heat upon impact. The electron beam is a highly concentrated, high-powered source of heat and acts similarly to an arc or plasma in making welds. For more information, refer to the AWS publication *Recommended Practice for Electron Beam Welding* [4].

The electron beam welding process was developed in the 1950s by the atomic energy industry. Commercial equipment has been available since 1960 and has been improved continuously. The use of electron beam welding is now widespread.

The electron beam is generated in a high-vacuum chamber known as an electron beam gun. There are several variations of the process. The original and still common method is to have the entire welding operation performed in a high-vacuum enclosure. In a variation of the process, the beam is transmitted through chambers maintained at successively weaker vacuum, and the weld is made in a chamber with a soft (low) vacuum. In another variation, the electron beam is generated by the electrode gun but is transmitted through chambers of intermediate pressure to the open air, where the weld is made.

The electron beam gun is a device for producing, accelerating, and forming the electron beam. It consists of a filament or cathode called an emitter, a grid cup, an anode, and focusing and deflection coils, and it is housed in a hard vacuum (0.001 torr) (Fig. 4.11). The emitter is either a tungsten filament or a tungsten rod heated by a filament. Electrons are freed from the tungsten when it

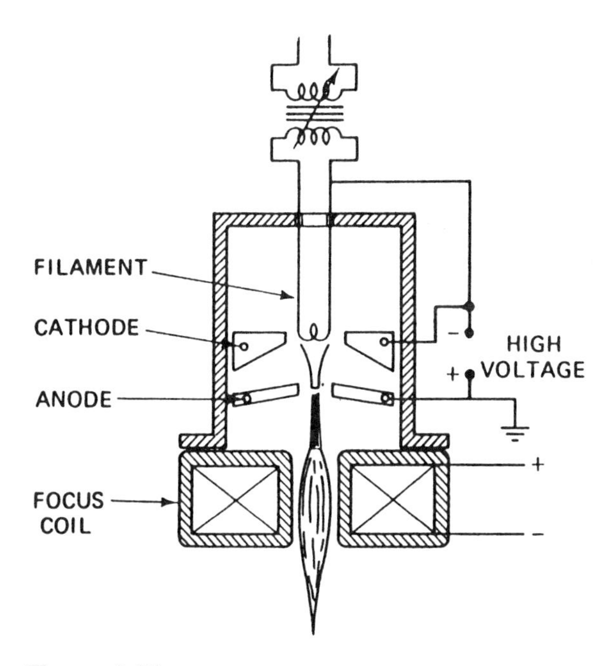

Figure 4.11 Electron beam gun.

is heated to a high temperature, a process known as thermionic emission. The freed electrons are attracted to the anode (the positive pole). The beam is collected and partially focused at the anode, which has a hole in the center. Beyond the anode the beam is further focused by means of magnetic forces generated by a focusing coil. Following this, the beam may be deflected by magnetic fields generated by deflection coils. It then leaves the electron beam gun through an exit port and impinges on the workpiece.

In the high-vacuum mode, the welding chamber is sufficiently large to hold the work to be welded and is at the same low pressure or vacuum as exists in the welding gun compartment. This chamber, which is evacuated to reduce the pressure, must be extremely strong so it will not crush under atmosphere pressure. All openings must be sealed for vacuum tightness. Precision work-handling equipment is used to move the weldment in the chamber. The vacuum chamber should be no larger than required due to the time and expense of evacuating it. Two pumps are required, a vacuum pump and a diffusion pump. The pump-down time can be lengthy. The chamber has interlocking doors so that work can be inserted or removed without repumping each time. Figure 4.12 shows an electron beam welding machine with the work compartment door open.

Figure 4.12 Electron beam welding machine.

Due to pump-down time and expense, a second EB variation was developed. This allows welding in a vacuum with a pressure of 0.1 torr (10^{-1} torr) known as a soft vacuum, so the work chamber can be much larger and there is less pump-down time because a diffusion pump is not required. A more powerful electron beam gun is needed, still enclosed in a hard vacuum. There are intermediate chambers between it and the work chamber to provide a pressure gradient. The hole through which the electron beam passes is sufficiently small that air does not leak into the chamber rapidly. The standoff distance between the electron beam gun and the workpiece is less in a soft vacuum chamber. In addition, metal approximately 4 in. (50 mm) thick can be welded, and standoff distance and travel speed are reduced.

A nonvacuum or open-air system requires a container to shield people from potential radiation. The electron beam gun is in a hard vacuum chamber here also, and there are more intermediate chambers. The welding capabilities are reduced to 2 in. (50 mm) thickness. Travel speed is slower, and standoff distance is reduced to 1½ in. (30 mm). The size of the workpiece can be much greater, and moving the work is much easier since the components are in air. Precision motion control and alignment are still required.

An electron beam welding machine consists of at least an electron beam gun, a power supply, work motion equipment, a welding chamber with vacuum pumps, an alignment and viewing system, and a computerized controller.

The power in the electron beam is controlled by four parameters, one being the accelerating voltage, which controls the electron speed at impact and is very high. There are two basic types of machines: low-voltage, with an accelerating voltage of 15–50 kV, and high-voltage, with an output of 100–200 kV. The depth of penetration is a function of the accelerating voltage. The second parameter is the number of electrons per second hitting the workpiece, which is related to beam current. The beam current ranges from 40 mA for the high-voltage machines to 500 mA for the low-voltage machines. The higher voltage machines produce a greater penetration depth-to-width ratio.

The third variable in the control of electron beam power is the diameter of the beam or the beam spot size, which depends on the power density. Beam power relates to the power density of the electron beam. Power densities can range from 100,000 to 10,000,000 W/in.^{2}. Temperatures are in the neighborhood of 25,000°F, which causes instantaneous vaporization of the surface. The fourth variable is the speed of travel of the beam.

The electron beam produces extremely deep penetration. When the electrons hit the base metal, they release the bulk of their kinetic energy, which turns to heat energy. This brings about a tremendous temperature increase. The electron beam travels through the vapors much more easily than through solid metal and thus penetrates more deeply. As the power density is increased, penetration increases. The high depth-to-width ratio produces a weld with almost parallel sides, which greatly minimizes distortion.

The weld width is narrow, and for this reason small misalignments between the beam and weld joint will allow the beam to miss the joint completely. Optical systems are sometimes used to align the beam with the joint. Travel and motion are always controlled by precision automatic systems. The entire welding operation is usually controlled by a computer.

Electron beam welding requires expensive equipment. Figure 4.13 shows an electron beam weld being made in air using the nonvacuum mode. Figure 4.14 shows electron beam welded parts.

The electron beam can be used to weld almost any metal. The metals most often welded are the superalloys, the refractory metals, the reactive metals, and stainless steel. Many combinations of dissimilar metals can also be welded.

Electron beam welding is never applied manually, and a semiautomatic variation has not been widely used. All electron beam welding machines are either mechanized or automated.

Figure 4.13 Making an electron beam weld in air.

Figure 4.14 Typical electron beam welded parts.

4.4 Laser Beam Welding

Laser beam welding is a welding process that produces coalescence with the heat from a laser beam impinging on the joint. The process is used without a shielding gas and without the application of pressure. The word *laser* is an acronym for *l*ight *a*mplification by *s*timulated *e*mission of *r*adiation. The laser beam is a highly concentrated source of energy very similar to the electron beam. The laser is a device that produces this concentrated coherent light beam by stimulating electronic or molecular transitions to lower energy levels. Its many uses include welding, cutting, cladding, piercing, and heat treating in the metalworking field. It can also be used for cutting many nonmetals. Lasers also have other uses including medical applications, communications, marking, compact disc players, bar code reading, and surveying.

The laser was conceived in 1951, but it was not until the 1960s that it was developed and demonstrated. The earliest lasers used solid synthetic ruby crystals excited by flash tubes and emitted short pulses of red coherent light. The CO_2 laser was developed in the mid-1960s and has become an industrial workhorse. The focused laser beam has a very high energy concentration, on the same order as that of an electron beam in a hard vacuum. It is a source of electromagnetic energy, or light, that can be projected with low divergence and can be concentrated to a very small spot. Light from an incandescent electric light bulb is "incoherent," or out of phase, and as a result has a high divergence and is radiated in all directions from the source. It contains a wide spectrum of wavelengths from short to long. The radiation from a laser is monochromatic, which means that it provides a single wavelength, which in turn provides for minimum beam divergence. The beam is also coherent in that the light is all in phase. Because the laser beam has a high energy content, when it impinges on a surface it creates heat, so it can be used exactly like an electron beam or a welding arc.

Two types of lasers are used commercially in metalworking. Solid-state lasers use a solid medium, and gas lasers use a mixture of helium, nitrogen, and CO_2 in a tube. In either case, when the medium is sufficiently excited it emits photons, which become the laser beam.

The most popular solid-state laser is the Nd:YAG laser, which is a crystal doped with neodymium (Nd) and made of yttrium, aluminum, and garnet. The Nd ions emit photons when their electrons are excited and then allowed to return to their original energy state. Figure 4.15 is a diagram of a solid-state laser. Solid-state lasers use a single crystal made into a round rod approximately ¾ in. (19 mm) in diameter and approximately 8 in. (200 mm) long. The end surfaces of the rod are ground flat and parallel and are polished to extreme smoothness.

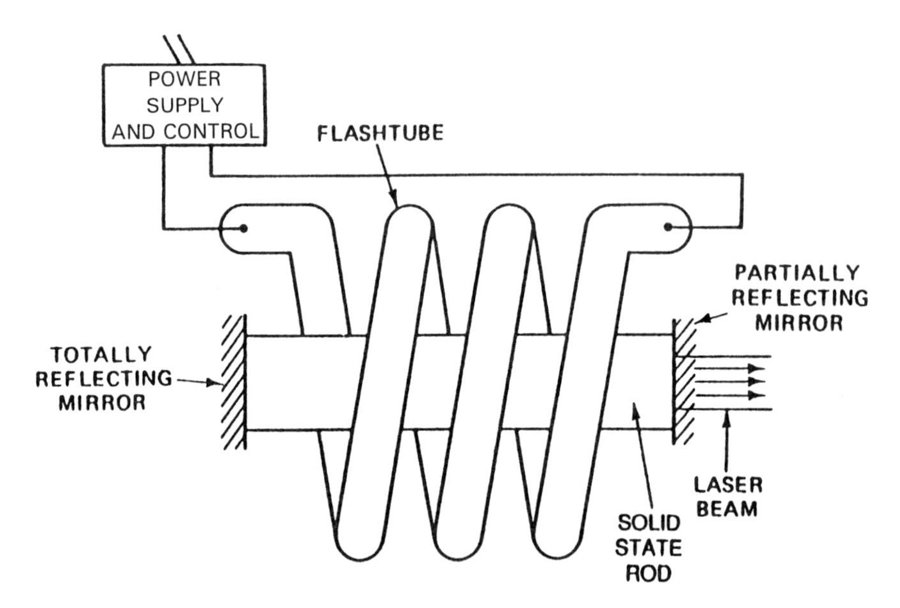

Figure 4.15 Solid-state laser.

Both flat ends are covered with silver to reflect light; however, a small area in one end is left uncovered to allow the laser beam to exit from the crystal rod. The solid-state rod is closely surrounded by a high-intensity light source, which is a flash tube with a xenon or krypton element. When the tube is flashed it emits an intense pulse of light that lasts for approximately 2 msec. A burst of laser beam light, which lasts the same length of time, occurs each time the flash tube is flashed. It is not possible to flash the crystal too often because of heat generated in the crystal and in the flash tube. It does not operate continuously because of heat buildup. The flash pulse durations are very short, and there is a relatively long period between pulses. This is known as the pulsed mode of operation. A recent development of the Nd:YAG laser allows for better cooling and continuous-wave operation. The power available for Nd:YAG lasers has also increased to 3000 W (3 kW). The power levels are expected to further increase in the future.

The wavelength of the Nd:YAG laser beam is 1.06 μm (micrometers). This wavelength is much shorter than that of CO_2 lasers. It presents an eye hazard. In general, safe use of lasers requires eye protection specifically blocking the wavelength of light that the laser produces. Unprotected eyes are at risk from both direct laser energy and reflected energy. Beam-reflecting goggles coated with material that blocks or reflects radiation prevent the beam from hitting the eyes and must be used when working around lasers.

For the Nd:YAG laser the beam diameter is 0.020 in. (0.05 mm) or more. The upper diameter limit is the usable diameter of a laser beam and is dependent on the power limitations.

The CO_2 laser uses gas that is a mixture of CO_2, helium, and nitrogen. Excitation of the gas laser is by means of high-voltage, low-current electric power. The electric discharge excites the CO_2 molecules, which on returning to their original energy state emit photons. Mirrors are placed on both ends of the tube, one entirely reflective and the other with a small hole in the reflector to allow the beam to exit. This forms a cavity in which photons build up. The freed photons travel between the mirrors and excite the CO_2 molecules, starting a chain reaction of photon emissions. A stream of photons, the laser beam exits through the unsilvered section in the one mirror. The CO_2 laser is shown in Fig. 4.16.

The wavelength of the CO_2 laser beam is 10.6 μm. This wavelength is 10 times as long as that of the solid-state laser and does not pose quite the eye hazard of the shorter wavelength. Conventional safety eyewear can give satisfactory protection. The CO_2 gas laser can be operated in the continuous-wave or pulse mode. The spot size can be made as small as 0.010 in. (0.2 mm) or as large as ½ in. (13 mm). The smaller focus spot size is used for welding and cutting, and the larger spot size is used for heat treating. CO_2 lasers are available up to 25 kW power levels. The laser beam is very intense and unidirectional but can be focused and reflected in the same way as an ordinary light beam. The focus size is controlled by the choice of lenses and mirrors and the distance to the workpiece.

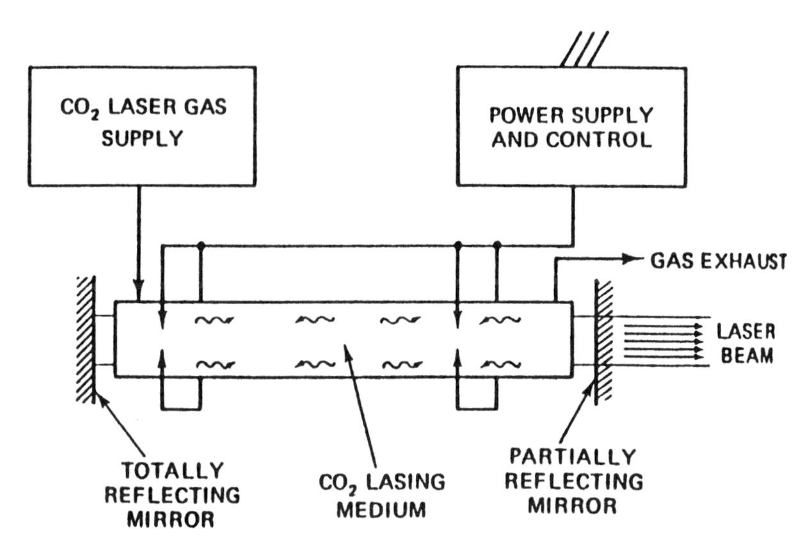

Figure 4.16 CO_2 laser.

The laser beam can be used in open air and can be transmitted long distances with only minimal loss of power.

The beam delivery system must match the type of laser used. Fiber optics are transparent to the shorter wavelength beam and are used to transmit the beam from solid-state lasers. The beam delivery system for longer wavelength lasers, specifically the CO_2 laser system, must use a lens and mirrors. These systems are more complex than a fiber-optics cable, and hence it is difficult to deliver the CO_2 laser by means of a robot. In addition, mirrors must be cleaned and readjusted frequently. The delivery system must also match the type of laser and particular application. The distance from the optical cavity to the workpiece has a minor effect on welding. This is because the laser can be focused to the proper spot size at the work surface without loss of energy whether it is close or far away.

A block diagram of a laser beam welding system is shown in Fig. 4.17. The major components are the laser beam source (sometimes called the oscillator), the power supply, the cooling system, the gas supply for the laser beam source, the beam delivery system, the beam output coupling to the workpiece, the motion system for moving the beam or the workpiece, the control system for the beam source, auxiliary systems, and the real-time monitor or feedback system. Workpiece motion, parts handling, and workpiece motion feedback or monitoring are similar to those of other automated welding systems except that the accuracy of movement must be very precise.

With laser welding, the molten metal takes on a radial configuration similar to that of the plasma arc welding known as "melt-in" or conduction mode

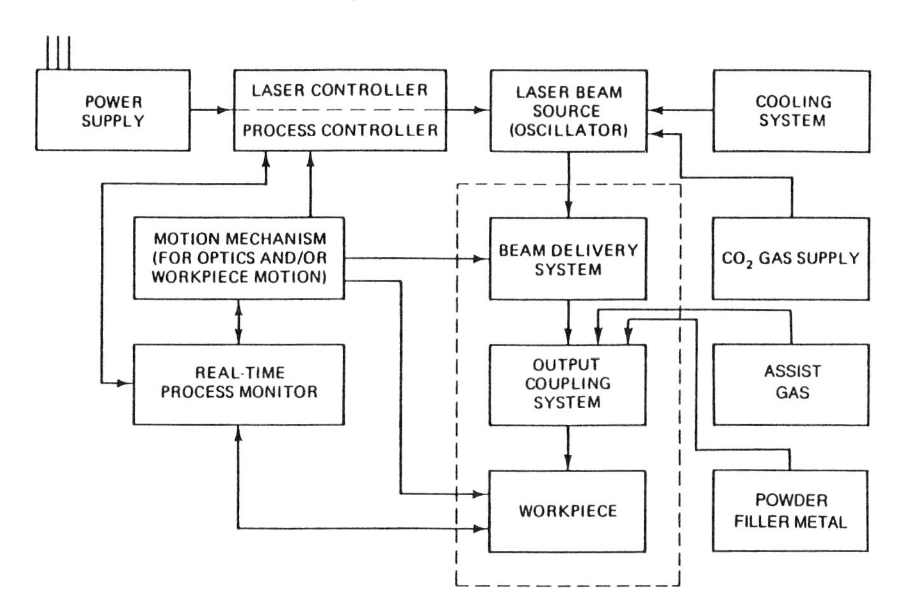

Figure 4.17 Block diagram of CO_2 laser system.

welding. When the power density rises above a certain threshold level, keyholing occurs the same as with plasma arc or electron beam welding. Keyholing provides for extremely deep penetration, which gives the weld a high depth-to-width ratio. Keyholing also minimizes the problem of beam reflection from the shiny molten metal surface because the keyhole behaves like a blackbody and absorbs the majority of the energy. For most applications, inert gas is used (the metal vapor in the weld area will ionize), and inert gas will minimize plasma formation. Plasma absorbs energy from the laser beam and can actually block the beam and reduce melting. This is overcome by using an inert gas jet directed along the metal surface, which blows away the plasma buildup. It also shields the weld from the atmosphere.

The welding characteristics of the laser are similar to those of the electron beam. Lasers can weld the same types of joints. The concentration of energy by both beams is similar, with the laser having a slightly lower power density than the electron beam.

The location of the focal point of the beam with respect to the surface of the workpiece is important. Maximum penetration occurs when the beam is focused slightly below the surface. Penetration is less when the beam is focused on the surface or deep within the workpiece. As power is increased, the depth of penetration increases.

The laser beam produces a tremendous temperature differential between the

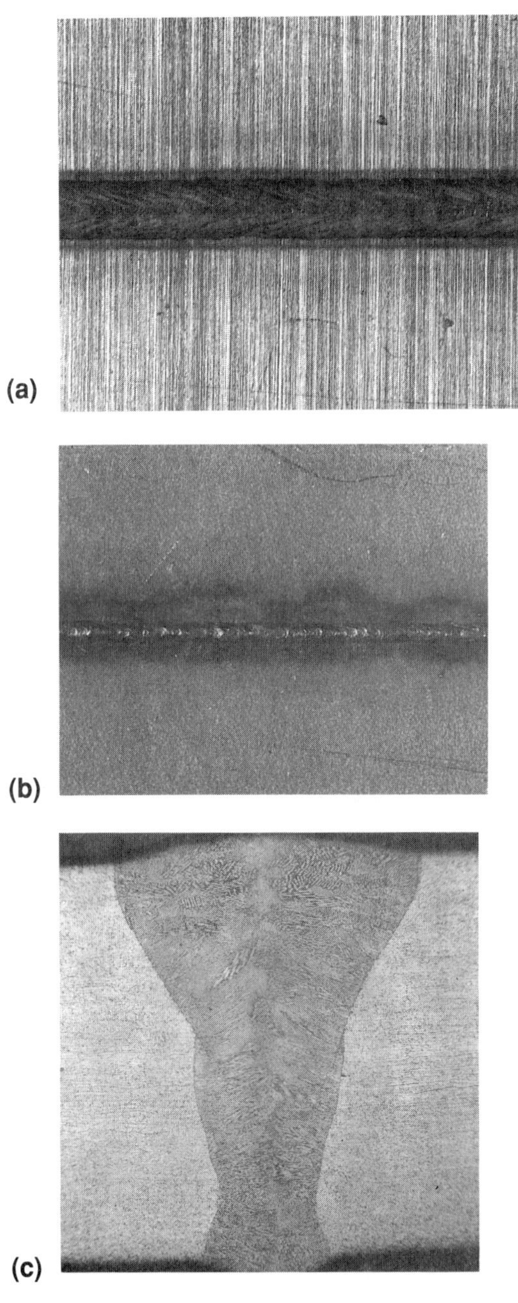

Figure 4.18 Top (a), underside (b), and cross section (c) of a weld made in stainless steel.

molten metal and the base metal immediately adjacent to the weld. Heating and cooling rates are much higher in laser beam welding than in arc welding, and the heat-affected zone is much smaller. Rapid cooling rates can create problems such as cracking in high-carbon steel. The laser beam has been used to weld carbon steels, high-strength low-alloy steels, aluminum, stainless steel, titanium, and some dissimilar metal combinations. Filler metal is used to weld metals that tend to be porous.

Materials ½ in. (12 mm) thick are welded at speeds of 30 in. (760 mm) per minute. Figure 4.18 shows the top, underside, and cross section of a laser weld made on stainless steel. This shows the characteristic cross section and depth-to-width ratio of a laser weld.

Laser welding is used extensively in the automotive industry. The laser's high speed is attractive for cost reduction. A major use is for autogenous welding of thin material. Figure 4.19 shows an autogenous weld being made on thin galvanized steel at very high travel speed.

The efficiency of laser beam welding equipment is relatively low but is being increased as the process becomes more popular. The master controller that controls the motion devices also controls the parameters of the welding procedure. Lasers are being continually improved and will find wider applications in welding.

The laser beam welding process has always been automated. The small beam spot size and the need to align it perfectly with the weld joint require precision movement of either the beam or the workpiece. Smooth constant travel and perfect alignment can be accomplished only by precise motion devices. For this reason, automated travel is required. Semiautomatic or manual operation of the laser beam is not used, also because of the safety problem.

4.5 Solid-State Welding

Solid-state welding is included in the AWS Master Chart of Welding and Allied Processes (Fig. 2.9). It is a group of welding processes that produce coalescence by the application of pressure at a temperature below the melting temperatures of the base and filler metals. This family of processes includes cold welding, diffusion welding, explosion welding, forge welding, friction welding, hot pressure welding, roll welding, and ultrasonic welding.

Forge welding is one of the oldest welding processes and is the one performed by blacksmiths. It is a manual process and is not automated.

Cold welding is a solid-state welding process in which pressure is used to produce a weld at room temperature with substantial deformation at the weld. Welding is accomplished by using extremely high pressures on extremely

Figure 4.19 Laser autogenous welding on auto muffler.

clean surfaces. For automatic cold welding, pressure is applied by an air or hydraulic system. The timing sequence is rather unimportant. This is a machine welding operation.

Diffusion welding is a solid-state welding process that produces a weld by the application of pressure at elevated temperature with no macroscopic deformation or relative motion of the workpieces. A filler metal may be inserted between the faying surfaces. Diffusion welding is performed in a pressurized vessel in a special atmosphere and with various means used to apply pressure. Heating is also accomplished by various methods such as induction or resistance.

Explosion welding is a solid-state welding process that produces a weld by high-velocity impact between the workpieces as the result of controlled detonation. This is an automatic operation; once the explosive material and holding fixtures are in place, the detonation progresses and creates the weld.

Friction welding or inertia welding is a solid-state welding process that produces a weld under compressive force in contact with the workpieces, which rotate or move relative to one another to produce heat and plastically displace material from the faying surfaces. The most common method uses rotational motion. One part is rotated, pressure is applied until the weld occurs, and rotation is stopped. This is an automatic process because a complex control system is required to regulate pressures and rotational speeds in proper sequence. It is accomplished on a machine similar to a metal-cutting lathe.

Hot pressure welding is a solid-state welding process that produces a weld with heat and the application of pressure sufficient to produce macrodeformation of the workpieces. It is used in making small-diameter pipe. Hot skelp is formed into a tube and pulled through a pressure die to make the weld. It is an automatic continuous process.

Roll welding is a solid-state welding process that produces a weld by the application of heat and sufficient pressure exerted with rolls to cause deformation at the faying surfaces. This is done in a machine. The parts being welded are fed through high-pressure rolls at a high temperature. Roll welding is used to clad alloy to plates. It is an automatic welding process because the pressure and rotating speed must be closely controlled.

Ultrasonic welding is a solid-state welding process that produces a weld by the local application of high-frequency vibratory energy as the workpieces are held together under pressure. It is an automatic welding process because the pressure and vibratory energy are closely controlled by sequence timers. It is normally used for thinner materials, both metals and nonmetals.

4.6 Other Welding Processes

The AWS Master Chart of Welding and Allied Processes shows a number of other welding processes in addition to those already discussed. These include induction welding, percussion welding, and thermit welding. They are mentioned briefly here to complete the coverage of the processes in the master chart.

Induction welding is any welding process that uses induced welding current. It uses high-frequency welding current to concentrate the welding heat at the desired location. Induction welding is an automatic process that uses a complex control system.

Percussion welding produces a weld with an arc resulting from a rapid discharge of electric energy. Pressure is applied progressively during or immediately following the electric discharge. It is an automatic process.

Thermit welding is an old process that produces welds with a superheated liquid metal formed by a chemical reaction between a metal oxide and aluminum. It is used for welding railroad rails.

References

1. *AWS Soldering Manual,* 2nd ed., American Welding Society, Miami, FL.
2. *AWS Brazing Manual,* 4th ed., American Welding Society, Miami, FL.
3. *Resistance Welding Manual,* Resistance Welder Manufacturers Association, Philadelphia, PA.
4. *Recommended Practice for Electron Beam Welding,* AWS C7-1, The American Welding Society, Miami, FL.

5

Automation of Related Processes

5.1 Oxygen Cutting

Thermal cutting, which involves heat, is a family of processes allied to welding shown on the AWS Master Chart of Welding and Allied Processes. The most widely used is oxygen cutting, which is discussed in this section. Another class is arc and plasma cutting processes, which is covered in Section 5.2, and other cutting processes, including laser cutting, covered in Section 5.3.

Oxygen cutting is a group of thermal cutting processes that sever or remove metal by means of the chemical reaction between oxygen and the base metal at elevated temperatures. The necessary temperature is maintained by the heat from an arc, an oxyfuel gas flame, or other source. The most popular is oxyfuel gas cutting (Fig. 5.1) with a specific fuel gas such as acetylene or natural gas. The necessary temperature is attained by means of the combustion of the fuel gas and oxygen. This process is used to cut ferrous materials and is based on bringing steel (iron) up to its kindling temperature, where it will ignite and burn in an atmosphere of pure oxygen.

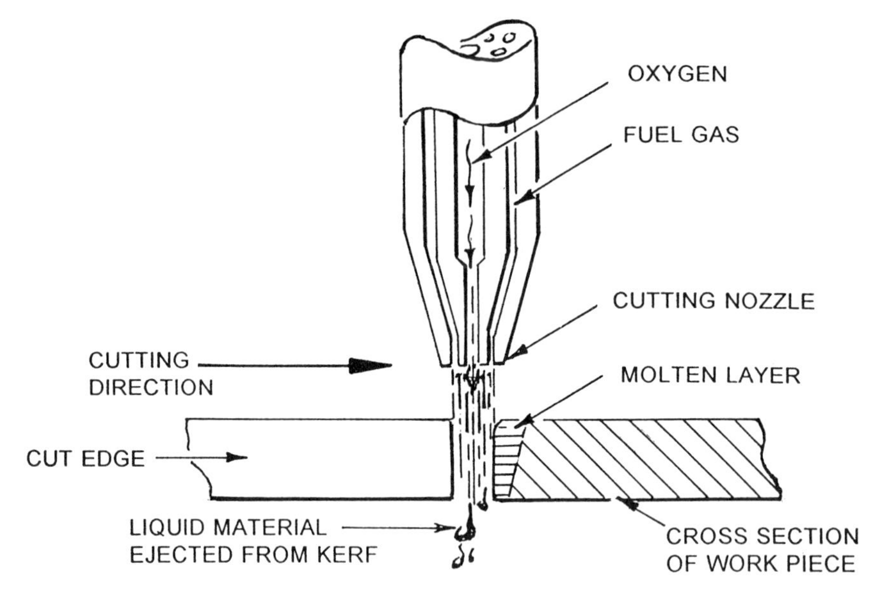

Figure 5.1 Oxyfuel gas cutting process.

Steel and a number of other metals are flame cut with the oxyfuel gas cutting process. The following conditions must apply:

1. The melting point of the material must be above its kindling temperature in oxygen.
2. The oxides of the metal should melt at a lower temperature than the metal itself and below the temperature that is developed by oxyfuel cutting.
3. The heat produced by the combustion of the metal with oxygen must be sufficient to maintain the oxygen cutting operation.
4. The thermal conductivity must be low enough that the material can be brought to its kindling temperature.
5. The oxides formed in cutting should be fluid when molten so as not to interrupt the cutting operation.

Iron and low-carbon and low-alloy steel fit all of these requirements and are readily cut with an oxygen flame.

Oxyfuel cutting was originally applied manually using a hand-held torch. This method was soon replaced by mounting the cutting torch on a small travel carriage or bug. This type of equipment used hand layout and was very useful for steel plate preparation for weldments.

A later development was the use of shape-cutting machines that followed specific contours and cut steel to these contours. The shapes were established by use of patterns. The first patterns used small strips of metal, usually aluminum bars approximately ¼ in. square, which were bent to shape and screwed to a sheet metal backing. A special mechanical tracing device would follow these bars around the contour of the design of the part. One or more torches would cut steel plate material to the exact pattern provided by the template. Templates were designed so that the final cut part would match the details of the pattern.

The metal shape cutting templates were followed by a more complex tracing method that employed an optical system using an electric eye to follow the interface between dark and light surfaces. Optical tracers became very popular in the 1940s. Based on the photocell tracer, the next improvement was the use of multiple patterns nested on the template to provide many pieces of plate chain continuously cut from the same piece of steel. Multiple torches were used, and this greatly enhanced the use of oxyfuel gas shape cutting.

The newest development in shape cutting is the use of microprocessor-based CNC controls that drive the cutting torches. A machine of this type is shown in Fig. 5.2. The controller provides x and y directions for cutting each part. The parts are nested so chain cutting can be accomplished. Computer software is used

Figure 5.2 Automatic oxyfuel flame cutting operation.

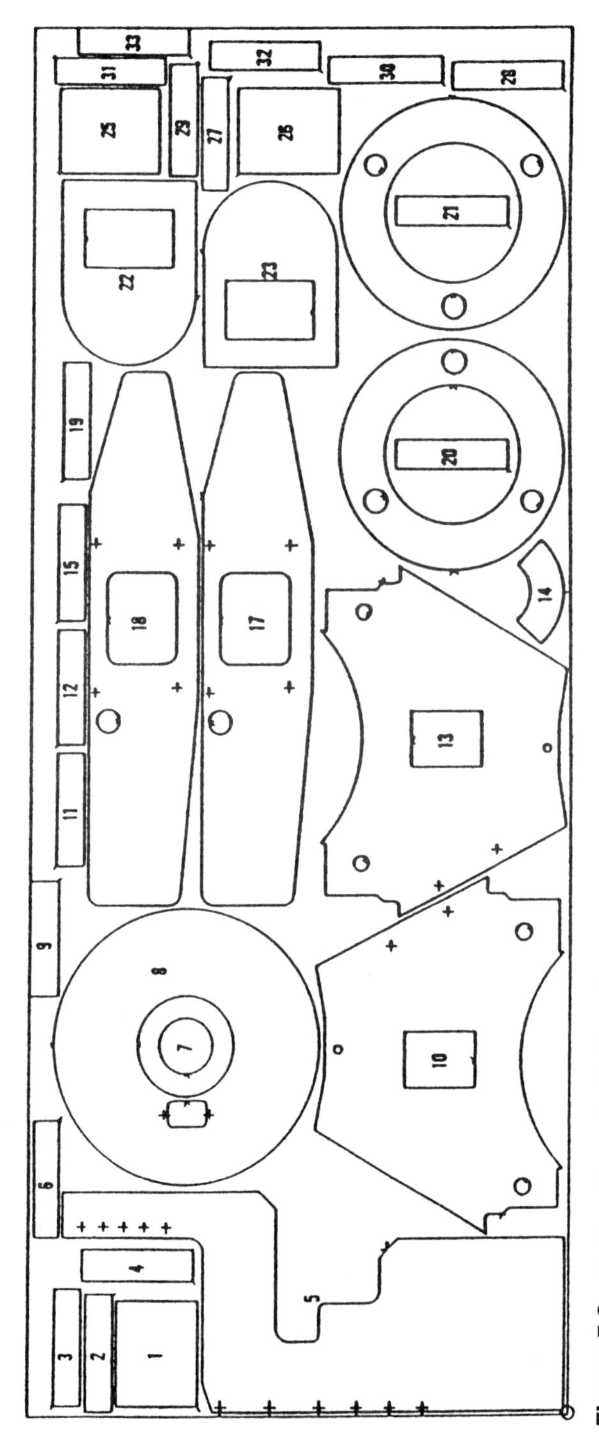

Figure 5.3 Nesting of parts for flame cutting.

to nest the pieces to maximize steel utilization. Figure 5.3 shows a layout of nested parts to be cut from a single thickness of steel plate material. The gantry or pantograph cutting machine can be equipped with a number of torches. This type of machine has become standard for producing contour-cut parts from steel plate. Most equipment of this type uses a pilot light for automatic torch lighting and sometimes a special water table to reduce distortion and smoke. Bevel attachments are available to prepare for vee or bevel joints. Cut surfaces that are smooth and with close tolerances are attainable with quality equipment. Manual cutting torches are used to cut the chain link holding the individual parts together.

The cutting machines are usually specified on the basis of the span of the gantry or pantograph frame that carries the torches. This enables the machine to accommodate the widest plates used in the manufacturing operation. Specifications include the number of torches, the fuel gas, and the type of computer control.

The software for computer nesting and control must be designed specifically for the type of equipment used and should accommodate the production control system in use. Another function of computer-controlled chain burning is production control. When the cutting program has been completed, a specific number of cut items are produced.

The cutting speeds for oxyfuel cutting on carbon steel are listed in Table 5.1. This schedule covers oxyfuel gas cutting of carbon steels from ⅛ in. (3.2 mm) to 8 in. (200 mm) thick. Special high-speed cutting tips can be used to increase cutting speeds. They are available for various fuel gases.

Table 5.1 Schedule for Oxyacetylene Flame Cutting

Material thickness		Cutting orifice diameter (center hole)			Approx. gas pressure (psi)		Travel speed[a] (in./min)
in.	mm	Drill size	in.	mm	Acetylene	Oxygen	
⅛	3.2	60	0.040	1.0	3	10	22
¼	6.4	60	0.040	1.0	3	15	20
⅜	9.5	55	0.052	1.3	3	20	19
½	12.7	55	0.052	1.3	3	25	17
¾	19.0	55	0.052	1.3	4	30	15
1	25.4	53	0.060	1.5	4	35	14
1½	38.1	53	0.060	1.5	4	40	12
2	50.8	49	0.073	1.9	4	45	10
3	76.2	49	0.073	1.9	5	50	8
4	101.6	49	0.073	1.9	5	55	7
5	127.0	45	0.082	2.1	5	60	6
6	152.4	45	0.082	2.1	6	70	5
8	203.2	45	0.082	2.1	6	75	4

[a]Mechanized.

Oxyfuel gas cutting is used for low-carbon and low-alloy steels. Stainless steels cannot be cut with the normal flame cutting process, and nonferrous metals such as aluminum and copper are not cut with oxyfuel flame cutting.

Robots are sometimes used but are not popular for this process. An oxyfuel gas flame cutting torch can be mounted on a robot. These are usually used for out-of-flat position cutting on special formed parts.

5.2 Arc and Plasma Cutting

Air carbon arc cutting is an arc cutting process in which metals to be cut are melted by the heat of an arc between a carbon electrode and the work. A high-velocity air jet traveling parallel to the electrode hits the molten pool under the arc to remove the molten metal. It is popular for gouging out defects, back-gouging root pass, and preparing joint edges. It is used to cut and gouge carbon steels, low-alloy steels, and stainless steels. Nonferrous metals may also be cut and gouged. The cut surface must be ground, normally manually. The welding equipment is the same as that used for shielded metal arc welding with the exception of a special electrode holder that holds the carbon electrode and directs the air stream and a compressed air supply. The process can be applied automatically and will produce fairly smooth cut surfaces. It is not popular as an automatic or automated process.

Plasma arc cutting is an arc cutting process that uses a constricted arc and removes the molten metal with a high-velocity jet of ionized gas issuing from the constricting orifice of the torch. It is used only to cut electrically conductive materials. For more information, see the AWS publication *Recommended Practices for Plasma Arc Cutting* [1]. This process is diagrammed in Fig. 5.4. There are two major variations. Low-current plasma cutting, or fine plasma cutting, is restricted to thin materials and usually manual application. The high-current plasma system will cut heavy materials and usually employs automated machine shape-cutting equipment.

Plasma cutting operates at a much higher speed than oxyfuel gas flame cutting. Figure 5.5 compares the cutting speeds of high-power plasma cutting, low-power plasma cutting, and oxyfuel gas cutting of steel. Shape-cutting machines similar to those used for oxyfuel gas cutting can be used for plasma arc cutting. However, travel speed must be higher.

The heavy-duty plasma torches are water-cooled, but they fit the same torch holders as those used for oxyfuel flame cutting machines. A water spray or water-shielded torch is often used to reduce smoke and noise. Water worktables that provide water in contact with the metal being cut also reduce smoke and noise. Table 5.2 is a schedule of cutting speeds using the high-current plasma system with a proprietary water injection system. Water-injected torches provide

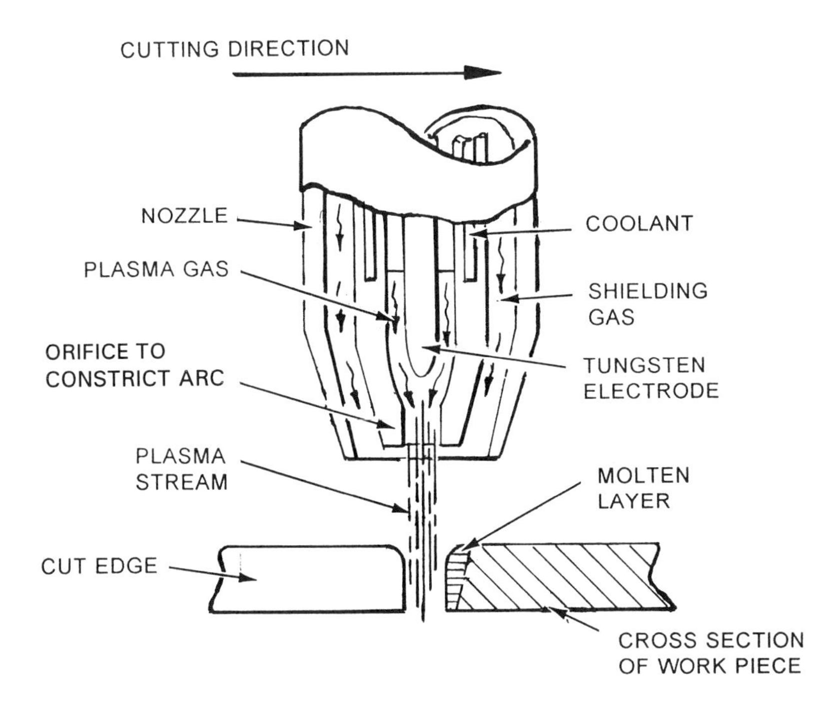

Figure 5.4 Plasma arc cutting—process diagram.

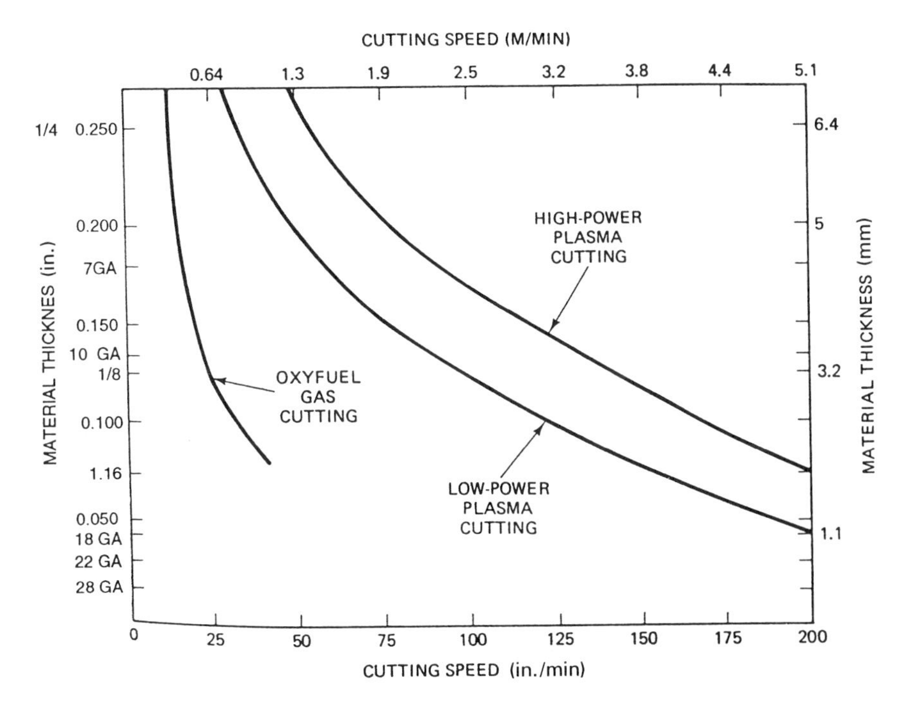

Figure 5.5 Plasma cutting schedule—mild steel.

Table 5.2 Plasma Arc Cutting Schedule—Mild Steel[a]

Material thickness		Travel speed		Arc current	Arc voltage	Plasma	Gas flow	Nozzle diameter
in.	mm	in./min	mm/min	(A)	(V)	gas	(cfm)	(in.)
1/8	3.2	175	600	300	140	N_2	165	0.166
1/4	6.4	150	300	350	150	N_2	165	0.166
3/8	9.6	125	150	350	150	N_2	165	0.166
1/2	12.8	100	100	400	160	N_2	165	0.166
3/4	19	70	50	500	165	N_2	165	0.187
1	25.4	60	30	550	165	N_2	165	0.187
1 1/2	38	30	10	600	170	N_2	165	0.187
2	50	25	8	700	190	N_2	260	0.220
3	75	15	5	900	210	Ar/He	270	0.250
4	100	10	3	1000	210	Ar/He	270	0.250
5	125	6	2	1000	210	Ar/He	300	0.250

[a]Based on water-injected plasma cutting.
Source: Hypertherm Inc.

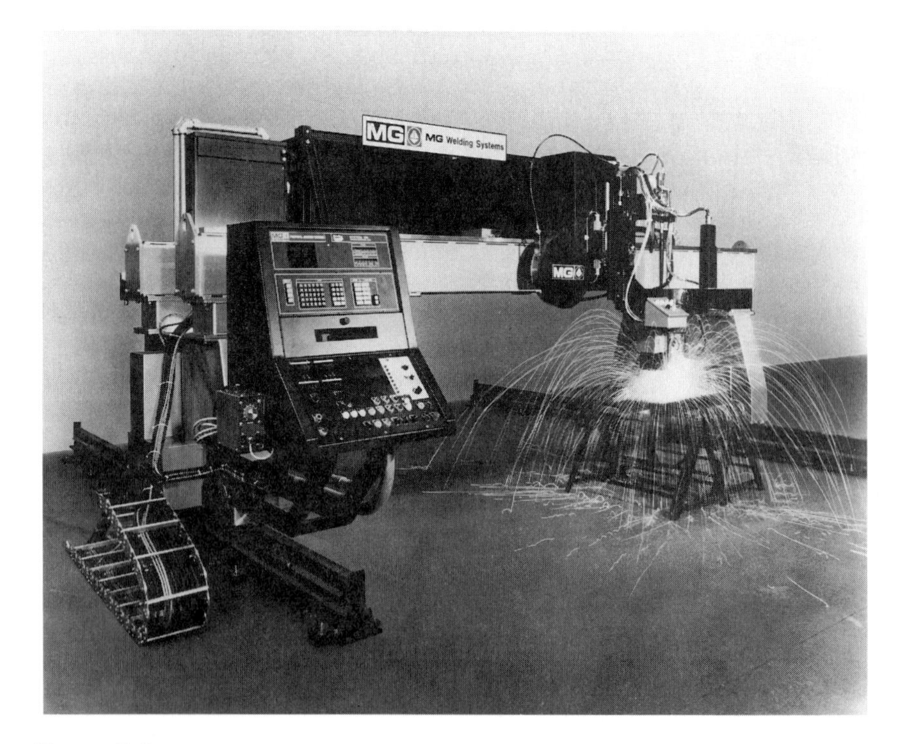

Figure 5.6 Automatic plasma arc cutting.

Figure 5.7 Robot doing plasma cutting.

higher performance. "Dry cut" plasma would have lower speed and lower cut quality. Nitrogen is used for the plasma gas. Figure 5.6 shows a gantry type of machine equipped with a heavy-duty plasma cutting torch.

A robot is sometimes used for plasma arc cutting. Figure 5.7 shows a typical application.

For the low-powered plasma variations, air is usually used for the cutting plasma. For the high-current variation, the plasma gas should be nitrogen; however, in some cases a mixture of argon and hydrogen is used. The plasma gas should be matched to the work being cut.

The other arc cutting processes are not suitable for automation and for this reason are not covered here.

5.3 Laser Beam Cutting

The basics of the laser beam were given in Section 4.4. Laser beam cutting is a thermal cutting process that severs metal by locally melting or vaporizing the metal with the heat from a laser beam. The process is used with or without assist gas to aid in the removal of molten and vaporized material. The laser beam cutting process is shown schematically in Fig. 5.8. There are several variations. In one case an inert gas jet assists in the removal of molten and vaporized material. Laser beam oxygen cutting, uses the heat from the chemical reaction between oxygen and the base metal at elevated temperatures. The necessary temperature is maintained with a laser beam. Other assist gases are compressed air and nitrogen.

The concentrated energy in a laser beam is only slightly less than the energy in an electron beam. The ability of either beam to cut materials is essentially the same, but laser beam cutting has many advantages over electron beam cutting. The laser beam can cut metal up to 1 in. (25 mm) in air. It can be used with automatic shape-cutting equipment at high travel speeds. The laser beam cut is narrower, and the angle of the cut is almost a perfect right angle. The quality of the cut surface is equal or superior to that of the best oxyfuel gas cut surface.

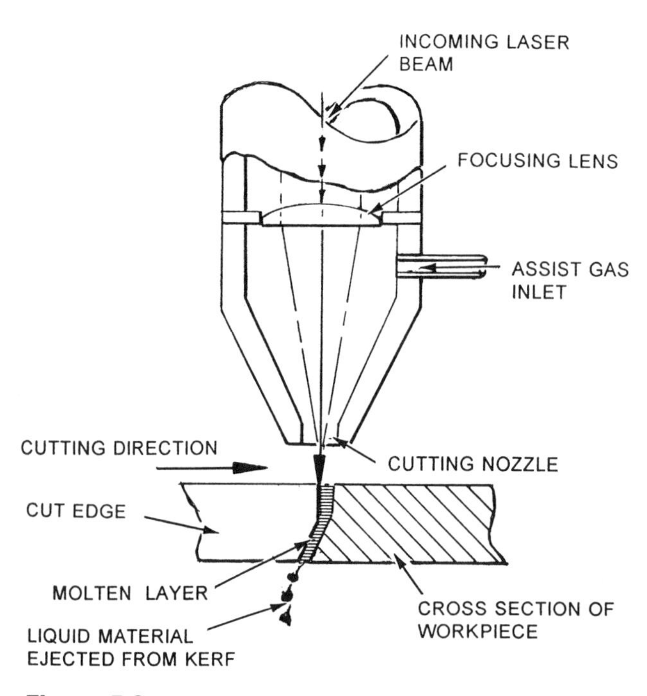

Figure 5.8 Laser cutting process.

Typical laser-cut parts are shown in Fig. 5.9. The dimensional accuracy is better than that of oxyfuel gas cutting. The edges of a laser cut are square and sufficiently smooth that additional finishing is not necessary, as shown by Fig. 5.10.

Precision laser cutting machines of various types are commercially available. The more common type uses a gantry frame that spans the workpiece. Others are similar to turret punches; in some case the table moves, in others the head moves. Laser cutting systems are often combined with NC turret punches with two axes of motion for sheet metal processing. Software is available for nesting parts on material for minimum scrap loss. Figure 5.11 shows a typical precision laser cutting machine.

Lasers are also used with robots, often for formed parts. Figure 5.12 shows a Nd:YAG laser cutting thin sheet metal for a pickup truck floor panel. In this case a proximity sensor is employed to maintain constant work-to-beam tip dimensions. The robot is finding wider use with solid-state lasers.

Table 5.3 lists the metals and nonmetals that can be cut with a CO_2 laser and the cutting speeds for specific thicknesses.

The power required for laser cutting is relatively low. In general, the

Figure 5.9 Typical parts cut with a laser.

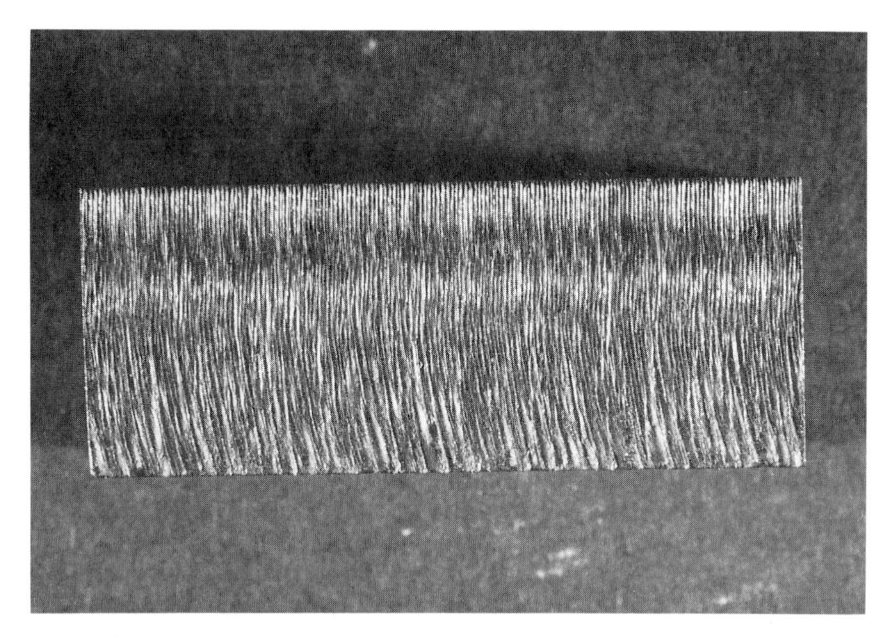

Figure 5.10 Surface of cut edge.

Figure 5.11 Laser cutting machine.

Figure 5.12 Robot laser cutting.

continuous-wave CO_2 laser with up to 1 kW power is sufficient to cut thin-gauge metals.

As mentioned previously, there are two commercial types of industrial lasers used for cutting: the CO_2 (carbon dioxide) laser and the Nd:YAG (neodymium-doped crystal, yttrium, aluminum, garnet) laser. They generate laser beams of different wavelengths. Both are invisible. The CO_2 laser emits a beam of 10.6 μm wavelength. The output light wavelength of the Nd:YAG is 1.06 μm. CO_2 and Nd:YAG lasers have different cutting abilities depending on the absorption of their wavelength. Table 5.4 lists the materials that can be cut with the two types of lasers and the quality of the cut. Some materials are transparent to a specific wavelength and transmit that wavelength. Optical grade fused silica glass

Table 5.3 CO_2 Laser Cutting Speeds

Material	Thickness		Cutting speed	
	mm	in.	m/min	in./min
ABS plastic	4	0.157	4.5	177.1
Acrylic	6	0.236	1.7	66.9
Cardboard	0.1	0.004	96.0	3779.5
Ceramic tile	6.3	0.248	0.3	11.8
Formica	1.6	0.063	7.8	307.1
Plywood	18	0.708	0.5	19.7
Wool suit material	—	—	48.0	1889.7
Galvanized steel	1	0.039	4.5	177.1
High-carbon steel	3	0.118	1.5	59.0
Mild steel	1	0.039	4.5	177.1
Stainless steel	2.8	0.110	1.2	47.2
Titanium	3	0.188	4.1	161.4

used in fiber optics is transparent to the Nd:YAG laser and is used to transmit the beam to the cutting nozzle. The laser beam from the CO_2 laser must be transmitted by mirrors. For this reason the CO_2 is popular for the two-dimensional automatic cutting machines. The fiber-optic Nd:YAG laser is more popular with the multiaxis robot delivery system.

Recent developments of the Nd:YAG laser have increased its power output

Table 5.4 Metals and Nonmetals—Cut Quality

Material	Laser type	
	1.06 μm Nd:YAG	10.6 μm CO_2
Mild steel	Excellent	Excellent
Stainless steel	Excellent	Excellent
Aluminum	Good	Good
Copper	Good	Difficult
Gold	Good	Not possible
Titanium	Good	Good
Ceramics	Fair	Good
Acrylics	Poor	Excellent
Polyethylene	Poor	Excellent
Polycarbonate	Poor	Good
Plywood	Poor	Excellent

and have provided for continuous-wave operation. The data given in Fig. 5.13 show the travel speed of lasers for cutting aluminum and steel of various thicknesses. Curves are shown for 600 W and 1500 W lasers used for steel; the higher power laser uses a jet of oxygen to improve cutting speed. Sharp corners, smooth surfaces, narrow cut width, minimum thermal damage, nonadherent dross, and 90° surfaces are all achieved with laser cutting. It is necessary to compare cutting ability and speeds for the material you plan to cut before deciding which type of laser to use, pulsed or continuous-wave CO_2 or Nd:YAG.

The parameters for laser cutting are the laser type and power, travel speed, assist gas type and pressure, and the cutting nozzle diameter and distance above the work. The focus of the laser beam is related to the nozzle parameters and affects the surface of the cut and the kerf and its angle. Cutting accuracy depends almost entirely on the accuracy of the motion system. The assist gas is usually an inert gas when a weld-ready edge is desired. The use of oxygen as the assist gas allows a higher travel speed.

Laser beam cutting has largely replaced electron beam cutting for two reasons. Electron beam cutting is done in a vacuum chamber, which complicates the introduction of material to be cut and the removal of cut parts. In addition, the material melted to make the cut "plates out" on the interior surface of the chamber.

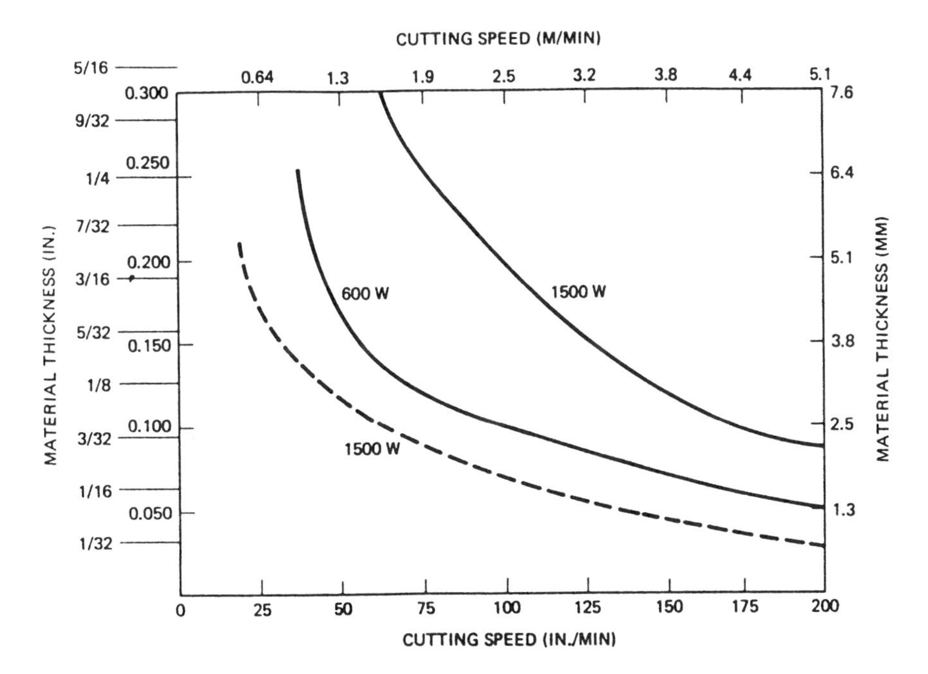

Figure 5.13 Cutting speed for (——) steel and (– – –) aluminum.

5.4 Thermal Spraying

The AWS Master Chart shows thermal spraying as a group of processes allied with welding. In all thermal spray processes a deposit of finely divided metallic or nonmetallic surfacing materials is sprayed on a substrate or base metal. There are three thermal spray processes: arc spraying, plasma spraying, and flame spraying. They differ considerably, and each has several variations. The surfacing material may be supplied in the form of powder, wire, or rod.

If the surfacing material is available in wire form, the electric arc spray or the flame spray process can be used. If it can be obtained only in powder form, the flame spray or plasma spraying process must be used.

Originally, spraying was a semiautomatic process, with the operator holding and manipulating the spray gun. Mechanized spray systems, and recently robotic spraying, have since become more popular.

Flame spraying was the original thermal spray process and was called metallizing. It uses an oxyfuel gas flame as the source of heat for melting the surfacing materials. Compressed gas is used to atomize and propel the molten metal to the surface of the substrate. There are two major variations, one using metal in wire form and the other using material in powder form. The material is melted in the gas flame and atomized by an air jet that propels the atomized particles to the workplace. The flame spraying process is slow but very flexible.

A recent variation of flame spraying is the high-velocity oxyfuel spraying system. It uses a high-speed jet that heats the powder particles and propels them to the workpiece. It operates at higher temperatures and with much higher particle velocities than the older methods. The density of the deposited coating is extremely high, and the bond with the workpiece is extremely good. This system is normally mechanized due to the noise and power involved.

Arc spraying uses an arc between two consumable electrodes of surfacing materials as the heat source. As the wires melt in the arc, a compressed gas jet atomizes the molten metal and propels the fine molten particles to the workpiece. A welding power source is used to maintain the arc. The gun is normally mounted in a holder with a movement mechanism. This system will deposit 15–100 lb of metal per hour, depending on the current level and the metal being sprayed. The deposition rate of arc spraying is three to five times greater than that of flame spraying, and the deposit is more dense and has a better mechanical bond. This process can be mechanized with motion devices or with a robot, as shown by Fig. 5.14.

Plasma spraying is the newest variation and uses a nontransferred arc to create an arc plasma for melting and propelling the surfacing material to the base

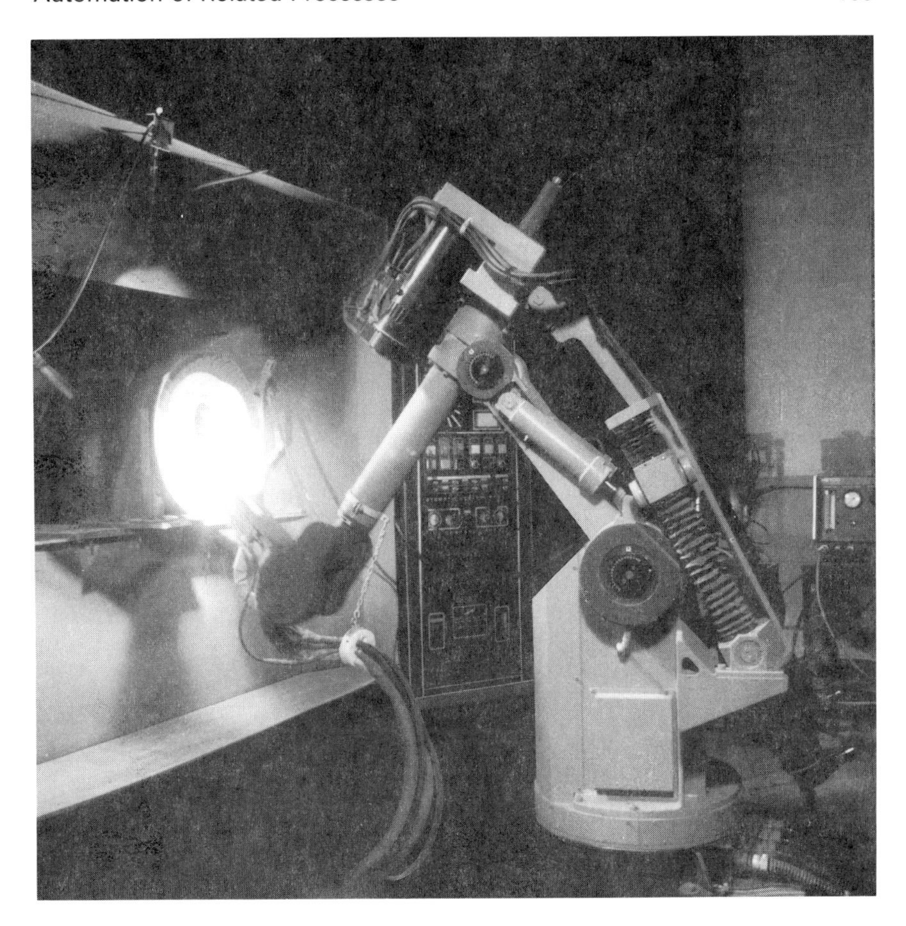

Figure 5.14 Arc spraying with a robot.

metal. The temperature is much higher than for either arc or flame spraying. The material to be sprayed must be in powder form. The high-temperature plasma melts the powdered material and propels it to the surface of the workpiece. Inert gas and extrahigh temperatures are used. The mechanical and metallurgical properties of the coating are superior to those achieved by either flame spraying or arc spraying.

The robot has become the popular way to automate thermal spraying operations. Figure 5.15 shows a robotic plasma spraying operation. The controller

Figure 5.15 Plasma spraying with a robot.

used for a thermal spray operation must control the spraying variables, including motion. This becomes extremely complex. Motion devices need not be quite as accurate as those used for arc welding but must be more accurate than those used for paint spraying.

The selection of a thermal spraying process, thermal spraying deposits, and materials are beyond the scope of this book.

Reference

1. *Recommended Practices for Plasma Arc Cutting,* AWS C5.2, American Welding Society, Miami, FL.

6

Procedures, Schedules, and Variables

6.1 Welding Procedure Development

A welding procedure is a document that describes how to make a particular weld. It provides a record of all welding variables. If the part is covered by a code or specification, the procedure must show the details of the weld joint design and record the welding process and the method of application. It includes all parameters—the welding current, arc voltage and electrode polarity, type and size of electrode, travel speed, shielding gas, electrode angle, etc.

The development of the welding procedure may require experimental work to obtain the optimum values of the various parameters. The most common way to develop a procedure is to consult welding procedure schedules that relate to the welding process, the metal being welded, the metal thickness, and the desired weld. This chapter provides schedules for welding with the more common processes used in mechanized welding. They are for low-carbon mild steels and for popular weld types. These schedules will produce quality welds under normal conditions. They do not guarantee specific properties, only that the strength should equal that of the basic mild steel. These schedules are starting points.

When developing a welding procedure, it is important to know the end use of the weldment being produced. The end use might require the highest weld quality for structural integrity or normal quality and maximum productivity. If maximum productivity is desired, the welding procedure schedules in this chapter can be modified to increase travel speed and welding current. Experimental welds should be made using higher travel speeds and higher welding currents to produce the weldment desired.

It is extremely important to understand the interaction of welding variables, the effect of changing one or the other, and the limit of specific variables. This information is presented in the next section.

Once the variables are selected, a test weldment should be made and tested to see that it meets the objective and the requirements. For code work, welding procedures must be qualified. To qualify a procedure, the weld is made in accordance with the procedure and then tested to determine whether the desired quality is obtained. The equipment and the operator must also be qualified. It is usually necessary to use the forms provided by the specific code. This is a convenient way to record all the relevant parameters and factors. With mechanized equipment, additional data may be required. Special forms may be developed to record all of the pertinent data such as the welding current waveform for pulsing current and ramping up or down the welding current or travel speed. For critical or repetitious work it is important to provide all welding procedure details that completely describe all of the relevant parameters. Upon the completion of the procedure development, the product must be tested in accordance with the code, and if the tests are acceptable the procedure is qualified.

6.2 Arc Welding Variables

Arc welding variables are those parameters that can be adjusted to control the weld. During manual welding, the welder can increase or decrease the speed of travel. The welder can also increase or decrease the arc length, which increases or decreases the arc voltage. With a conventional power source this also changes the welding current. The welder can change the angle of the electrode, which affects the penetration and shape of the weld. This must also be accomplished in automatic welding. A good welding procedure has the proper balance to produce a smooth-running arc that will deposit high quality weld metal. Many welding variables interrelate, and some are more easily changed and are more useful for the control of the weld. These variables and the changes produced are essentially the same for all of the arc welding processes in which the weld metal crosses the arc.

Arc welding variables can be divided into three classifications: preselected or distinct level variables, primary adjustable variables, and secondary adjustable variables.

The distinct level variables cannot be changed while welding. They are changed in increments and are therefore fixed during a weld. They affect the weld and are included in a welding procedure. Distinct level variables include electrode size and type, welding current type and polarity, shielding gas composition. These variables are selected on the basis of the type of metal being welded, its thickness, the joint design, welding position, desired deposition rate, and required weld appearance. The starting point for preselecting these factors can be determined by reviewing the weld schedules in this chapter.

The primary adjustable variables are those that are most commonly used to change the characteristics of the weld: travel speed, arc voltage, and welding current. They can be easily measured and continually adjusted over a wide range. These primary variables control the formation of the weld by influencing the depth of penetration, the bead width, and the bead reinforcement (or height) (see Fig. 6.1). They also affect deposition rate, arc stability, and spatter level. Specific values are assigned to these primary adjustable variables in the welding schedule.

The relationship between penetration and primary variables is shown by Fig. 6.2. Weld penetration is very important and must be sufficient to provide the joint strength required. Weld penetration is affected by all three classes of arc welding variables—primary, secondary, and preselected. The simplest way to increase welding penetration is to increase welding current. If the welding current exceeds the maximum current allowable for the electrode size selected, the next larger electrode should be used. This is the preselected variable. Reducing travel speed will also increase penetration, but this is less desirable than increasing current. Arc voltage will also, up to a point, increase penetration.

Weld bead width is important for groove welding, and a wider bead can be used to compensate for minor changes in joint location. The bead width relationship to the primary variables is shown by Fig. 6.3. The most effective way to

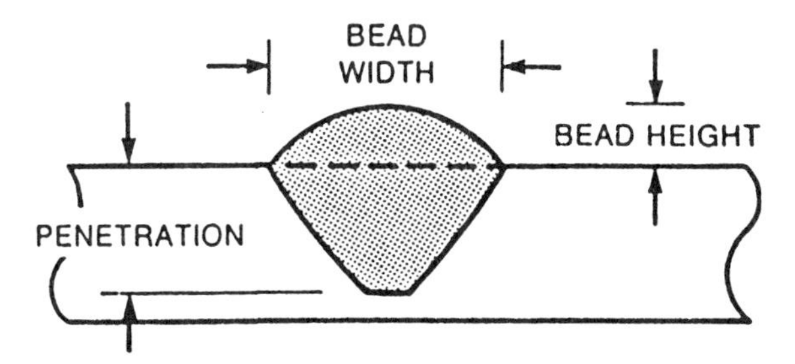

Figure 6.1 Bead height, width, and penetration.

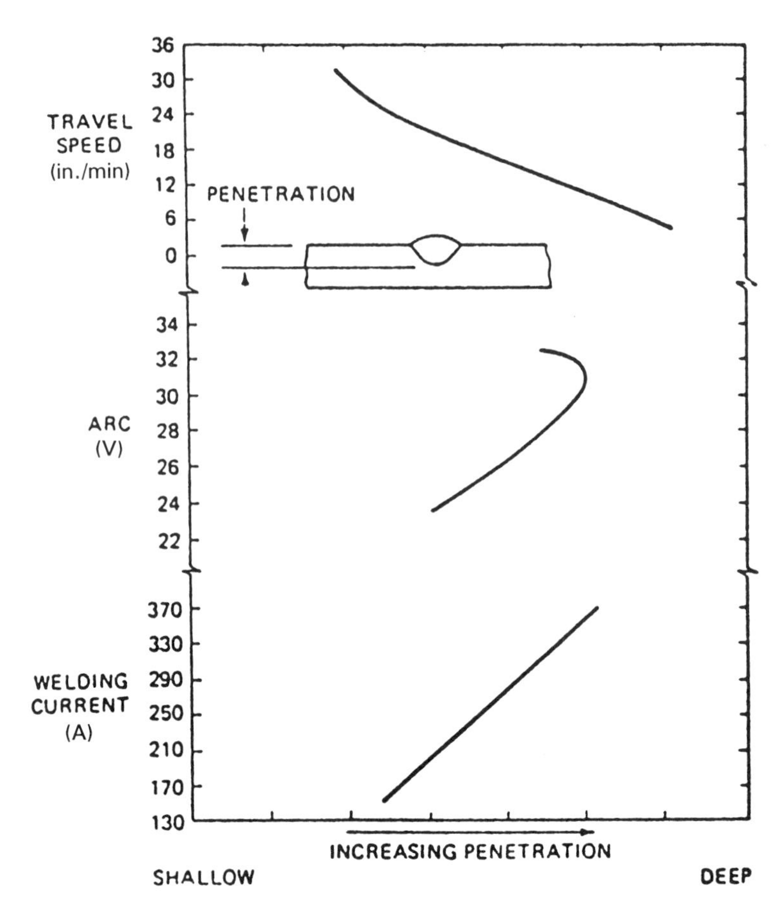

Figure 6.2 Weld penetration related to primary variables.

control bead width is by adjusting the arc voltage. Figure 6.4 helps explain the arc length–weld bead width relationship. It is a straight-line relationship, but arc voltage can be changed only over a narrow range. Reducing travel speed widens the bead, and increasing current widens the bead up to a limit.

The relationships between weld bead reinforcement or height and the three primary variables are shown by the curves in Fig. 6.5. Weld reinforcement is important when considering the requirements for filling a groove with the proper amount of metal using the desired number of passes. The weld bead height is most effectively controlled by travel speed, which is a relatively straight-line relationship. This is the first choice. Weld bead height versus welding current is also a relatively straight-line relationship, and welding voltage or arc length would be a third choice for controlling weld bead reinforcement.

If either welding voltage or welding current is changed excessively, the arc will become unstable. This is because there is a specific relationship between arc

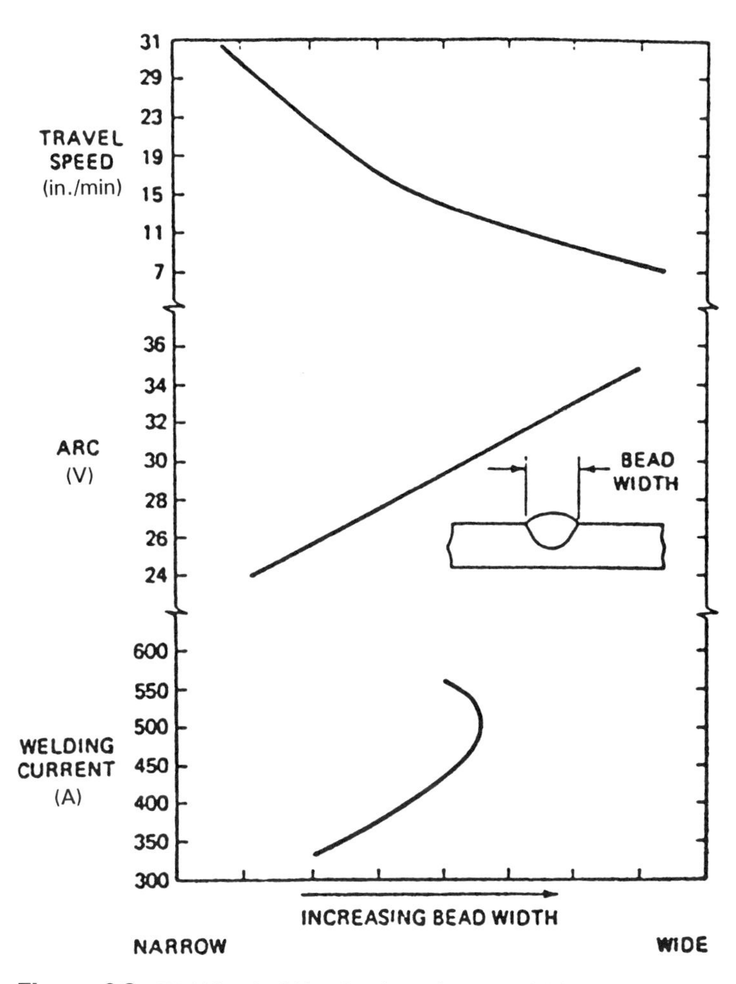

Figure 6.3 Weld bead width related to primary variables.

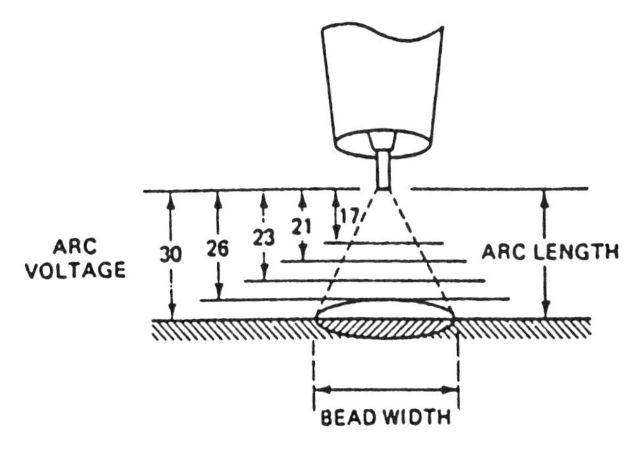

Figure 6.4 Arc length–weld bead width relationship.

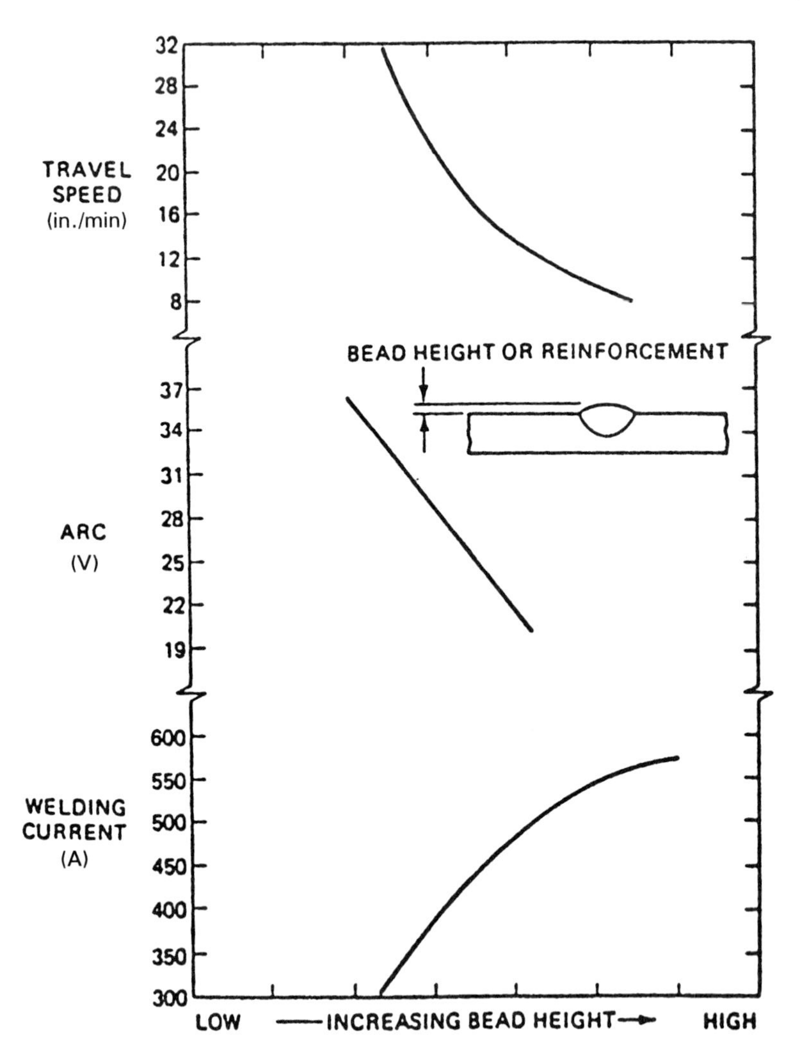

Figure 6.5 Weld bead reinforcement related to primary variable.

voltage and welding current to maintain the stable operating range. This relationship changes for different processes, shielding gas atmospheres, and electrode sizes. The relationship shown by Fig. 6.6 is typical, even though the values of current and voltage will change for the different processes, shielding gases, electrode sizes, etc.

The secondary adjustable variables include tip-to-work distance (stickout) and electrode or nozzle angle. These variables can be adjusted during the welding operation and can also be changed continuously over a fairly wide range. They do not directly affect weld bead formation; instead, they cause a change in a

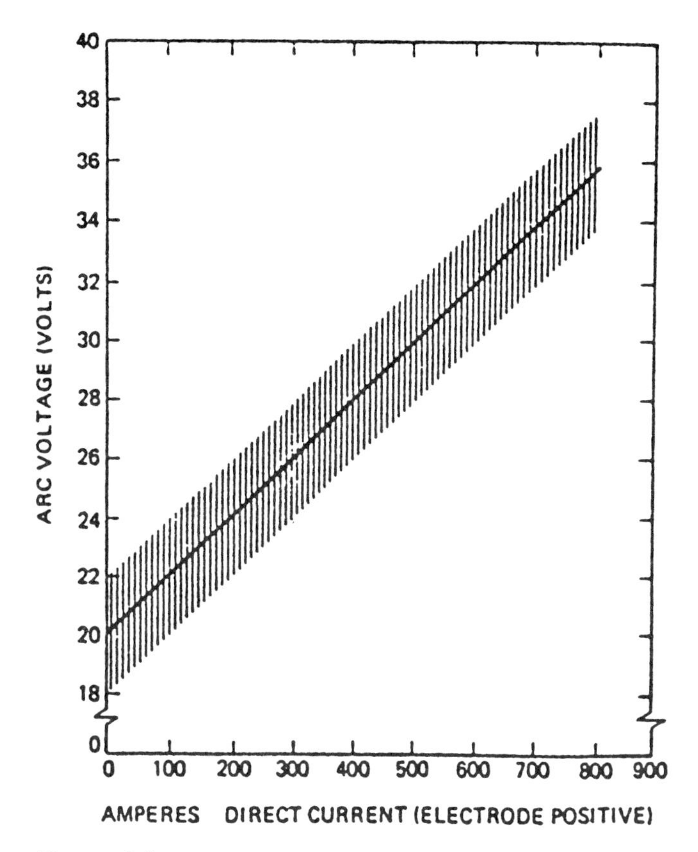

Figure 6.6 Welding voltage–current relationship.

primary variable, which in turn causes a change in bead formation. Secondary adjustable variables are more difficult to measure and accurately control than the primary adjustable variables. They are assigned values and are usually included in the welding schedules.

The welding current and electrode wire feed rate have a direct relationship with each other. This relationship can be changed with changes in *stickout,* also known as electrode extension. This is shown in detail by Fig. 6.7. In the stickout area, preheating of the electrode occurs that is sometimes called I^2R heating. The electrode wire extending from the current pick-up tip to the arc is heated by the tremendous amount of current being carried by the relatively small electrode wire. This preheats the electrode wire so that when it enters the arc it is at an elevated temperature, which increases the melt-off rate. Increasing stickout increases the deposition rate only if the wire feed speed is increased sufficiently to maintain the current at a constant value. The relationship between stickout and welding current is shown by Fig. 6.8. Increasing the stickout will reduce the welding

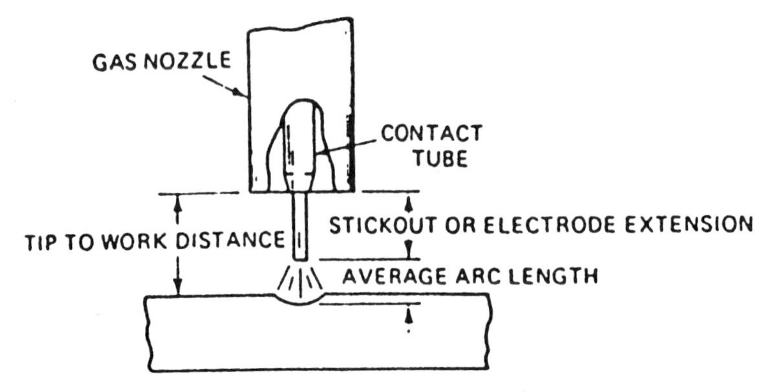

Figure 6.7 Stickout or electrical extension.

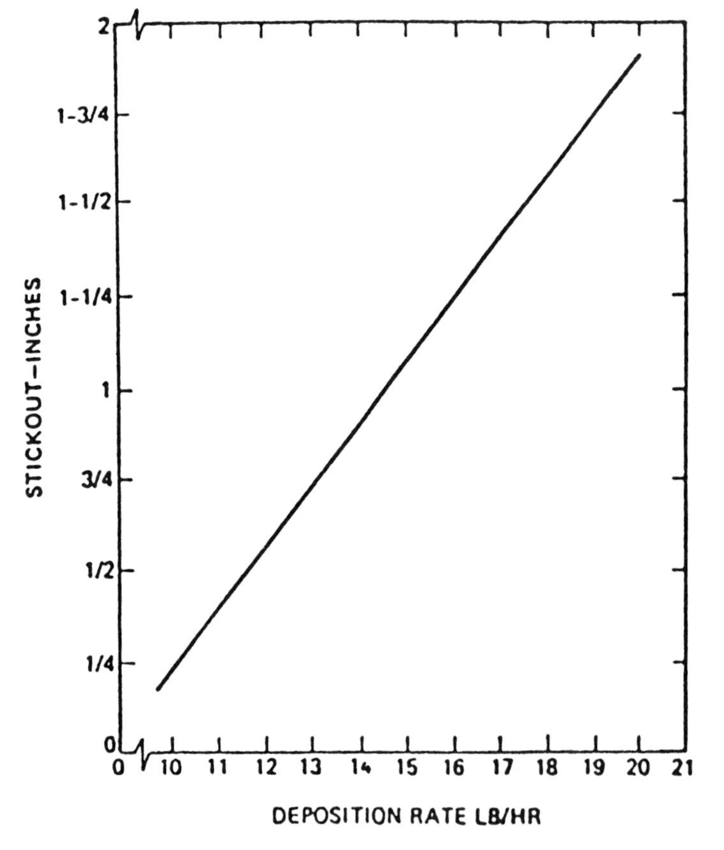

Figure 6.8 Stickout versus deposition rate.

current in the arc by a fairly large amount when the wire feed speed rate is not changed. This reduces penetration a proportional amount.

Another secondary adjustable variable is electrode or nozzle travel angle, which affects penetration as shown by Fig. 6.9. The travel angle is the angle that the electrode or the centerline of the welding torch makes with a reference line perpendicular to the axis of the weld. The travel angle is sometimes described as either a drag angle or a push angle. The drag angle points backward from the direction of travel. The push angle points forward in the direction of the welding.

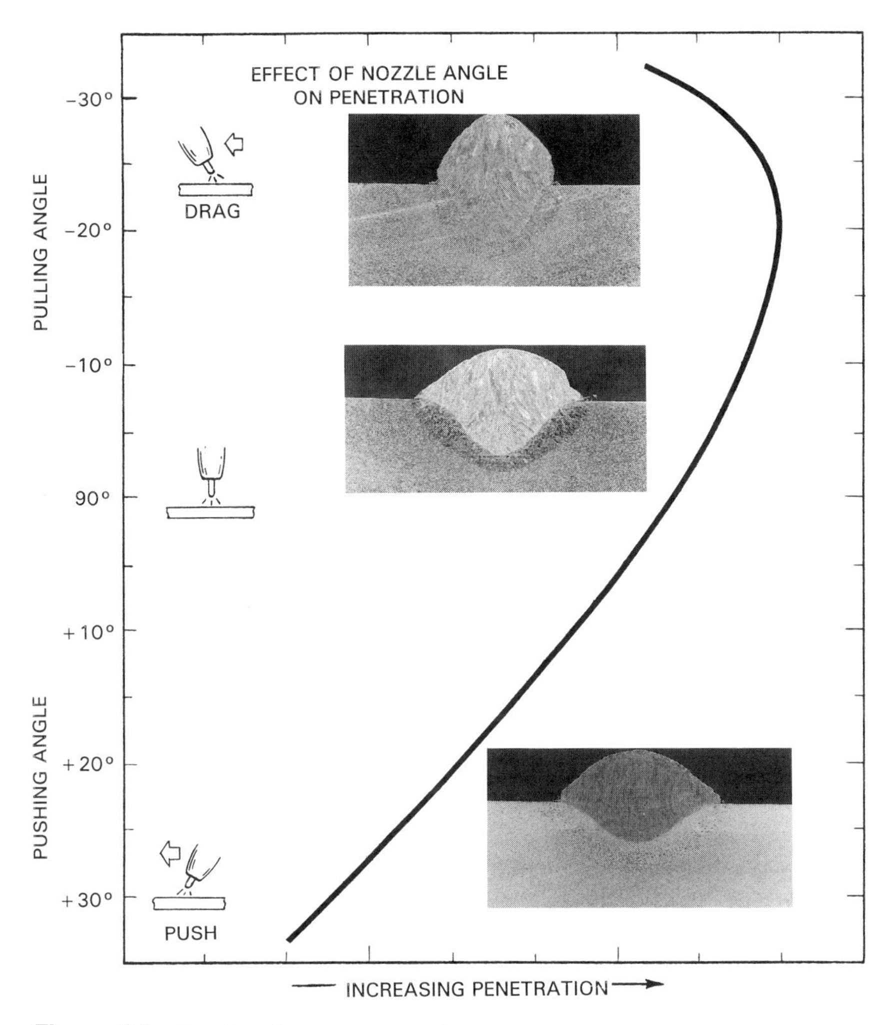

Figure 6.9 Travel angle versus penetration.

Table 6.1 Summary of Variable Adjustments to Change the Weld

	Primary welding variable			Secondary welding variable		Distinct welding variable	
Change required	Arc voltage (arc length)	Welding current (wire feed speed)	Travel speed	Travel angle	Stickout or tip-to-work distance	Electrode size	Shielding gas
Deeper penetration		1. Increase		2. Trailing max. 25°	2. Decrease	5. Smaller[a]	4. CO_2
Shallower penetration		1. Decrease		3. Leading	2. Increase	5. Larger[a]	4. $Ar + CO_2$
Bead height and bead width							
Larger bead		1. Increase	2. Decrease		3. Increase[a]		
Smaller bead		1. Decrease	2. Increase		3. Decrease[a]		
Higher, narrower bead	1. Decrease			2. Trailing	3. Increase		
Flatter, wider bead	1. Increase			2. 90° or leading	3. Decrease		
Higher deposition rate		1. Increase			2. Increase	3. Smaller[a]	
Lower deposition rate		1. Decrease			2. Decrease[a]	3. Larger[a]	

Key: 1, First choice; 2, second choice; 3, third choice; 4, fourth choice; 5, fifth choice.

[a]When these variables are changed, the wire feed speed must be adjusted so that the welding current remains constant.

Maximum penetration results when a drag angle of 15–20° is used. Torch angle can be continually changed with robot welding, and it can also be changed during welding with most automated systems. If the torch travel angle is altered from the optimum condition, penetration decreases. From a drag angle of 15° to a push angle of 30°, the relationship between penetration and travel angle is almost a straight line. A drag angle of greater than 25° should not be used.

The adjustments just mentioned are summarized in Table 6.1. Making experimental welds as you change variables will soon help you understand these relationships.

6.3 Metal Transfer and GMAW Procedure Schedules

Gas metal arc welding (GMAW) is the most popular of the arc welding processes used for automated and robotic arc welding. There are at least four variations that relate to the mode of metal transfer across the arc. The mode of metal transfer has a strong influence on the welding operation and depends on the metal being welded, the type and size of the electrode, and the shielding gas. It also affects the type of weld that can be produced. Once the requirements for a particular application are known, the appropriate metal transfer mode can be selected. Table 6.2 provides information to help you select the metal transfer mode. The proper selection of metal transfer mode will contribute to the quality and appearance of the finished weld.

The four major types of metal transfer are

Spray transfer
Globular transfer
Short-circuiting transfer
Pulsed spray metal transfer

They are distinguished by the size, frequency, and characteristics of the metal drops crossing the arc. These types or modes of metal transfer are fairly well defined; however, in the transition zones between modes, two types of transfer may occur at the same time.

Spray transfer (Fig. 6.10) is a very smooth mode of transfer. The droplets crossing the arc are smaller in diameter than the electrode. The welding current density is relatively high, and the deposition efficiency approaches 100%; there is very little spatter, and that which occurs is of a very fine variety. Spray transfer occurs primarily with a relatively small diameter electrode wire and an arc atmosphere of argon or one with a predominance of argon. The transition amperage is around 250 A, depending on the size of the electrode wire. Below this transition point the droplets increase in size, approaching the diameter of the electrode wire, and more spatter occurs.

Table 6.2 Variations of Gas Metal Arc Welding Relating to Metal Transfer

	Metal transfer			
	Globular	Short-circuiting	Spray	Pulsed spray
Shielding gas	CO_2	CO_2 or CO_2 + argon (C25: 75% Ar, 25% CO_2)	Argon + oxygen (1–5% O_2)	Argon + oxygen (1–5% O_2)
Metals to be welded	Low-carbon and medium-carbon steel; low-alloy, high-strength steels	Low-carbon and medium-carbon steels; low-alloy, high-strength steels; some stainless steels	Low-carbon and medium-carbon steels; low-alloy, high-strength steels	Aluminum, nickel, steels, nickel alloys
Metal thickness	10 gauge (0.140 in.); up to ½ in. without bevel preparation	20 gauge (0.038 in.) to ¼ in.; economical in heavier metals for vertical and overhead welding	¼–½ in. with no preparation; max. thickness practically unlimited	Thin to unlimited thickness
Welding position	Flat and horizontal	All positions (also pipe welding)	Flat and horizontal with small electrode wire, all positions	All positions
Major advantages	Low-cost gas; high travel speed, deep penetration, high deposition	Thin material—will bridge gaps; minimum cleanup	Smooth surface; deep penetration; high travel speed	Uses larger electrode
Limitations	Spatter removal sometimes required; high heat	Uneconomical in heavy thickness, except out of position	Position; minimum thickness	Special power source
Appearance of weld	Relatively smooth; some spatter	Smooth surface; minor spatter	Smooth surface; minimum spatter	Smooth surface; minimum spatter
Travel speed	Up to 250 in./min	Max. 50 in./min	Up to 150 in./min	Up to 100 in./min
Electrode diameter (in.)	0.045, 1/16, 5/64, 3/32	0.030, 0.035, 0.045	0.035, 0.045, 1/16, 3/32	1/16, 5/64, 3/32, 1/8

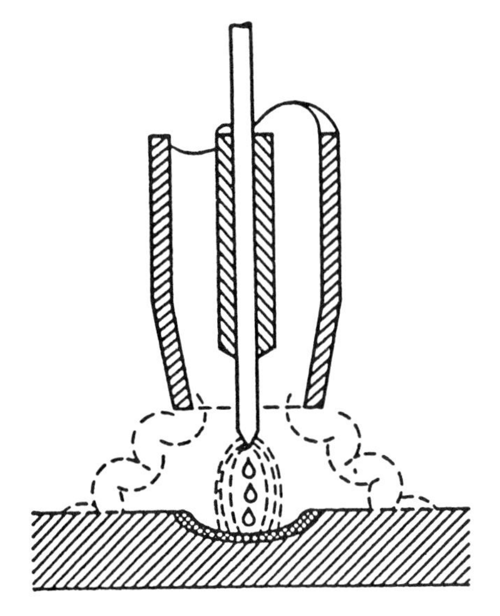

Figure 6.10 Spray transfer.

Since a relatively high current density is required for spray transfer, the welding current is relatively high. This creates a large molten weld pool and deep penetration. Normal spray transfer is limited to the flat and horizontal positions and cannot be used on thin materials. If fit-up is poor, the spray transfer mode is difficult to use because burn-through will occur. The spray transfer mode is used primarily for welding steel, aluminum, and stainless steel.

Globular transfer, illustrated in Fig. 6.11, is carried out below the transition range, which is below the true spray transfer mode. In this transition range the drops of molten metal are approximately the same size as the electrode wire diameter. Globular transfer occurs when CO_2 gas or CO_2-rich gas with a relatively small amount of argon is used as the shielding atmosphere. Welding arcs in a CO_2 atmosphere are longer than those in an argon atmosphere, and the voltage is higher. This type of metal transfer results from the "cathode jet" when the electrode is positive (i.e., the anode). The cathode jet originates from the workpiece and supports the molten ball of metal on the tip of the electrode. As melting continues, the molten globule grows in size until its diameter reaches approximately twice the diameter of the electrode wire. As it grows it takes on unusual shapes and is finally separated from the electrode and transferred across the arc by electromagnetic and gravity forces. As the globule transfers across the arc, its irregular shape and direction change, and it sometimes reconnects with the electrode while it is in contact with the work. This causes a short circuit, which momentarily extinguishes the arc.

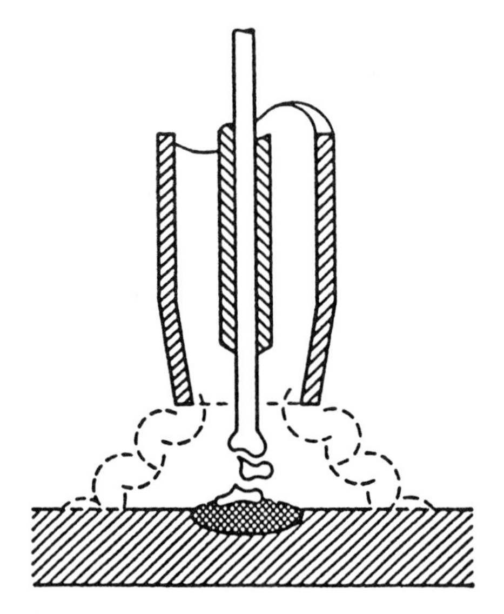

Figure 6.11 Globular transfer.

Globular transfer produces more spatter than spray transfer. The spatter comes from the molten puddle as well as from the metal transferring across the arc. The frequency of globular detachment and flight is random but relatively low. Globular transfer takes place at a lower current density than spray transfer, and the resulting weld deposit is not as smooth. The welding parameters that produce globular transfer are usually used for welding in the flat and horizontal positions.

The short-circuiting mode of metal transfer (Fig. 6.12) allows all-position welding on thinner materials and can be used when fit-up is poor. The mechanism of short-circuiting transfer is illustrated by Fig. 6.13. In this transfer mode the molten tip of the electrode wire is supported by the cathode jet; however, the electrode wire is feeding at a high rate of speed so the molten tip will occasionally come into contact with the molten weld pool. A short circuit occurs, and the surface tension of the molten pool draws the molten metal off the tip of the electrode into the pool. This creates a bridge across the arc gap and momentarily shorts or extinguishes the arc. The molten metal on the end of the electrode will then separate from the electrode wire and reestablish the arc. These conditions continue at a random but relatively low frequency. The short-circuit arc outages occur so rapidly that they are not noticed while welding. This mode of metal transfer, which allows for all-position welding, occurs with low arc voltage (25 V maximum) and low welding current (200 A maximum).

Short-circuiting transfer is normally used with small-diameter electrode wires and with a CV (constant-voltage) power source. It is generally limited to the welding of steels, since CO_2 as a shielding atmosphere cannot be used on nonferrous metals. There is a small amount of spatter involved. The molten pool

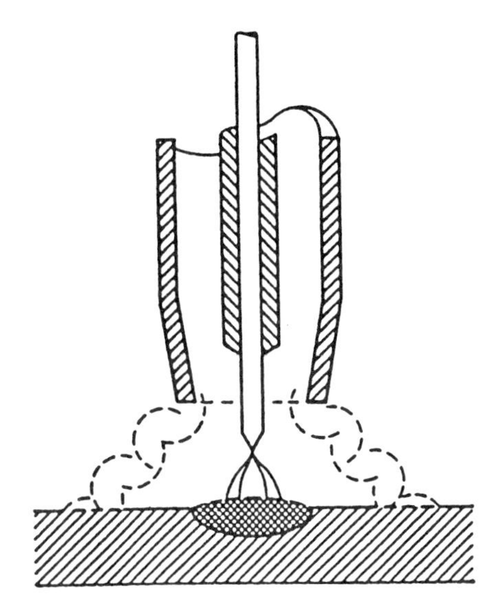

Figure 6.12 Short-circuiting transfer.

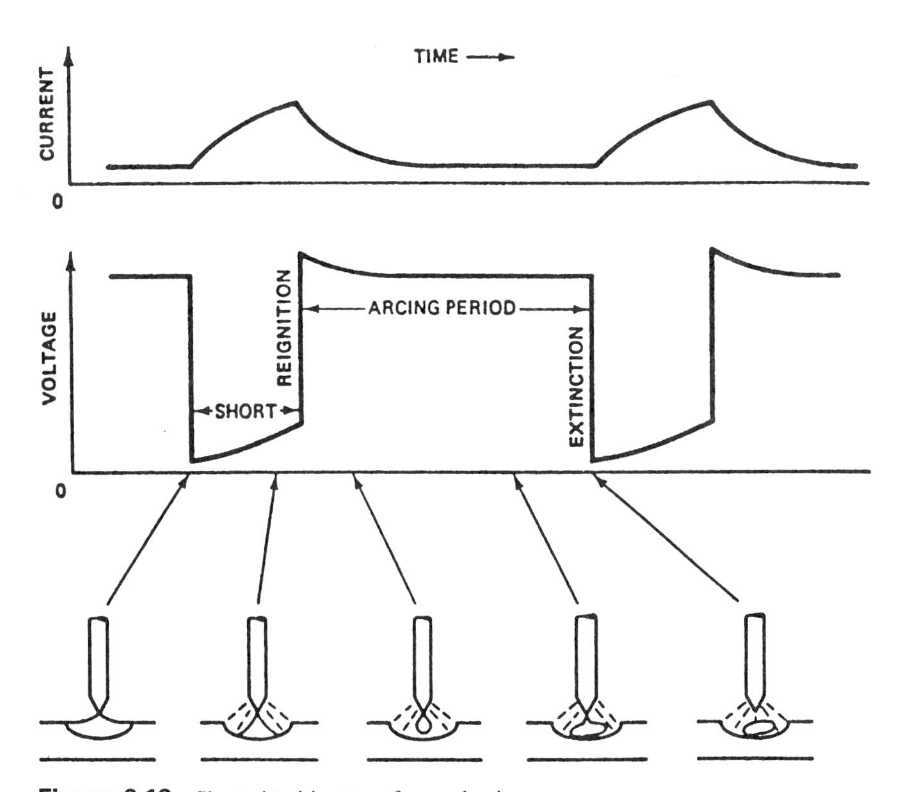

Figure 6.13 Short-circuiting transfer mechanism.

is relatively small and easily controlled. Weld penetration can be controlled by travel speed. This metal transfer mode of welding will bridge gaps much wider than the thickness of the material. It is extremely popular for welding thin materials.

The pulsed spray mode was developed to overcome the limitations of the three modes of metal transfer mentioned above. In this mode, discrete droplets of molten metal are transferred across the arc in a regular pattern. A special power source is used to provide for pulsed spray transfer. The welding current produced by the pulsing power source varies between a low or background level and a high or pulsed level, shown by Fig. 6.14. The high (peak) current level is above the transition point, and the low (background) current level has sufficient energy to sustain the arc but not sufficient to transfer metal across it. Each peak current pulse supplies enough energy to transfer the molten droplet across the arc to the puddle, with one drop of metal being transferred across the arc at each high pulse of the current. In this way pulsed current power allows the spray transfer mode to be used at a lower average current than normal. This means a lower heat input and a smaller controllable weld pool, which allows welding in all positions and on thinner materials. The pulsed spray transfer mode produces droplets of a size approximately equal to the diameter of the electrode wire. Weld spatter is much less than with spray or globular transfer. A welding atmosphere rich in argon gas with 1–5% oxygen or with 5–20% CO_2 allows welding on ferrous alloys. Pure argon or argon–helium mixtures are generally used for the pulsed spray welding of aluminum. Pulsed spray welding produces very smooth welds that require a minimum of metal finishing.

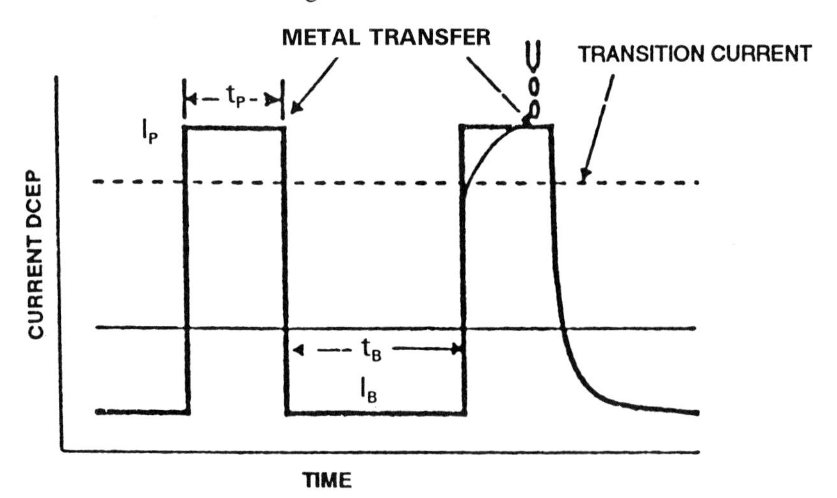

Figure 6.14 Machine output waveform. I_P, peak current (pulse amplitude); t_P, time 0 peak (pulse width); t_B, time 0 background; I_B, background current.

Pulsed spray metal transfer can use larger than normal electrode wires, which is an economic advantage because larger wires are less expensive. Originally, the power sources developed for pulsed spray welding were designed so that the high current pulse was at the same frequency as the incoming utility line current. Today's advanced power source produces an almost infinite number of current frequencies and allows the operator to change the waveform of the high pulse. This provides the synergic mode of welding, which produces one droplet per pulse. This type of machine can be tuned for different types of welding according to the gas mixture, base material, and welding position. It allows the use of large electrode wires and can be used for almost any weldable metal in any welding position. The resulting welds are extremely smooth with a minimum of weld spatter.

Each of the following schedules (Tables 6.3–6.5) is for welding with a specific mode of metal transfer. The mode of metal transfer is noted for each schedule, but not the base metal type. The tables show the base metal thickness or fillet size, the weld type, electrode diameter, welding current, wire feed speed, arc voltage, gas type and flow rate, and welding travel speed. All of the schedules are based on using direct current, electrode positive (DCEP). Both the welding current and electrode wire feed speed are listed, since it is sometimes more convenient to establish the welding current without exactly knowing the wire feed speed.

Table 6.3 is the spray arc metal transfer schedule. Note that this schedule is for larger welds so it shows the number of passes to be used. Table 6.4 is the globular metal transfer schedule. This schedule is for steel and uses CO_2 for shielding. It can be used at high currents. Table 6.5 is the short-circuiting arc transfer schedule. This is the most common schedule and is used on thin carbon steel. It can use CO_2 or CO_2—argon mixtures for gas shielding.

For pulsed spray transfer schedules, variable-frequency power sources are used. Consult the equipment manufacturer for schedules that match the output of the machine.

New shielding gases are constantly being introduced; many are premixed and may be promoted heavily. It is impossible to provide information concerning each of these gases; however, Table 6.6 is a guide to shielding gases and their application in gas metal arc welding. The applications are listed for each gas, and the remarks column describes the type of metal transfer that will result.

It is essential that welds be made on the production part using the parameters selected. It is possible that more than one combination of parameters or type of metal transfer may be involved. If this is true, the weld should be examined from the appearance and strength point of view and also from the productivity point of view. After the decision has been made, welds should be produced and tested to qualify the procedure.

Table 6.3 Spray Arc Transfer Schedules

Material thickness				Electrode diameter		Welding current	Arc voltage	Wire feed	Travel speed	Shielding gas[b] flow
in.[a]	mm	Type of weld	Number of passes	in.	mm	(A, dc)	elec. pos.	(in./min)	(in./min)	(cfh)
1/8	3.2	Fillet or square groove	1	1/16	1.6	300	24	165	35	40–50
3/16	4.8	Fillet or square groove	1	1/16	1.6	350	25	230	32	40–50
1/4	6.4	Vee groove	2	1/16	1.6	325–375	24–25	210–260	30	40–50
1/4	6.4	Vee groove	2	3/32	2.4	400–450	26–29	100–120	35	40–50
1/4	6.4	Fillet	1	1/16	1.6	350	25	230	32	40–50
1/4	6.4	Fillet	1	3/32	2.4	400	26	100	32	40–50
3/8	9.5	Vee groove	2	1/16	1.6	325–375	24–25	210–260	24	40–50
3/8	9.5	Vee groove	2	3/32	2.4	400–450	26–29	100–120	28	40–50
3/8	9.5	Fillet	2	1/16	1.6	350	25	230	20	40–50
3/8	9.5	Fillet	1	3/32	2.4	425	27	110	20	40–50
						325–375	24–26	210–260		

½	12.7	Vee groove	3	1/16	1.6	375	26	250	24	40–50
						400–450	26–29	100–120		
½	12.7	Vee groove	3	3/32	2.4	425	27	110	30	40–50
½	12.7	Fillet	3	1/16	1.6	350	25	230	24	40–50
½	12.7	Fillet	3	3/32	2.4	425	27	105–110	26	40–50
		Double	4	1/16	1.6	325–375	24–26	210–260	24	40–50
¾	19.1	Vee groove	4	1/16	1.6	350	25	230	24	40–50
¾	19.1	Double vee	4	3/32	2.4	400–450	26–29	100–120	24	40–50
		Double	4	3/32	2.4	425	27	110	24	40–50
¾	19.1	Fillet	5	1/16	1.6	350	25	230	24	40–50
¾	19.1	Fillet	4	3/32	2.4	425	27	110	26	40–50
1	24.1	Fillet	7	1/16	1.6	350	25	230	24	40–50
1	24.1	Fillet	6	3/32	2.4	425	27	110	26	40–50

Use only in flat and horizontal fillet position.
[a]For fillet welds, material thickness indicates fillet weld size.
[b]Shielding gas is argon plus 1 to 5% oxygen.

Table 6.4 Globular Transfer (CO_2) Schedule

Material thickness				Electrode diameter						
gauge	in.	mm	Type of weld[a]	in.	mm	Welding current (A, dc)	Arc voltage elec. pos.	Wire feed (in./min)	Travel speed (in./min)	CO_2 gas flow (cfh)
18	0.050	1.3	Fillet	0.045	1.1	280	26	350	190	20–25
			Square groove	0.045	1.1	270	25	340	180	20–25
16	0.063	1.6	Fillet	0.045	1.1	325	26	360	150	30–35
			Square groove	0.045	1.1	300	28	350	140	30–35
14	0.078	2.0	Fillet	0.045	1.1	325	27	360	130	30–35
			Square groove	0.045	1.1	325	29	360	110	30–35
			Square groove	0.045	1.1	330	29	350	105	30–35
11	0.125	3.2	Fillet	1/16	1.6	380	28	210	85	30–35
			Square groove	0.045	1.1	350	29	380	100	30–35
3/16	0.188	4.8	Fillet	1/16	1.6	425	31	260	75	30–35
			Square groove	1/16	1.6	425	30	320	75	30–35
			Square groove	1/16	1.6	375	31	260	70	30–35
1/4	0.250	6.4	Fillet	5/64	2.0	500	32	185	40	30–35
			Square groove	1/16	1.6	475	32	340	55	30–35
3/8	0.375	9.5	Fillet	3/32	2.4	550	34	200	25	30–35
			Square groove	3/32	2.4	575	34	160	40	30–35
1/2	0.500	12.7	Fillet	3/32	2.4	625	36	160	23	30–35
			Square groove	3/32	2.4	625	35	200	33	30–35

[a]For mild carbon and low alloy steels on square groove welds, backing is required.

Table 6.5 Short-Circuiting Transfer Schedule[a]

Material thickness[b]			Electrode diameter		Welding current	Arc voltage,	Wire feed speed	Travel speed	Shielding gas
Fraction	in.	mm	in.	mm	(A, dc)	electrode positive	(in./min)	(in./min)	flow[c] (cfh)
24 ga.	0.025	0.6	0.030	0.8	30–50	15–17	85–100	12–20	15–20
22 ga.	0.031	0.8	0.030	0.8	40–60	15–17	90–130	18–22	15–20
20 ga.	0.037	0.9	0.025	0.9	55–85	15–17	70–120	35–40	15–20
18 ga.	0.050	1.3	0.035	0.9	70–100	16–19	100–160	35–40	15–20
1/16	0.063	1.6	0.035	0.9	80–110	17–20	120–180	30–35	20–25
5/64	0.078	2.0	0.035	0.9	100–130	18–20	160–220	25–30	20–25
1/8	0.125	3.2	0.035	0.9	120–160	19–22	210–290	20–25	20–25
1/8	0.125	3.2	0.045	1.1	180–200	20–24	210–240	27–32	20–25
3/16	0.187	4.7	0.035	0.9	140–160	19–22	210–290	14–19	20–25
3/16	0.187	4.7	0.045	1.1	180–205	20–24	210–245	18–22	20–25
1/4	0.250	6.4	0.035	0.9	140–160	19–22	240–290	11–15	20–25
1/4	0.250	6.4	0.045	1.1	180–225	20–24	210–290	12–18	20–25

[a]Single-pass flat and horizontal fillet position. Reduce current 10–15% for vertical and overhead welding.
[b]For fillet and groove welds. For fillet welds, size equals metal thickness. For square groove welds, the root opening should equal half the metal thickness.
[c]Shielding gas is CO_2 or mixture of 75% Ar + 25% CO_2.

Table 6.6 Shielding Gases and Their Applications for Gas Metal Arc Welding

Shielding gas	Gas composition	Gas type	Applications	Remarks
Argon	Ar	Inert	Nonferrous metals	Least expensive inert gas; provides spray transfer
Argon + helium	50% Ar, 50% He	Inert	Al, Mg, Cu, and their alloys	Higher head in arc use on heavier thickness; less porosity; provides spray transfer
Argon + oxygen	Ar + 3–5% O_2	Oxidizing	Mild and low-alloy steels	Provides spray transfer
Argon + carbon dioxide	75% Ar, 25% CO_2	Slightly oxidizing	Mild and low-alloy steels (also some stainless with GMAW)	Smooth weld surface; reduces penetration; short-circuiting transfer
Helium + argon + carbon dioxide	90% He, 7.5% Ar, 2.5% CO_2	Essentially inert	Stainless steel and some alloy steels	Provides arc stability; helpful in out-of-position welding; short-circuiting transfer
Carbon dioxide	CO_2	Oxidizing	Mild and low-alloy steels (also on some stainless steels)	Deep penetration; short-circuiting or globular, depending on wire size and current

6.4 Flux-Cored Arc Welding Procedure Schedules

The welding procedure schedules for flux-cored arc welding are given in a different manner than for GMAW. The type of metal transfer is much less important than with GMAW. Flux-cored welding electrode wires can be run with either straight or reverse polarity, that is, with the electrode negative or positive. There are two basic types of flux-cored wires: self-shielded and gas-shielded. The welding parameters are considerably different for the two types. The schedules presented are based on welding plain carbon steel and present data for different electrode wire sizes. The smaller diameter electrode wires are used for out-of-position welding. Since most automatic welding is in the flat position, the larger electrode wires can often be used. In general, the electrode wire size for flux-cored wire can be matched to the procedure used for gas metal arc welding. Table 6.7 is the flux-cored arc welding schedule for gas-shielded electrode wires run with the electrode positive, and Table 6.8 is the schedule for the self-shielded flux-cored electrode wire, which is run with the electrode negative.

The current ranges for flux-cored electrode wire are broader than for solid wire. Table 6.9 shows the welding range for gas-shielded electrode wire run with the electrode positive. These provide the deepest penetration and usually a smooth bead surface. An automatic system can use higher currents. Table 6.10 shows the welding range for self-shielded flux-cored electrode wires run with the electrode negative. Electrical stickout is used for these electrodes. The welding range is not as wide as with the gas-shielded version. The arc voltage must be in accordance with the schedule.

In these schedules the welding current and the wire feed speed values are both given even though the welding current is set by the wire feed speed. It is sometimes more convenient to establish the welding current directly without exactly knowing the wire feed speed. As mentioned, test welds should be produced using the selected parameters to ensure that the welds meet the requirements. If they do, the procedure can be considered qualified.

6.5 Submerged Arc Welding Procedure Schedules

The submerged arc welding process applied automatically is popular for manufacturing construction equipment, pressure vessels, and tanks and in shipbuilding. When the welds can be applied in the flat position, the submerged arc welding process gives a high deposition rate.

The joint designs for submerged arc welding are a little different from those

Table 6.7 Flux-Cored Arc Welding—CO_2 Gas-Shielded Electrode with Reverse Polarity, Electrode Positive (DCEP)[a]

Electrode diameter		Weld position	Amperage	Current[b]	Voltage	Wire feed speed		Deposition rate		Stickout
in.	mm					in./min	mm/min	lb/hr	kg/hr	± 1/4 in.
0.045	1.1	Flat, horiz.	150	DCEP	25	225	75,715	3.5	1.58	3/4
0.045	1.1	Flat, horiz.	180	DCEP	27	280	7,112	5.3	2.40	3/4
0.045	1.1	Flat	250	DCEP	29	450	11,430	8.0	3.63	3/4
1/16	1.6	Flat, horiz.	200	DCEP	25	138	3,505	4.7	2.13	3/4
1/16	1.6	Flat, horiz.	250	DCEP	26	177	4,495	6.0	2.72	3/4
1/16	1.6	Flat, horiz.	300	DCEP	27	230	5,842	8.4	3.81	3/4
1/16	1.6	Flat	350	DCEP	28	280	7,112	10.9	4.94	3/4
1/16	1.6	Flat	375	DCEP	29	311	7,899	11.6	5.26	3/4
5/64	2.0	Flat, horiz.	250	DCEP	26	119	3,040	6.6	2.99	1
5/64	2.0	Flat, horiz.	300	DCEP	29	145	3,683	8.4	3.81	1
5/64	2.0	Flat, horiz.	350	DCEP	31	181	4,597	10.2	4.63	1
5/64	2.0	Flat	400	DCEP	33	226	5,740	12.1	5.49	1

3/32	2.4	Flat, horiz.	350	DCEP	26	120	3,048	9.2	4.17	1
3/32	2.4	Flat, horiz.	400	DCEP	29	142	3,606	11.5	5.22	1
3/32	2.4	Flat	450	DCEP	32	174	4,419	13.7	6.21	1
3/32	2.4	Flat	500	DCEP	34	201	5,105	15.2	6.89	1
3/32	2.4	Flat	550	DCEP	36	234	5,943	18.1	8.21	1
7/64	2.8	Flat	500	DCEP	30	125	3,175	13.4	6.08	1
7/64	2.8	Flat	550	DCEP	32	145	3,683	15.5	7.03	1
7/64	2.8	Flat	600	DCEP	34	176	4,470	18.5	8.39	1
7/64	2.8	Flat	650	DCEP	36	196	4,978	20.6	9.34	1
7/64	2.8	Flat	700	DCEP	36	221	5,613	23.6	10.70	1
1/8	3.2	Flat	600	DCEP	32	120	3,048	17.8	8.07	1
1/8	3.2	Flat	650	DCEP	34	130	3,302	19.7	8.93	1
1/8	3.2	Flat	700	DCEP	36	143	3,632	21.4	9.70	1
1/8	3.2	Flat	750	DCEP	38	155	3,937	22.0	9.97	1
1/8	3.2	Flat	800	DCEP	38	166	4,216	24.6	10.88	1

[a] Use CO_2 shielding gas at 30–40 ft^3/hr (16–19 liters/min).
[b] DCEP = direct current; electrode positive.

Table 6.8 Flux-Cored Arc Welding—Self-Shielding Electrode on Straight Polarity, Electrode Negative (DCEN)[a]

Electrode diameter						Wire feed speed		Deposition rate		
in.	mm	Weld position	Amperage	Current[b]	Voltage	in./min	mm/min	lb/hr	kg/hr	Stickout ± 1/4 in.
0.045	1.1	Flat	130	DCEN	15	105	2667	1.80	0.82	1/2
0.045	1.1	Flat fillet	160	DCEN	17	170	4318	2.76	1.25	1/2
0.045	1.1	Horiz. fillet	160	DCEN	16.5	170	4318	2.76	1.25	1/2
0.045	1.1	Vert. up fillet	130	DCEN	16	125	3175	1.80	0.82	1/2
0.045	1.1	Ovhd. fillet	130	DCEN	16	125	3175	1.80	0.82	1/2
1/16	1.6	Flat	150	DCEN	18	70	1778	2.40	1.09	3/4
1/16	1.6	Horiz. fillet	200	DCEN	19	99	2514	3.54	1.61	3/4
1/16	1.6	Flat	250	DCEN	20	144	3675	5.88	2.67	3/4
0.068	1.7	Flat	175	DCEN	18.5	49	1244	1.92	0.87	3/4
0.068	1.7	Horiz. fillet	200	DCEN	20	94	2387	3.54	1.61	3/4
0.068	1.7	Flat	200	DCEN	20	94	2387	3.54	1.61	3/4
0.068	1.7	Flat	225	DCEN	21	111	2819	5.16	2.34	3/4
0.068	1.7	Flat	275	DCEN	22	159	4038	7.38	3.35	3/4
5/64	2.0	Horiz. fillet	200	DCEN	19	67	1701	3.30	1.50	3/4
5/64	2.0	Horiz. fillet	250	DCEN	21	90	2286	5.52	2.50	3/4
5/64	2.0	Flat	300	DCEN	22.5	124	3170	7.50	3.40	3/4
3/32	2.4	Flat	250	DCEN	18.5	57	1447	4.98	2.26	3/4
3/32	2.4	Flat	300	DCEN	20	75	1905	6.48	2.94	3/4
3/32	2.4	Horiz. fillet	300	DCEN	21.5	75	1905	6.48	2.94	3/4
3/32	2.4	Flat	300	DCEN	21.5	75	1905	6.48	2.94	3/4

[a]No external shielding gas used.
[b]DCEN = direct current, electrode negative.

Table 6.9 Flux-Cored Arc Welding Range for E7OT-1 with CO_2 Shielding: Direct Current, Electrode Positive (DCEP)

Electrode diameter		Minimum				Maximum			
				Wire feed speed				Wire feed speed	
in.	mm	A	Volts	in./min	mm/min	A	Volts	in./min	mm/min
0.045	1.2	120	21	168	4267	300	30	625	15,875
1/16	1.6	150	24	100	2540	425	31	400	10,160
5/64	2.0	200	26	95	2413	450	33	270	6,858
3/32	2.4	300	26	95	2413	600	36	255	6,477
7/64	2.8	450	30	110	2794	750	38	237	6,019
1/8	3.2	550	32	98	2489	850	39	175	4,445

for the other processes. Figure 6.15 illustrates joint designs for submerged arc welding. Note that backup is usually required because of the deep penetration achieved with submerged arc welding.

The welding schedules shown by Table 6.11 are based on using a single electrode on mild or low alloy steel.

Tests should be made to see that the chosen procedure produces the desired weldment. If so, the weldment should be tested; if it is acceptable, the procedure is qualified.

Table 6.10 Flux-Cored Arc Welding Range for E71T-11 Self-Shielding: Direct Current, Electrode Negative (DCEN)

Electrode diameter		Minimum				Maximum			
				Wire feed speed				Wire feed speed	
in.	mm	A	Volts	in./min	mm/min	A	Volts	in./min	mm/min
0.045	1.2	95	13	65	1651	180	18.5	200	5080
1/16	1.6	100	15	47	1193	300	22	189	4800
0.068	1.7	125	17	49	1245	300	23	184	4673
5/64	2.0	150	18	47	1193	300	22.5	124	3149
3/32	2.4	200	17	40	1016	350	22	93	2410

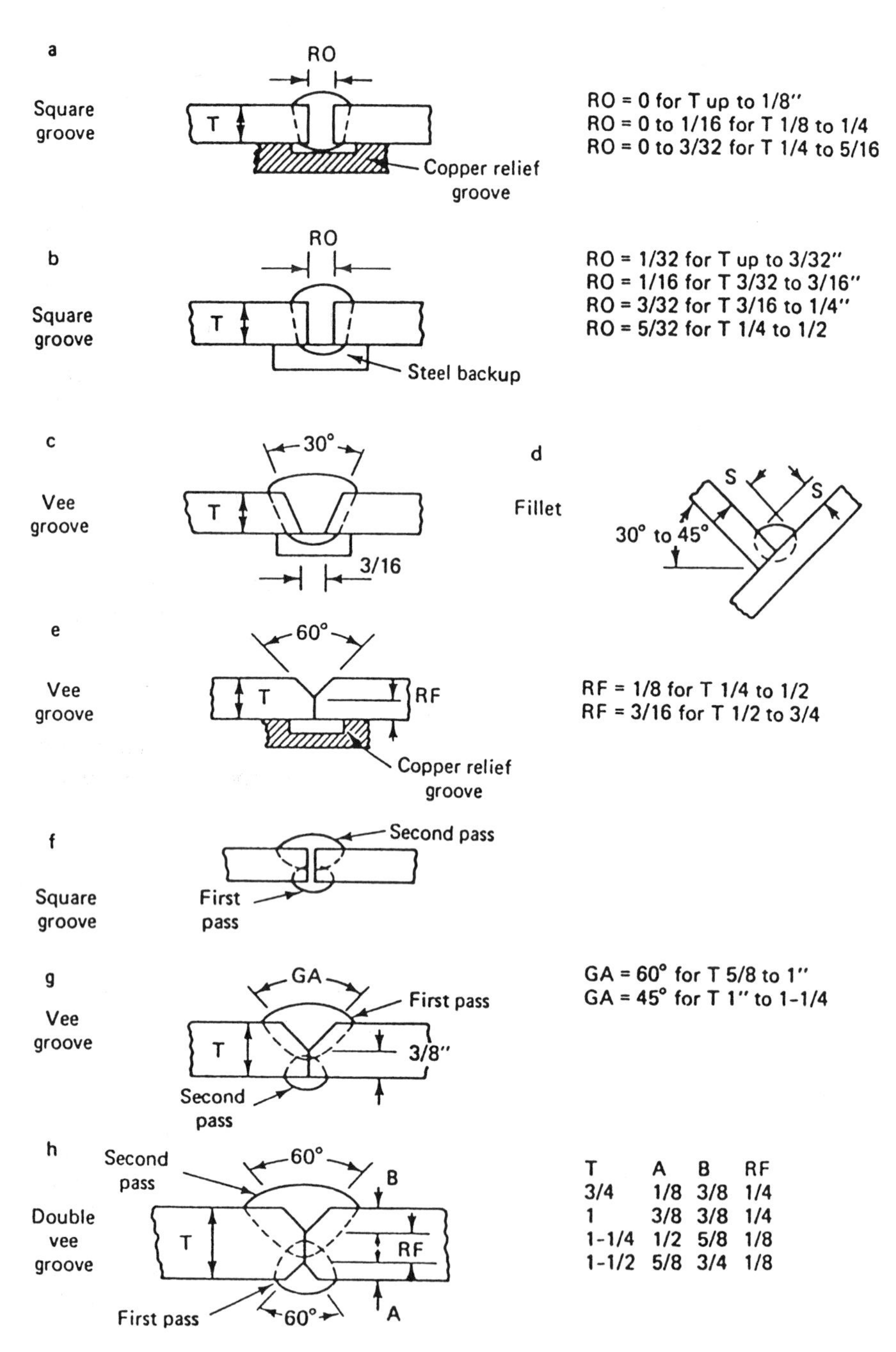

Figure 6.15 Joint designs for submerged arc welding.

6.6 Gas Tungsten Arc Welding Procedure Schedules

Gas tungsten arc welding uses a nonconsumable electrode. The procedure must be developed for a particular metal, and the metal dictates the welding current that should be used. Alternating current is often used for welding aluminum; however, when the part is cleaned immediately before being automatically welded, straight polarity, or dc electrode negative, current can be used. For more critical applications, the square-wave, pulsed current, or variable polarity welding variation would be used. In these cases, specific research is required to establish the appropriate welding procedure.

The preselected variables must be chosen with care. This includes the tungsten electrode type and size as well as the shielding gas composition, which depends on the metal being welded and the type of power to be used. The American Welding Society specifies seven different compositions of tungsten electrodes. These range from pure tungsten (EWP) through tungsten alloyed with cerium oxide, lanthanum oxide, 1% or 2% thorium, and zirconium oxide. The pure tungsten electrode is the most commonly used and least expensive and works on most metals. The tungsten alloys containing oxides are better electron emitters, so they have greater current-carrying capacity and provide for better arc starting and a more stable arc. They usually have a longer electrode life.

Table 6.12 shows the current-carrying ranges for the various types of electrodes that use dc electrode negative, dc electrode positive, or alternating current with unbalanced or balanced waveform.

Table 6.12 provides guidance in determining the most suitable electrode size. However, when making welding tests it will become evident if the size is inappropriate. If the tungsten electrode tends to overheat or have a wet surface appearance, the electrode is too small. A larger size should be selected, or a tungsten alloy electrode of the same size may be used. If the current is too low or the electrode is too large, the arc will wander erratically over the end of the electrode. Grinding the electrode to a point will alleviate this problem. The electrode should remain shiny after use and should never be allowed to touch the molten metal.

The correct choice of shielding gas depends on the metal being welded and the resultant weld. Only inert gas should be used. Since gas tungsten arc welding is often used on unusual metals, research may be required to select the gas for the metal being welded. However, gas tungsten arc welding is most commonly used for welding aluminum, and argon is the usual shielding gas. Argon is good for arc starting and operates at a lower arc voltage. Helium is more expensive and must be used at a higher flow rate. It allows a higher voltage, and it is possible to weld at a higher speed with helium than with argon. In some cases, argon and helium are mixed for optimum shielding for a particular metal.

Table 6.11 Submerged Arc Welding Schedules

Material thickness								
gauge	in.	Type of weld[a]	Electrode diameter	Welding current (A, dc)	Arc voltage (electrode positive)	Wire feed (in./min)	Travel speed (in./min)
16	0.063	Square groove (a)	3/32	300	22	68	100–140
		Square groove (b)	1/8	425	26	53	95–120
14	0.078	Square groove (a)	3/32	375	23	85	100–140
		Square groove (b)	1/8	500	27	65	75–85
12	0.109	Square groove (a)	1/8	400	23	51	70–90
		Square groove (b)	1/8	550	27	65	50–60
		Fillet (d)	1/8	400	25	51	40–60
10	0.140	Square groove (a)	1/8	425	26	53	50–80
		Square groove (b)	5/32	650	27	55	40–45
3/16	0.188	Square groove (a)	5/32	600	26	50	40–75
		Square groove (b)	3/16	875	31	55	35–40
		Fillet (d)	1/8	525	26	67	35–40
1/4	0.250	Square groove (a)	3/16	800	28	50	30–35
		Square groove (b)	3/16	875	31	56	22–25
		Fillet (d)	5/32	650	28	56	30–35
		Vee groove (e)	3/16	750	30	47	25–40
3/8	0.375	Square groove (b)	3/16	950	32	61	20–25
		Square groove (f)	3/16	1st pass 500	32	27	30
				2nd pass 750	33	47	30

		Vee groove (e)	3/16	900	33	57	23–25
		Fillet (d)	3/16	950	31	61	30–35
½	0.500	Vee groove (c)	3/16	975	33	63	12–17
		Square groove (f)	3/16	1st pass 650	34	40	25
				2nd pass 850	35	54	23–27
		Vee groove (e)	3/16	950	35	61	18–20
		Fillet (d)	3/16	950	33	61	14–17
¾	0.750	Vee groove (c)	7/32	1000	35	49	68
		Square groove (f)	3/16	1st pass 925	37	59	12
				2nd pass 1000	40	65	11
		Vee groove (e)	7/32	950	36	46	10–12
		Fillet (d)	7/32	1000	35	49	6–8
		Vee groove (g)	7/32	1st pass 950	34	46	15
				2nd pass 750	34	25	22
		Double vee groove (h)	3/16	1st pass 700	35	42	20–22
				2nd pass 1000	36	65	14–16
1	1.000	Vee groove (g)	7/32	1st pass 1150	36	58	11
				2nd pass 850	36	40	20
		Double vee groove (h)	7/32	1st pass 900	36	42	13–15
				2nd pass 1075	36	52	12–14
1¼	1.25	Double vee groove (h)	7/32	1st pass 1000	36	50	13
				2nd pass 1125	37	56	8
1½	1.50	Double vee groove (h)	7/32	1st pass 1050	36	51	9
				2nd pass 1125	37	56	7

[a]Letters refer to parts of Fig. 6.15.

Table 6.12 Current Ranges for Tungsten Electrodes

Tungsten electrode diameter		Typical current ranges for tungsten electrodes[a]					
				Alternating current, unbalanced wave		Alternating current, balanced wave	
in.	mm	DCEN, EWX-X	DCEP, EWX-X	EWP	EWX-X	EWP	EWX-X
0.010	0.30	Up to 15	na	Up to 15	Up to 15	Up to 15	Up to 15
0.020	0.50	5–20	na	5–15	5–20	10–20	5–20
0.040	1.00	15–80	na	10–60	15–80	20–30	20–60
0.060	1.60	70–150	10–20	50–100	70–150	30–80	60–120
0.093	2.40	150–250	15–30	100–160	140–235	60–130	100–180
0.125	3.20	250–400	25–40	150–200	225–325	100–180	160–250
0.156	4.00	400–500	40–55	200–275	300–400	160–240	200–320
0.187	5.00	500–750	55–80	250–350	400–500	190–300	290–390
0.250	6.40	750–1000	80–125	325–450	500–630	250–400	340–525

na = not applicable.
[a]All values based on the use of argon gas.

There are several variations of the gas tungsten arc welding process that often apply to automatic or automated welding. Pulsed current GTAW can be used to control heat input. It also gives deeper penetration and is advantageous for welding thick to thin materials. The pulsed current versus time relationship is shown by Fig. 6.16, which gives the waveforms for high and low pulses with long and short pulse time. The long pulse time is sufficient to provide the deep penetration desired. A short pulse time allows for the weld to partially solidify so

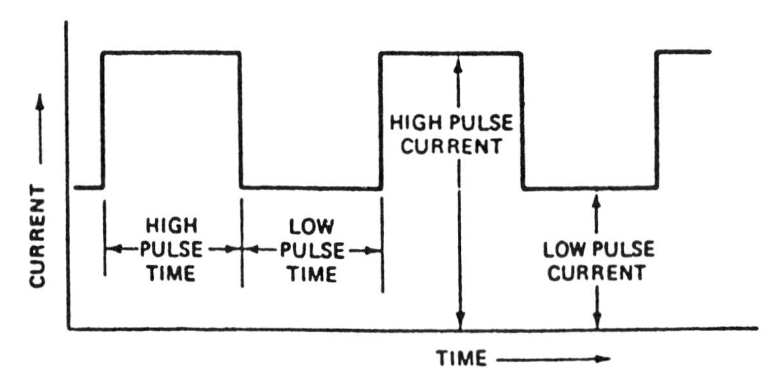

Figure 6.16 Current–time relationship for pulsed current.

that the weld pool is under control. Pulsing can be controlled by most automated welding system controllers.

No GTAW procedure schedules are provided in this section, because they vary considerably depending on the metal being welded, the joint detail, and the shielding gas composition. In general, arc length is related to arc voltage. The arc length should approximate the diameter of the tungsten electrode. Arc voltage depends on arc length and gas atmosphere. The welding current must be based on the joint metal thickness, and test welds may be needed to determine the proper current. For square groove joint welds, the welding current is the current required to melt the full thickness of the material. In some cases of automated welding, backing bars, normally water-cooled, are employed. The current required depends on the physical characteristics, thermal conductivity, etc., of the metal being welded. Shielding gas flow rates are usually on the order of 10–15 cubic feet per hour (3–7 liters/min) and can be adjusted to improve the weld results. If the weld surface shows oxidation or contamination from the atmosphere, the gas flow rate should be increased. The inside diameter of the nozzle should be between ¼ in. (6.4 mm) and ¾ in. (19 mm) and is related to the diameter of the electrode. Travel speed for gas tungsten arc welding is relatively slow, 10–20 in./min (250–500 mm/min). Travel speeds may be extended up or down depending on the weld achieved during procedure development tests.

Once the procedure has been developed, tests should be made to make sure that it gives a weld of acceptable quality. If so, the procedure is considered qualified.

Programmed gas tungsten arc welding is also very common, particularly with mechanized and automated welding systems. For example, on a small tube head, the welding current must be ramped to the maximum penetration at the starting point, but as the metal heats up, the current should be reduced to compensate for heating of the part. At the termination of the weld, it should be returned to zero. At the same time, the flow of shielding gas should begin before the weld starts and end when the weld is completed.

The productivity of gas tungsten arc welding can be increased by using the "hot wire" system. This uses an automatically fed filler wire that is electrically "hot." The current in the filler wire increases its temperature and therefore the melt-off rate, thereby increasing productivity. This system is used only for automatic systems. It requires an extra wire feed system and must be programmed for optimum results.

6.7 Plasma Arc Welding Procedure Schedules

Procedure schedules for the plasma arc welding process are more complex than those for gas tungsten arc welding because of the additional variables. The process is very similar to gas tungsten arc welding and can be used to weld the same metals. However, the tungsten is enclosed within the welding torch, and the torch orifice emits the plasma stream that does the heating and welding. There are two variations. In the nontransferred mode the current flow is from the electrode inside the torch to the nozzle containing the orifice of the torch and back to the power supply. This mode is normally used for welding thin materials and for using powdered filler metal.

The more common method of plasma welding uses a transferred arc; the current flows from the tungsten electrode inside the welding torch through the orifice to the workpiece and back to the power supply. The plasma has a higher temperature than the arc of the GTAW process and is therefore faster. In addition, the plasma is stiffer than the tungsten arc and makes the torch-to-work distance slightly less critical. Using the transferred arc mode, there are two ways to weld. The melt-in mode of operation is quite similar to gas tungsten arc welding but faster. In the "keyhole" welding method the plasma jet penetrates through the workpiece and forms a hole known as the keyhole. Surface tension forces the molten base metal to flow around the keyhole to form the weld. The keyhole method can be used only for joints the plasma can pass through. It is used for welding materials from 1⁄16 in. (1.6 mm) to 1⁄2 in. (12 mm) in thickness, depending on base metal composition, current, and welding gas. The keyhole method provides for full-penetration single-pass welds.

The weld joint design is based on the thickness of the metal welded and the mode of operation. The preferred joint design is the square groove with a minimum root opening that allows for full keyhole penetration. When using the melt-in mode for thicker materials, the same general joint details as those used for the gas tungsten arc can be employed. All joint detail designs can be welded.

The apparatus for plasma arc welding is very similar to that of gas tungsten arc welding, with the exception of the torch and the complexity of the control system. The torches must be water-cooled because the arc is generated inside the torch. The control circuits are more complex because of the need to start the arc within the torch, which normally uses a high-frequency pilot arc, and the need to be able to switch between the nontransferred arc mode and the transferred arc mode. For automated welding, filler wire may or may not be required, depending on the material thickness, the joint detail, and the productivity requirements. When it is required, a wire feeder is used, which must be under the control of the control circuit.

Argon gas is usually used both for shielding the weld area from the atmosphere and for generating the plasma. Active gases are not recommended for plasma welding.

Many of the same factors mentioned for GTAW procedure development apply to plasma arc welding. However, Table 6.13 provides welding procedure

Table 6.13 Plasma Arc Welding Procedure Schedule

Material thickness (in.)	Type of weld	Orifice diameter (in.)	Filler wire diameter (in.)	Shield gas at 20 cfh	Plasma gas flow (cfh) Argon	Weld current[b] (A)	No. of passes	Travel speed (in./min)
Stainless steel[a]								
0.008	Edge butt	0.093	—	A	0.5	12 DCEN	1	7
0.008	Edge butt	0.093	—	A-5H_2	0.5	10 DCEN	1	13
0.020	Square groove	0.046	—	A-5H_2	0.5	12 DCEN	1	21
0.030	Square groove	0.046	—	A-5H_2	0.5	34 DCEN	1	17
0.062	Square groove	0.081	—	A-5H_2	0.7	65 DCEN	1	14
0.093	Square groove	0.081	—	A	2.0	85 DCEN	1	12
0.093	Square groove	0.081	—	A-5H_2	2.0	85 DCEN	1	16
0.125	Square groove	0.081	—	A	2.5	100 DCEN	1	10
0.125	Square groove	0.081	—	A-5H_2	2.5	100 DCEN	1	16
0.187	Square groove	0.081	—	A-5H_2	3.5	100 DCEN	1	7
0.250	Vee groove	0.081	—	A-5H_2	3.0	100 DCEN	First	5
0.250	Vee groove	0.081	3/32	A-5H_2	1.4	100 DCEN	Second	2
Mild steel								
0.030	Square groove	0.081	—	A	0.5	45 DCEN	1	26
0.080	Square groove	0.081	—	A	1.0	55 DCEN	1	17
Copper[a]								
0.016	Edge butt	0.093	—	He	0.5	18 DCEN	1	24
Aluminum								
0.036	Square groove	0.081	1/16	He	0.05	47 DCEP	1	24
0.050	Edge joint	0.081	—	He	0.6	48 DCEP	1	22
0.090	Fillet	0.081	3/32	He	1.4	34 DCEP	1	4

cfh = cubic feet per hour.
[a]Backing gas argon at 5–10 cfh.
[b]DCEN = direct current, electrode negative; DCEP = direct current, electrode positive.

schedules for keyhole welding of stainless steel, mild steel, copper, and aluminum. Mild steel is rarely welded with the plasma process.

Several variations of plasma arc welding should be considered. The pulsed current method similar to that used in gas tungsten arc welding is often used with plasma arc welding. In addition, the hot-wire PAW system is also used in the automatic mode. Another system developed for welding aluminum uses the variable polarity plasma arc, which provides only a very small amount of reverse polarity current with the maximum amount of straight polarity current. This is used for specialized applications.

7

Welding Problems and Quality Control

7.1 Weldment Integrity

The user of a welded product needs assurance that the weldment will meet its service requirements. This is particularly important for welded products manufactured with automated welding equipment. The purpose of an automated welding station is to produce a high quality weld on each and every weldment.

The welding procedure (Chapter 6) describes how a particular weld or weldment is made. It is a record of all the different elements, variables, and factors that enter into production of the weldment. Once the procedure is established and approved, weldments must be produced strictly following the procedure and using the appropriate equipment. The resulting weldments must be thoroughly tested in every way. If the weldment passes the tests, the welding procedure is qualified. Qualification means that specific tests were made on a completed weldment produced according to the procedure and the weldment met the expected service requirements. Normally prototypes are tested to destruction, but such tests ensure that the design of the product is acceptable.

In automatic or automated welding we should eliminate, or at least reduce, the amount of actual testing or inspection required. In other words, if the qualified welding procedure is explicitly followed each and every time, the quality of the weldment will be acceptable. There are four major factors that affect the quality of a weldment:

1. Piecepart accuracy
2. Joint cleanliness
3. Weld procedure consistency
4. Tooling quality

If these factors are strictly controlled, a high quality weldment will result.

In making the transfer from manual or semiautomatic welding to automatic welding, the intent is to reduce welding time. The time needed for inspecting the product should also be reduced. The method of application determines the human involvement and affects the method of ensuring quality. For example, in mechanized welding the equipment requires manual adjustment in response to visual observations during welding. The welder is constantly observing the welding operation and in a sense is visually inspecting the weld even though he is not physically making the weld.

For automatic welding, the equipment requires only occasional operator assistance, if any. No manual adjustments are made during the welding operation. The operator is not observing the welding while it is being performed. There is no human involvement. Visual inspection may not occur even when the operator unloads the machine.

Robotic welding is much like automatic welding in that human observation is not involved. All parameters are programmed, and if they change beyond specific limits the welding operation will shut down. With adaptive control, sensing devices compensate for variations and ensure quality even if they do occur.

In adaptive control welding, the operator is not involved other than possibly loading and unloading the welding machine. Quality is assumed because there are sensing devices to determine if any factor falls outside its normal limits; if it does, the machine automatically adjusts the parameters.

Various techniques are used to ensure a continuing high level of quality. One is to set up an automatic inspection station following the welding operation. This inspection station can test each weldment or randomly test specific weldments, or it can perform tests after a large number of articles have been produced. Some products may be required to be leaktight. If so, an automatic leak-testing inspection station can be made part of the system. In other cases, destructive testing may be required after the production of a large number of units, and statistical quality control records may be kept to ensure acceptable quality.

7.2 Visual Inspection

The most widely used inspection technique is visual inspection. It is very popular for noncritical welding production. It requires the least time and is the least expensive inspection method. It references the welding procedures and can eliminate problems that create weldment quality defects. It is very effective. It can be used in automated or robotic welding, possibly when parts are unloaded from the welding station.

Any flaw or abnormality that is different from the expected results can be considered a defect. Defects have a degree of seriousness that may or may not affect the service life of the weldment. It is therefore necessary to establish limits of acceptability for defects beyond which they are cause for rejection. The establishment of such limits depends on the product, its intended service, and similar factors. If the product is made to a specification, the specification will establish the limits of defects.

One technique used by the volume production industries is the use of a workmanship specimen. This is an actual piecepart made by the automatic weld station and posted at that station. A workmanship specimen provides the welder, the inspector, and the supervisor an example of what is expected from that particular workstation. Often the weld is highlighted with colored paint to make it stand out. Welding schedules and weld size information can be posted adjacent to the weldment specimen. In some cases, welds that do not meet requirements are also posted as examples of unacceptable welded parts that must be repaired or rejected.

Visual inspection throughout the welding operation has another advantage. It can catch errors at various steps of the production process and have them corrected. Defects can be caused by such factors as faulty, oily, or dirty joints; poor piecepart preparation, which causes inconsistent fit-up; and improper joint preparation. Early detection of these problems greatly reduces the expense of making repairs.

The fillet weld is the most widely used type of weld. It is easily inspected, but inspection requires the use of a fillet weld gauge. Figure 7.1 illustrates the use of the fillet gauge. Figure 7.2 illustrates the acceptance criteria for a fillet weld, which can be undersized in any of three ways. Visual inspection also determines the presence of poor weld starts and stops or tie-ins when a weld overlaps. It also can determine the smoothness of a weld, determine whether the weld properly follows the weld joint, and detect surface cracks and weld porosity. Excessive spatter, poor bead shape, undercutting, or excessively oversized welds are also detected. Visual inspection cannot, of course, find subsurface defects. Tiny or fine cracks can be easily missed if the weld is not cleaned.

Tools used by the welding inspector include the pocket magnifier, weld

Figure 7.1 Measuring a fillet weld.

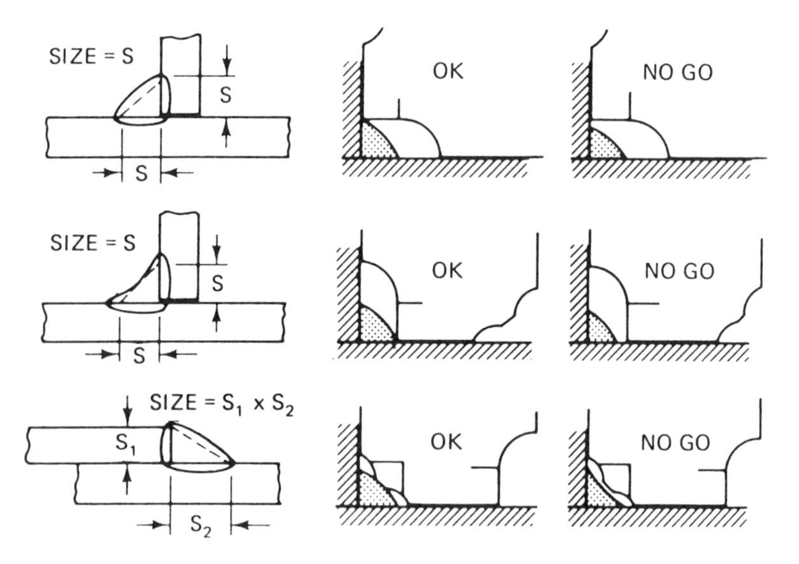

Figure 7.2 Fillet weld size.

gauge, scale and straight edge, borescope, small mirror, flashlight, and weld or procedure standards. A high level of illumination is required at the point of visual inspection.

7.3 Nondestructive Evaluation

There are a number of nondestructive evaluation techniques that can be used to ensure quality welds. The more popular ones are briefly described here.

Magnetic particle inspection (magnetic testing, MT) is a nondestructive method for detecting cracks, porosity, inclusions, lack of fusion, and other discontinuities in ferromagnetic materials. Surface and shallow subsurface discontinuities can be detected. Although there is no restriction as to the size and shape of the parts to be inspected, only ferromagnetic materials can be inspected by magnetic particle inspection. This method consists of establishing a magnetic field in the test part, applying magnetic particles to the surface of the weld, and examining the surface for the accumulation of the particles, which indicates a defect. Magnetic particle inspection is shown in Fig. 7.3. Magnetic particle

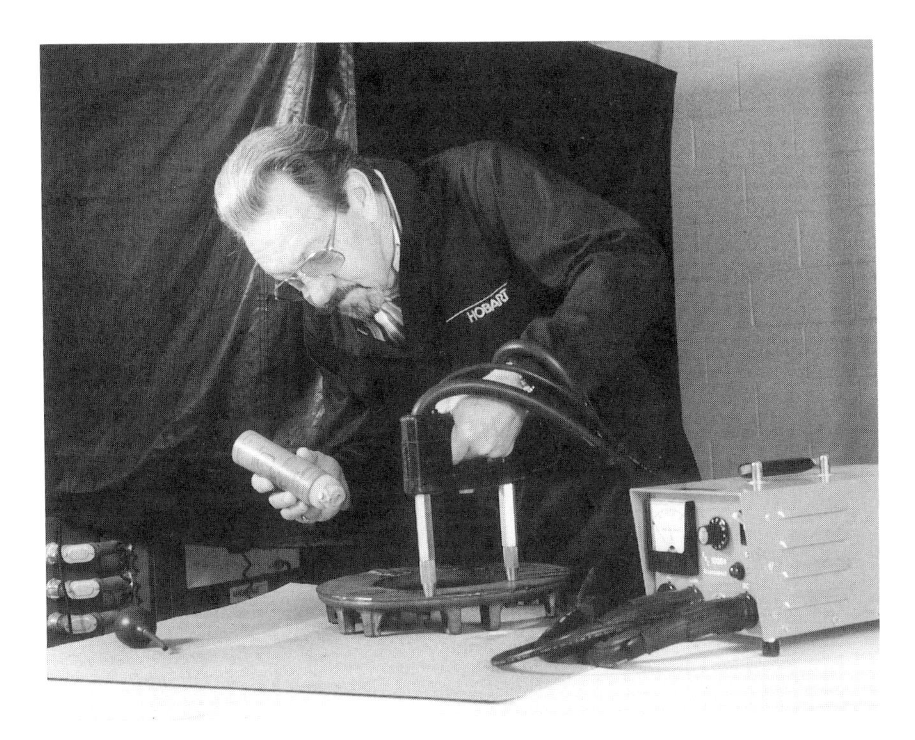

Figure 7.3 Magnetic particle inspection.

inspection can use a dry or wet method. In the dry method, powder is applied to the surface and collects at the indications. In the wet method, the magnetic particles are suspended in a liquid that flows over the part being tested and, again, the particles collect at the defect location. This process is often automated.

Another popular techniques is liquid penetrant inspection (PT), which is a highly sensitive nondestructive method for detecting small surface discontinuities (flaws) such as cracks and pores. The PT method can be used for most materials, including ferrous and nonferrous metals and nonmetals. Although there are several types of penetrants and developers, all of them follow the same fundamental principles. There are variants that use fluorescent materials for higher sensitivity. Liquid penetrant inspection is shown by Fig. 7.4. The penetrant is sprayed onto the surface of the weld. It is then wiped clean, and a developer is sprayed onto the same surface. If any of the penetrant was retained, for instance, in a crack, it will show up on the developer with a contrasting color, which is easily seen by the naked eye, as shown in Fig. 7.5. In other cases, the penetrant is fluorescent and will be visible under special lights. This inspection technique can be mechanized for volume production.

Another popular nondestructive testing technique is ultrasonic testing (UT),

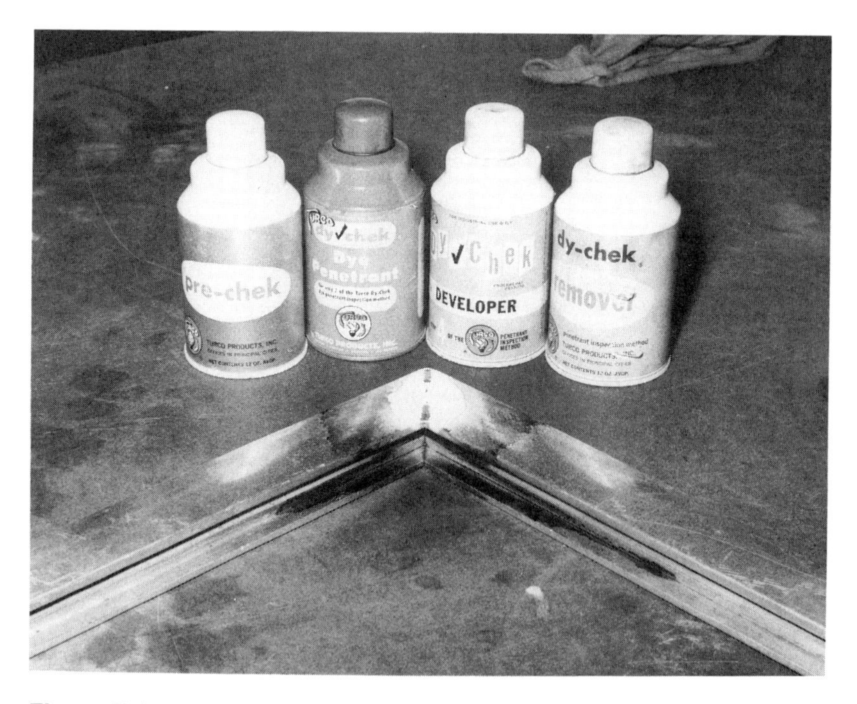

Figure 7.4 Liquid penetrant inspection.

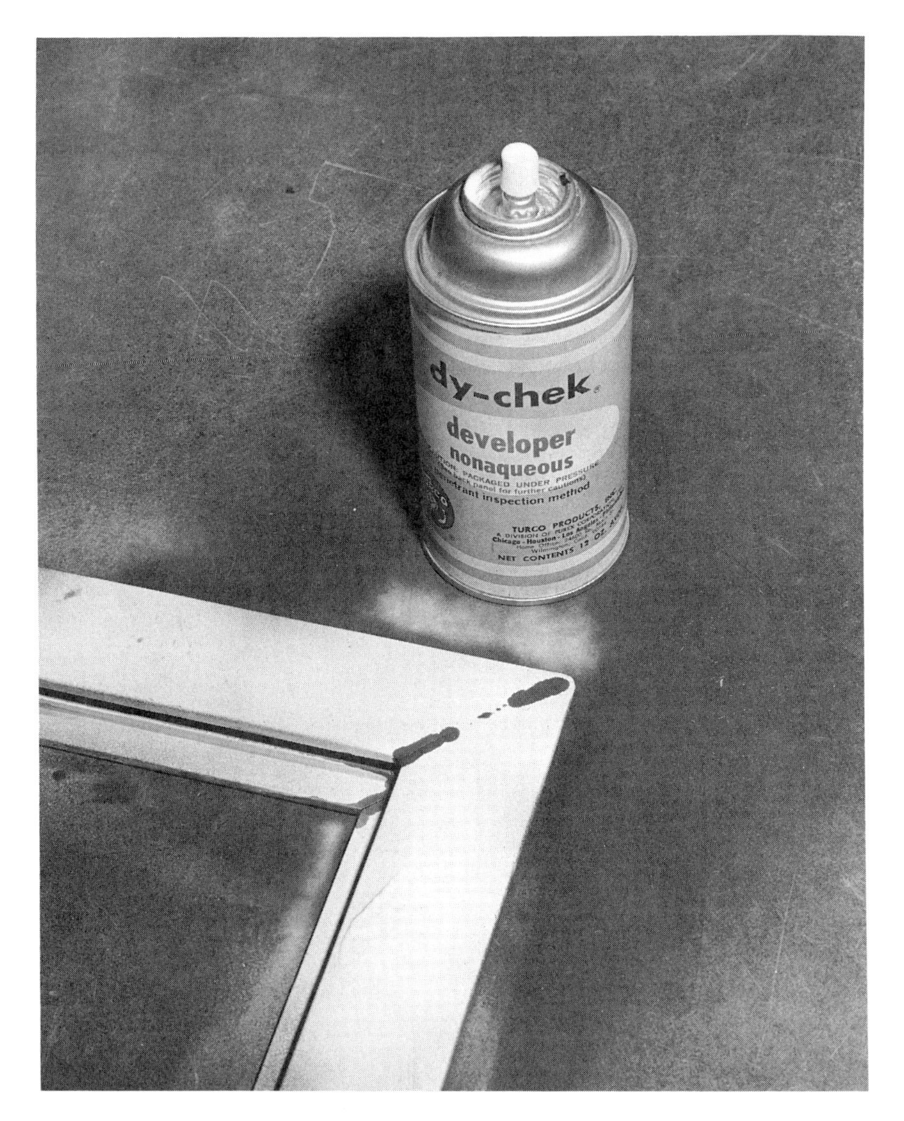

Figure 7.5 Liquid penetrant indication.

which employs mechanical vibration of high-frequency sound waves. A beam of ultrasonic energy is directed into the specimen being tested. This beam travels through the material with only a small loss. However, when it is intercepted by a discontinuity (defect or flaw), the ultrasonic energy is reflected and makes an indication on the face of the cathode ray tube. Ultrasonic testing is capable of finding subsurface discontinuities. In the ultrasonic contact pulse reflection technique, an electromechanical transducer is excited by a high-frequency voltage that causes the crystal to vibrate mechanically. This crystal probe becomes the source of ultrasonic mechanical vibrations that are transmitted into the test piece through a film of oil or other suitable couplant. When the pulse of ultrasonic waves strikes a discontinuity, the pulse is partially reflected, and part of its energy returns to the transducer, which now serves as a receiver for the return pulse from the flaw. The initial signal, the echoes from the discontinuities, and the echo of the rear surface of the test material are displayed on the screen of a cathode ray oscilloscope (Fig. 7.6). Permanent records can be made of UT inspection images.

Radiographic inspection (RT), one of the oldest nondestructive methods, uses X-rays or gamma radiation to examine the interior of welds. Radiographic examination provides a permanent film record of defects that, under normal conditions, is easy to interpret. This is a relatively slow and expensive method of nondestructive testing; however, it can be used to detect porosity, inclusions, cracks, and voids in the interior of castings, welds, and other structures. X-rays generated by electronic sources and gamma rays emitted by radioactive elements are penetrating radiation whose intensity is modified by passage through a material. The amount of energy absorbed by a material depends on its thickness

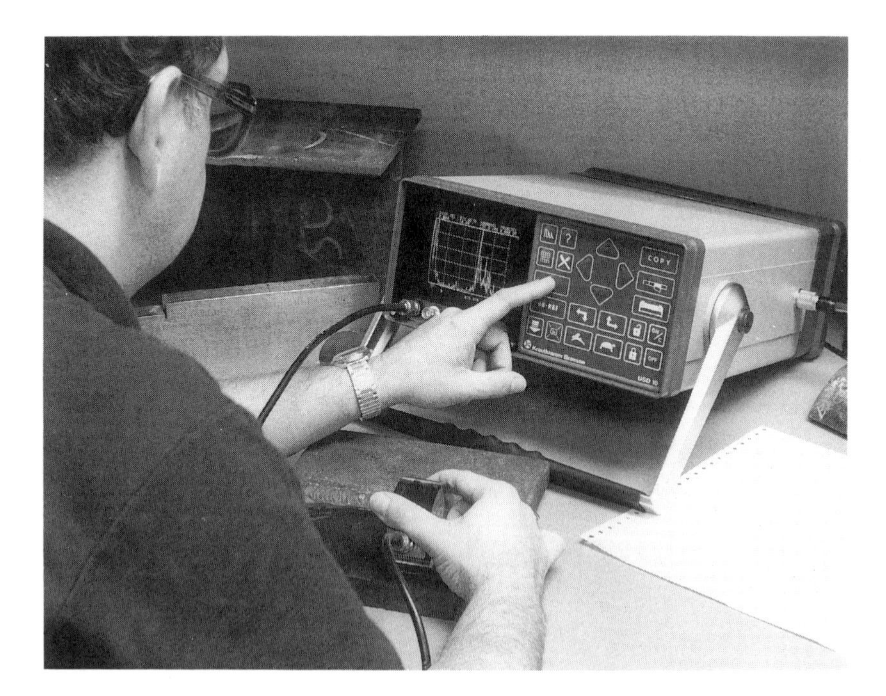

Figure 7.6 Ultrasonic inspection.

and density. Energy not absorbed by the material will cause exposure of the radiographic film. Exposed areas will be dark when the film is developed. Areas of the film exposed to less energy remain lighter. Therefore, areas of the material where the thickness has been changed by the discontinuities, such as pores or cracks, will appear as dark areas on the film, while inclusions of high density, such as tungsten, will appear as light areas. All discontinuities are detected by viewing variations in the density of the processed film. The X-ray or gamma-ray source and the penetrameter are placed on one side of the piece to be radiographed, and the film is placed on the opposite side of the part. Figure 7.7

Figure 7.7 Making a radiograph.

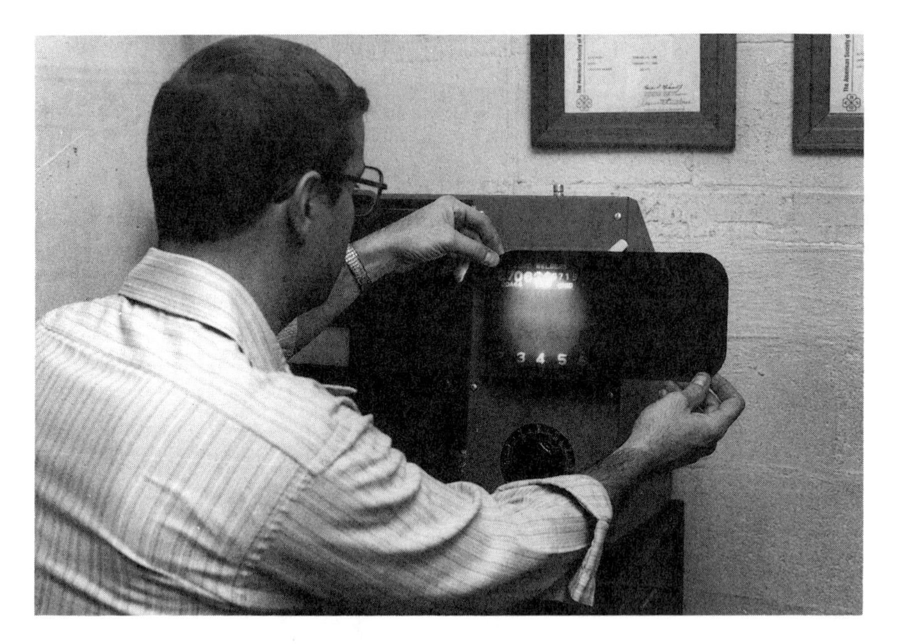

Figure 7.8 Viewing a radiograph.

shows the setup for taking a radiograph using X-rays. Figure 7.8 shows a processed X-ray film being read and interpreted. Radiography is used as a spot check in many cases as it is relatively slow and expensive for making continuous examinations.

Table 7.1 is a summary of nondestructive testing techniques for welds. It can be used as a guide for selecting the correct technique to be used for a particular application.

Specialized inspection methods can be used for specific products. They may be used to determine specific qualities such as dimension, which can then be statistically analyzed to determine production quality trends.

7.4 Potential Weld Defects

Gas metal arc welding is most often used for automated welding. Defects that can occur in GMAW are inclusions, porosity, undercutting, incomplete fusion, excessive melt-through, poor weld contour, excessive weld spatter, and craters. Defects are caused by the use of improper base metal, filler metal, or shielding gas or the improper choice of welding parameters. The base metal at the weld and the filler metal must be clean to avoid creation of a defect. Other occurrences that reduce the quality of the weld are arc blow, loss of shielding gas coverage, defective electrical contact between the contact tube and the electrode, wire feed stoppages, and defective electrical connection to the weldment. Many of these defects can be detected only by means of destructive evaluation. All of these potential defects are illustrated in Fig. 7.9.

Inclusions

The two basic types of inclusions (Fig. 7.9A) that can occur in a weld are slag inclusions and oxide inclusions. Inclusions cause a weakening of the weld and often serve as crack initiation points.

Some GMAW electrodes leave small glassy slag islands on the surface of the weld. Slag inclusions can be caused by welding over these in multiple-pass welds. The best way to avoid this problem is to clean the surface of the weld bead, especially the toes of the weld, where any slag can be easily trapped.

Oxide inclusions can be detected by radiographic testing. They often occur when excessively high travel speeds are used or during the welding of metals, such as aluminum, magnesium, or stainless steel, that have heavy oxide coatings. The oxide coating on the surface of the metal becomes mixed into the weld pool. Methods of preventing or correcting this problem include reducing the travel speed, increasing the welding voltage, using a more highly deoxidized type of electrode, and cleaning the thick oxide coatings from the surface of metals such as aluminum, magnesium, and stainless steel.

Porosity

Porosity is caused by gas pockets in the weld metal (Fig. 7.9B). They may be scattered in small clusters or occur along the entire length of the weld. These voids weaken the weld. This defect is caused by one or more of the following:

1. Inadequate flow of shielding gas
2. Wind drafts that deflect the shielding gas coverage
3. Blockage of the shielding gas flow due to spatter buildup on the nozzle or some other cause
4. Contaminated or wet shielding gas
5. Excessive welding current
6. Excessive welding voltage
7. Excessive electrode extension
8. Excessive travel speed, which causes the weld puddle to freeze before gases can escape
9. Rust, grease, oil, moisture, or dirt on the surface of the base metal
10. Impurities in the base metal, such as sulfur and phosphorus in steel
11. The use of filler metal with moisture trapped on the surface oxide coating

Table 7.1 Summary of Nondestructive Testing Techniques

Examination technique	Equipment	Defects detected	Advantages	Disadvantages	Other considerations
Visual (VT)	Pocket magnifier, welding viewer, flashlight, weld gauge, scale, etc.	Weld preparation, fit-up, cleanliness, roughness, spatter, undercuts, overlaps, weld contour and size; welding procedures	Easy to use; fast, inexpensive, usable at all stages of production	For surface conditions only; dependent on subjective opinion of inspector	Most universally used examination method
Dye penetrant (DPT) or fluorescent (FPT)	Fluorescent or visible penetrating liquids and developers; ultraviolet light for FPT	Defects open to the surface only; good for leak detection	Detects very small, tight, surface imperfections; easy to apply and interpret; inexpensive; used on magnetic or nonmagnetic materials	Time-consuming in the various steps of the process; normally, not permanent record	Often used on root pass of highly critical pipe welds; if material improperly cleaned, some indications may be misleading

Magnetic particle (MT)	Iron particles, wet or dry, or fluorescent; special power source; ultraviolet light for the fluorescent type	Surface and near-surface discontinuities, cracks, etc.; porosity, slag, etc.	Indicates discontinuities not visible to the naked eye; useful in checking edges prior to welding, also repairs; no size restriction	Used on magnetic materials only; surface roughness may distort magnetic field; normally, no permanent record	Examination should be from two perpendicular directions to catch discontinuities that may be parallel to one set of magnetic lines of force
Radiographic (RT)	X-ray or gamma ray; source: film processing equipment, film viewing equipment, penetrameters	Most internal discontinuities and flaws; limited by direction of discontinuity	Provides permanent record; indicates both surface and internal flaws; applicable on all materials	Usually not suitable for fillet weld inspection; film exposure and processing critical; slow and expensive	Most popular technique for subsurface inspection; required by some codes and specifications
Ultrasonic (UT)	Ultrasonic units and probes; reference and comparison patterns	Can locate all internal flaws located by other methods with the addition of exceptionally small flaws	Extremely sensitive; use restricted only by very small weldments; can be used on all materials	Demands highly developed interpretation skill	Required by some codes and specifications

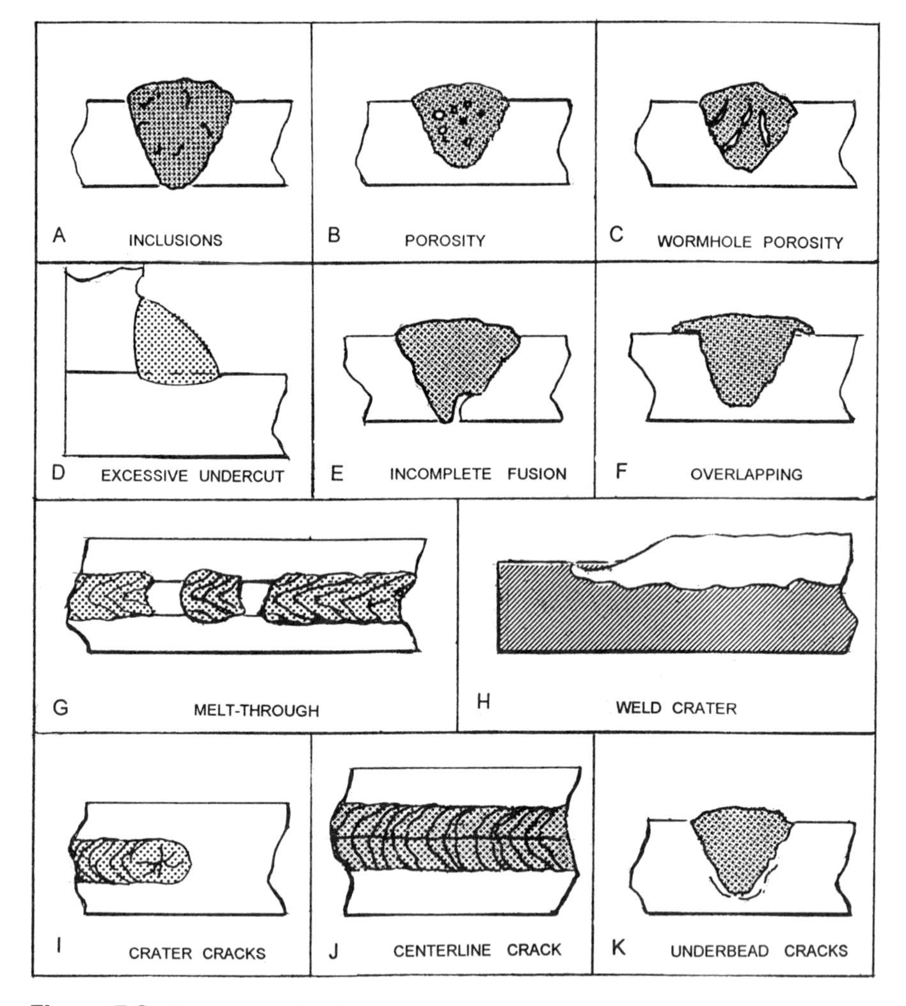

Figure 7.9 Potential weld defects.

Porosity can be corrected or avoided by

1. Increasing the shielding gas flow rate
2. Setting up wind shields
3. Cleaning the nozzle of the welding gun
4. Replacing the gas cylinder or replacing a defective system
5. Lowering the welding current (reducing the wire feed speed)
6. Decreasing the voltage
7. Decreasing the electrode extension

8. Reducing the travel speed
9. Cleaning the surface of the base metal and filler metal
10. Changing to a different base metal with a different composition

Wormhole porosity

Wormhole porosity is the presence of elongated gas pockets (Fig. 7.9C) and is usually caused by sulfur in the steel base metal or moisture on the surface of the base metal. The best methods for preventing wormhole porosity are to clean the surfaces of the joint and preheat the workpiece to remove moisture. If sulfur in the steel is thc problem, a more weldable grade of steel should be selected.

Undercutting

Undercutting is a groove melted in the base metal next to the toe or root of a weld that is not filled by the weld metal (Fig. 7.9D). This is particularly a problem with fillet welds. Undercutting causes a joint to be weaker at the toe of the weld, which may result in weldment failure. It is caused by one or more of the following:

1. Excessive welding current
2. Arc voltage too high
3. Excessive travel speed, which does not allow enough filler metal to be added
4. Incorrect electrode angle

Undercutting can be prevented by

1. Reducing the welding current
2. Reducing the welding voltage
3. Using a travel speed slow enough that the weld metal can completely fill all of the melted-out areas of the base metal
4. Correcting the electrode angles being used

Incomplete fusion

Incomplete fusion occurs when the weld metal is not completely fused to the base metal (see Fig. 7.9E). Incomplete fusion between the weld metal and the base metal is usually due to inadequate penetration. This is often a major problem with the short-circuiting mode of metal transfer. When short-circuiting welding is done, wider root openings are used to allow better penetration. Short-circuiting welding has the least penetration of the various modes of gas metal arc welding.

Causes of incomplete fusion are

1. Excessive travel speed, which causes an excessively convex weld bead or does not allow adequate penetration

2. Welding current too low
3. Letting the weld metal get ahead of the arc or letting the weld layer become too thick, which keeps the arc away from the base metal

Incomplete fusion can be corrected by reducing the travel speed, increasing the welding current, using proper electrode angles, or increasing the travel speed.

Overlapping

Overlapping is the protrusion of the weld metal over the edge or toe of the weld (Fig. 7.9F). This can cause an area of incomplete fusion, which creates a notch and can initiate a crack. Overlapping is produced by one or more of the following:

1. Too slow a travel speed, which permits the weld pool to get ahead of the electrode
2. Welding current too low
3. An incorrect electrode angle that allows the force of the arc to push the molten weld metal over unfused sections of the base metal

Overlapping can be corrected by using a higher travel speed, using a higher welding current, and/or using the correct electrode angle.

Melt-through

Melt-through occurs when the arc melts through the bottom of the weld and creates holes (Fig. 7.9G). This can be caused by the use of excessive welding current, a travel speed that is too slow, and/or a root opening that is too wide or a root face that is too small.

Melt-through can be prevented by reducing the welding current, increasing the travel speed, and by reducing the width of the root opening, using a slight weaving motion, or increasing the electrode extension.

Excessive weld spatter

Excessive weld spatter creates a poor weld appearance, wastes electrodes, and makes slag removal difficult. Excessive spatter can also block the flow of shielding gas from the nozzle, which causes porosity. The amount of spatter produced in gas metal arc welding depends on the type of metal transfer and the composition of the shielding gas. Globular transfer with carbon dioxide shielding creates high levels of spatter. It also is caused by excessive welding current, high arc voltage, or long electrode extension.

Methods of reducing the amount of spatter include reducing the welding current, reducing the arc voltage, and reducing the amount of stickout. Another method of reducing weld spatter is to change to an argon–carbon dioxide mixture, which, in many cases, produces spray transfer and less spatter.

Craters

Weld craters are depressions of the weld surface at the point where the arc was broken (Fig. 7.9H). The weld crater often contains cracks and can serve as an origin for further cracking. There are two common ways to prevent craters. For automatic welding, gradually reducing the welding current at the end of the weld gradually reduces the size of the molten weld pool. The second method is to stop the travel long enough to fill the crater before stopping the arc.

Cracking

Weldment cracking can be caused by an improper welding procedure or materials (Fig. 7.9I–K). Cracking can be classified as either hot or cold cracking and may be transverse or parallel to the weld axis. Cracks are often caused by high joint restraint and a high cooling rate.

Hot cracking occurs at elevated temperatures and generally happens when the weld metal starts to solidify. It is often caused by excessive sulfur, phosphorus, and lead contents in the steel base metal. In nonferrous metals, it is often caused by sulfur or zinc. Hot cracking may be prevented by

1. Preheating to reduce shrinkage stresses in the weld
2. Using clean or uncontaminated shielding gas
3. Increasing the cross-sectional area of the weld bead
4. Changing the contour of the weld bead
5. In steel, using a filler metal that is high in manganese

Crater cracks (Fig. 7.9I) are shallow hot cracks that are caused by improperly breaking the arc. They can be prevented in the same way as craters are, by gradually reducing the welding current at the end of the weld or by breaking the travel before stopping the arc.

Cold cracking occurs after the weld metal has completely solidified. It may occur several days after welding and is generally caused by hydrogen embrittlement, excessive joint restraint, and rapid cooling. Preheating and the use of a dry, high-purity shielding gas help reduce this problem.

Centerline cracks (Fig. 7.9J) are cold cracks that often occur in single-pass concave fillet welds. A centerline crack is a longitudinal crack that runs down the center of the weld as shown in the illustration. They may be caused by one or more of the following:

1. A weld bead that is too small for the thickness of the base metal
2. Poor fit-up
3. High joint restraint
4. Extension of a crater crack

Centerline cracks can be prevented by increasing the bead size, decreasing the gap width, or preheating, or by preventing the formation of weld craters.

Base metal and underbead cracks are cold cracks that form in the heat-affected zone of the base metal. Underbead cracks occur underneath the weld bead as shown in Fig. 7.9K. Base metal cracks originate in the heat-affected zone of the weld. They are caused by excessive joint restraint, entrapped hydrogen, and a brittle microstructure. A brittle microstructure is caused by rapid cooling or excessive hydrogen and a hardenable microstructure. Underbead and base metal cracking can be reduced or eliminated by using preheat.

Other problems

Arc Blow

The electric current that flows through the electrode, workpiece, and cables sets up magnetic fields in a circular path perpendicular to the direction of the current. When the magnetic fields around the arc are unbalanced, the arc tends to bend away from the greatest concentration of the magnetic field. This deflection of the arc is called *arc blow.* Arc blow can result in an irregular weld bead and incomplete fusion.

Direct current is highly susceptible to arc blow, especially when welding is being done in corners and near the end of a joint. The arc blow occurs because the induced magnetic field is in one direction. Arc blow is illustrated in Fig. 7.10. It is often encountered when welding magnetized metal or near a magnetized fixture or on complex structures. Forward arc blow is encountered when welding away from the work connection or at the beginning of a weld joint. Backward arc blow occurs during welding toward the work connection, into a corner, or toward the end of a weld joint. There are several corrective methods that can be used:

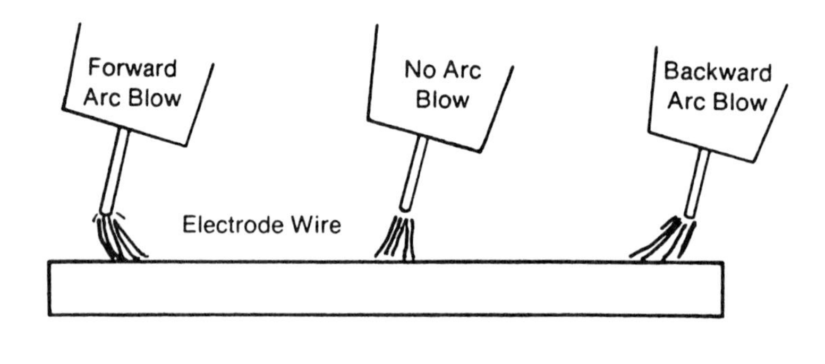

Figure 7.10 Arc blow.

1. Weld toward an existing weld or tack weld.
2. Reduce the welding current and the arc voltage.
3. Place the work connection as far as possible from the weld, at the end of the weld or at the start of the weld, and weld toward the heavy tack weld.
4. Change the position of the fixture or demagnetize the base metal or the fixture.

Inadequate Shielding

Many defects are caused by inadequate flow or a blockage of flow of shielding gas to the arc area. Inadequate gas coverage can cause oxidation of the weld pool and causes porosity in the weld bead, which usually appears as surface porosity. This is easily detected because the arc changes color, the weld bead is discolored, and the arc becomes unstable and difficult to control. The most common causes of this problem are

1. Blockage of gas flow in the torch or hoses or freezing of the regulator when using carbon dioxide
2. A leak in the gas system or an empty gas cylinder
3. Weld spatter blocking the nozzle of the welding torch
4. A very high travel speed
5. Improper flow rate
6. Wind or drafts
7. Too great a distance between nozzle and work

The torch and hoses should be routinely checked to make sure that the shielding gas flows freely and the connections are not leaking. The nozzle and contact tube should be cleaned of spatter regularly. A very high travel speed may leave the weld pool or part of it exposed to the atmosphere. This can be corrected by inclining the gun in the direction of travel, using a nozzle that directs shielding gas back over the heated area, or increasing the gas flow rate. The best method is to slow the travel speed. An improper flow rate may occasionally be the problem. Too high a flow rate can cause excessive turbulence in the arc area. When winds or air drafts are present, corrective steps should be taken. Setting up screens around the operation is the most effective. An excessive distance between the end of the nozzle and the molten weld pool will also create a problem in providing adequate shielding, which can be corrected by shortening this distance.

Clogged or Dirty Contact Tube

The power delivered to the arc in gas metal arc welding depends on the transfer of current from the contact tube to the electrode. A clogged, dirty, or worn contact tube can cause fluctuations in the amount of power transferred to the electrode, which can have an effect on the arc characteristics. It can also cause an irregular weld bead and possibly incomplete fusion because of the power fluctuations. A clogged contact tube can stop the feed of the electrode wire, which stops the welding arc. A contact tube can become dirty or clogged by spatter from the arc, by rust, scale, copper wire coating, drawing compounds left on the surface of the electrode, or metal chips created by overly tight wire feed rolls. These problems can be prevented by making sure that the electrode wire is clean and the wire feed rolls are tight enough to feed the wire without creating chips. A wire wipe made of cloth is often attached to the wire feeder to clean the electrode wire as it is fed.

Wire Feed Stoppages

Wire feed stoppages cause the arc to be extinguished and can create an irregular weld bead because of the stops and starts. Wire feed stoppages can be caused by

1. A clogged contact tube
2. A clogged conduit in the welding gun assembly
3. Sharp bends or kinks in the wire feed conduit
4. Excessive pressure on the wire feed rolls, which can cause breakage of the wire
5. Inadequate pressure on the wire feed rolls
6. Attempting to feed the wire over excessively long distances
7. Spool of wire clamped too tightly to the wire reel support

Problems such as sharp bends or kinks in the wire feed system, excessive pressure on the wire feed rolls, or attempting to feed the wire over an excessively long distance are particularly troublesome when using soft electrode wires such as aluminum, magnesium, and copper.

Wire feed stoppages must be corrected by taking the torch assembly apart and cutting and removing the electrode wire. Wire stoppages can be prevented by

1. Cleaning the contact tube
2. Straightening or replacing the wire feed conduit
3. Adjusting the pressure on the wire feed rolls to prevent breakage or to provide adequate driving force
4. Shortening the distance between the wire feeder and the gun or between the wire feeder and the electrode wire source
5. Adjusting pressure on the spool of wire or on the dereeler

7.5 Other Quality Assurance Methods

Customer demand for high quality products is continually increasing. Postweld inspection techniques find defects too late in the production cycle to be much use.

There is a need for "real-time" inspection of the manufacturing process to ensure that the quality requirements of the part are being met. Statistical process control techniques can be used to ensure quality weldments. Real-time analysis is possible with adaptive control welding systems that use dedicated fixtures or robots and can eliminate the need for human monitoring of the process.

A key to meeting statistical process control requirements is the controller operating the welding cell. Controllers must control all of the parameters involved in making the weld, which include arc voltage, current, wire feed speed, gas flow rate, gas analysis, travel speed, fit-up analysis, seam tracking, contact tube standoff, and other factors. The heart of this type of operation is a multiaxis computer controller that has specific sensing devices with feedback to evaluate the parameters and to make necessary adjustments in them.

The quality control program must have management's commitment to produce the highest quality products possible. The system must include equipment suitable for providing the high quality welded product and a multiaxis adaptive computer controller with feedback sensors to monitor critical parameters. The secret of attaining this "real-time" inspection is the ability to gather and report data obtained by a variety of sensors interfaced with the weld. These sensors monitor the key welding variables and provide data to be compared with established parameters. The data are recorded by a printer connected to the computer controller that prints out each of the variables. Depending on the product, the printouts can be made as often as several times a second or only once during a manufacturing cycle. They can be produced in the form of the standard charts and reports required by the statistical process control system. All of the collected data must be manipulated to provide the desired statistical analysis.

An arc data monitor system flow chart is presented in Fig. 7.11. This system will provide trend analysis and can be made to calculate "heat input." Precise calibration of the measuring sensors is essential, and they should be independent of the settings of the welding equipment. Instruments must be calibrated periodically. If pulsed current welding is used, special attention should be given to pulse, shape, and frequency.

Although the data are usually printed out by a printer connected to the computer controller, they can also be presented on a display screen. The data reporting system should sound an alarm to alert personnel when an out-of-limit condition is detected. Out-of-limit conditions are welding parameter values that fall outside the range determined to be acceptable.

Arc data monitors are available that can be attached to complex weld cells. Figure 7.12 shows a portable unit used to collect data and to print out the measured values. Monitors must be matched to the welding process. Instruments of this type can be used to provide a permanent record for each weld.

Real-time monitoring systems must be matched to the welding cell and can

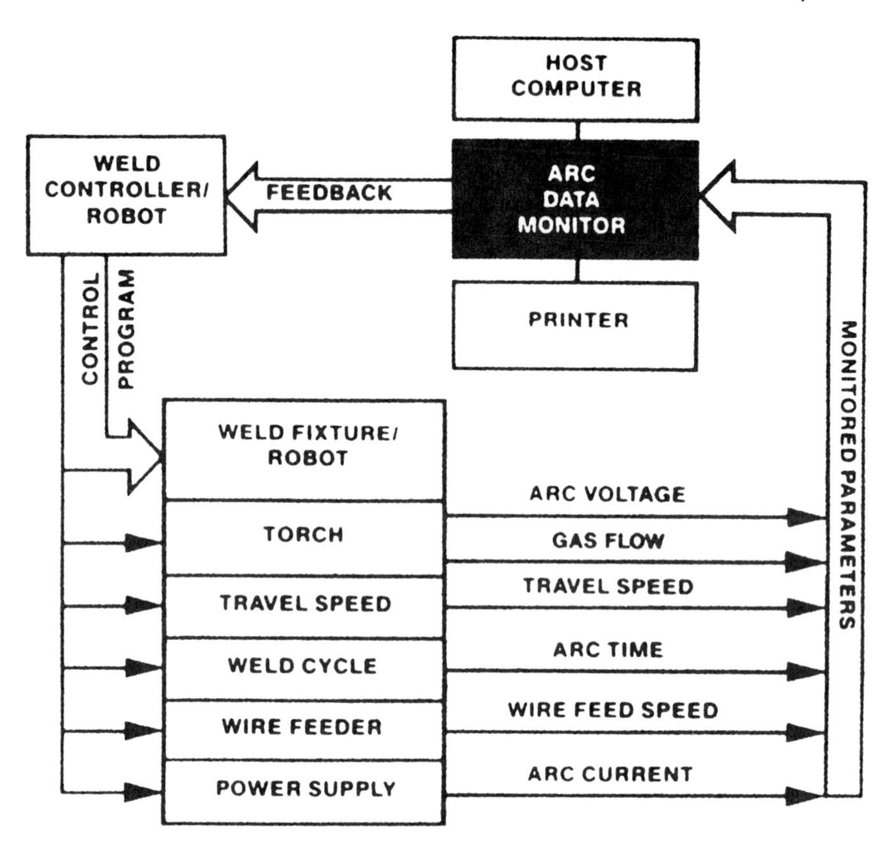

Figure 7.11 Flow chart of data monitor system.

be completely independent of the controller. However, they can also be designed to work with the system computer controller. They can be used to document welding data required for critical weldments.

Statistical analysis must be used to qualify the parameters and to establish the limits of each variable. For high-volume, high-quality products, this is a necessity prior to actual production. It can be used to quantify the welding procedure.

Arc data monitoring and the recording of all weld parameters will prevent defects from occurring. Variations in a parameter will be reported long before the parameter is out of limits and allow necessary adjustments or part replacement. This is the method of assured high quality required by automated or robotic weld cells.

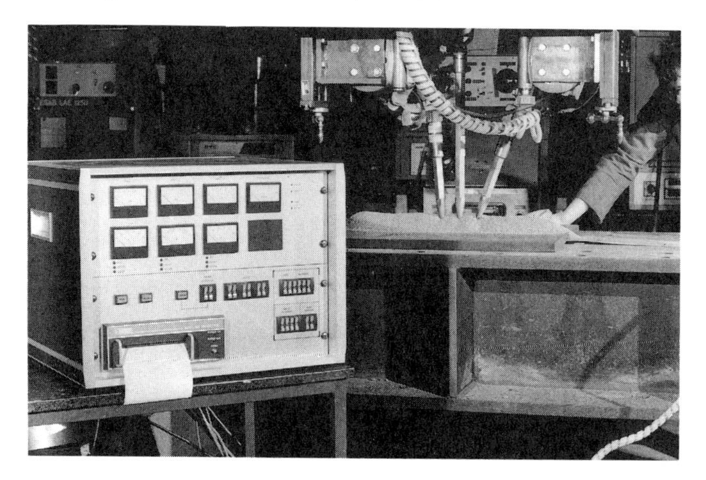

Figure 7.12 Portable arc data monitor.

7.6 Nondestructive Examination Symbols

An examination symbol consists of the following elements [1]:

1. Reference line
2. Arrow
3. Examination method letter designation
4. Extent and number of examinations
5. Supplementary symbols
6. Tail (specifications, codes, or other references)

Nondestructive examination methods are specified by use of the letter designation shown in Table 7.2. Supplementary symbols to be used in nondestructive examination symbols are shown by Fig. 7.13.

The elements of a nondestructive examination symbol have standard locations with respect to each other as shown in Fig. 7.14 and discussed in the following list, which is based on AWS 2.4 [1].

The arrow must connect the reference line to the part to be examined. The side of the part to which the arrow points is considered the *arrow side* of the part. The side opposite the arrow side of the part is referred to as *the other side.*

Examinations to be made on the arrow side of the part are specified by placing the letter designation for the selected examination method below the reference line.

Examinations to be made on the other side of the part are specified by

Table 7.2 Nondestructive Examination Methods

Examination method	Letter designation
Acoustic emission	AET
Electromagnetic	ET
Leak	LT
Magnetic particle	MT
Neutron radiographic	NRT
Penetrant	PT
Proof	PRT
Radiographic	RT
Ultrasonic	UT
Visual	VT

Source: Ref. 1.

placing the letter designation for the selected examination method above the reference line.

Examinations to be made on both sides of the part are specified by placing the letter designation for the selected examination method on both sides of the reference line.

When the letter designation has no arrow-side or other-side significance or there is no preference as to the side from which the examination is to be made, the letter designation is centered on the reference line.

More than one examination method may be specified for the same part by placing the combined letter designations of the selected examination methods in the appropriate positions relative to the reference line. Letter designations for two or more examination methods to be placed on the same side of the reference line or centered on the reference line must be separated by a plus sign.

Nondestructive examination symbols and welding symbols may be combined.

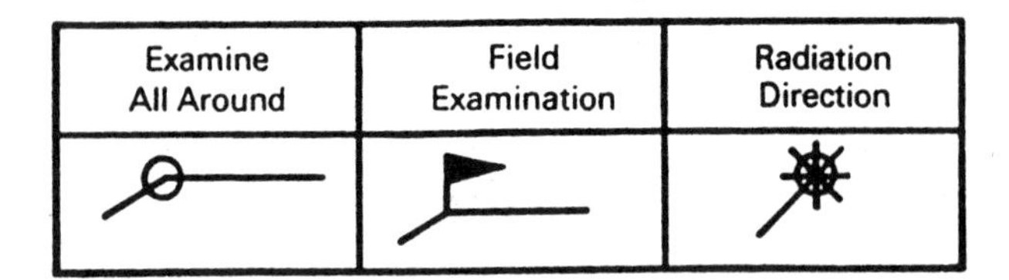

Figure 7.13 Supplementary nondestructive examination symbols.

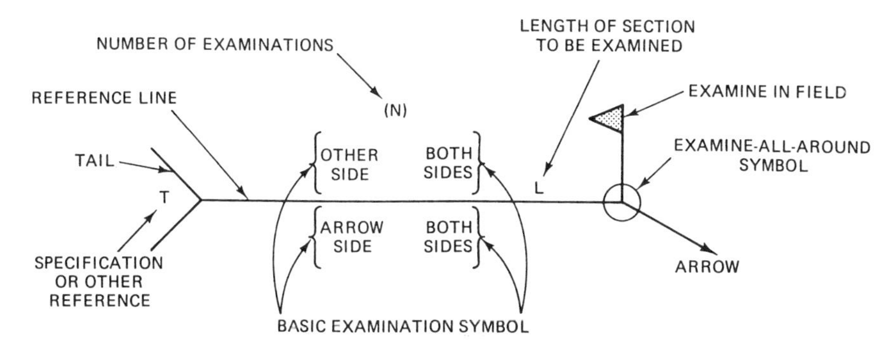

Figure 7.14 Location of elements of nondestructive examination symbols.

When it is required to specify dimensions with nondestructive examination symbols, the same system of units that is standard for the drawing is used. Dual dimensioning must not be used on nondestructive examination symbols. If conversions from metric to U.S. customary, or vice versa, are required, then a table of conversions may be included on the drawing.

Examinations required all around a weld, joint, or part must be specified by placing the "examine-all-around" symbol at the junction of the arrow and reference lines.

Examinations that must be conducted in the field (not in a shop or at the place of initial construction) must be specified by placing the field examination symbol at the junction of the arrow and reference lines.

The direction of penetrating radiation may be specified by use of the radiation direction symbol drawn at the required angle on the drawing, with the angle indicated, in degrees, to ensure no misunderstanding.

Information applicable to the examination specified and not otherwise provided may be placed in the tail of the nondestructive examination symbol. See the AWS standard [1] for more complete information.

Reference

1. *Symbols of Welding and Non-Destructive Testing,* AWS 2.4, The American Welding Society, Miami, FL.

8

Equipment Used for Automated Arc Welding

8.1 Welding Power Sources

A welding machine or power source for arc welding is designed to provide electric power of the proper values and characteristics to maintain a stable arc suitable for welding. The power required for arc welding ranges from less than 5 W to over 40 kW using 10–50 V and 5–1000 A. The electric power for arc mechanized or automatic welding is normally converted from utility line power. For mechanized or automated welding, static machines are preferred.

There are three types of arc welding power sources, distinguished according to their static characteristics output curve. The *constant-power* (CP) welding machine with the drooping volt–ampere curve is the conventional type of power source that has been used for many years for shielded metal arc welding using stick electrodes. It can be used for submerged arc welding, gas tungsten arc welding, and plasma arc welding. The *constant-voltage* (CV) welding machine with a fairly flat volt–ampere curve is the type normally used for gas metal arc and flux-cored arc welding using small-diameter electrode wire. The *constant-*

current (CC) welding machine with a fairly vertical volt–ampere curve is normally used for gas tungsten arc and plasma arc welding.

The output volt–ampere curves for these three types are shown in Fig. 8.1. They can be best understood by comparing their static characteristic output curves. The curves are obtained by loading the welding machine with variable resistance and plotting the voltage at the terminal for different amounts of current output. These curves provide the volt–ampere relationship for steady-state arc welding conditions.

The conventional or drooping characteristic welding machine, also known as the variable-voltage welding machine, is really a constant-power machine. It is used for manual covered (stick) welding (SMAW), gas tungsten arc (GTAW or TIG) welding, mechanized submerged arc welding (SAW), plasma arc welding (PAW), carbon arc welding (CAW), and arc stud welding (SW). The constant-power welding machine produces a drooping volt-ampere static output curve as shown in Fig. 8.2. These curves show that the welding machine produces maximum output voltage with no load, that is, zero current, and as the load increases, the output voltage decreases. Under normal welding conditions, the output voltage is between 20 and 40 V. The open circuit, or no-load voltage, is between 60 and 85 V. The desired current output is achieved by adjusting the electrical coupling within the machine. These types of machines produce either alternating (sinusoidal) current or steady direct current welding power, or in some cases both alternating and direct current.

When used for shielded metal arc welding (stick welding), the arc voltage is controlled by the human welder and is directly related to the arc length and thus indirectly related to welding current. The output curve shows that when the arc voltage increases, the welding current decreases, and when the arc voltage

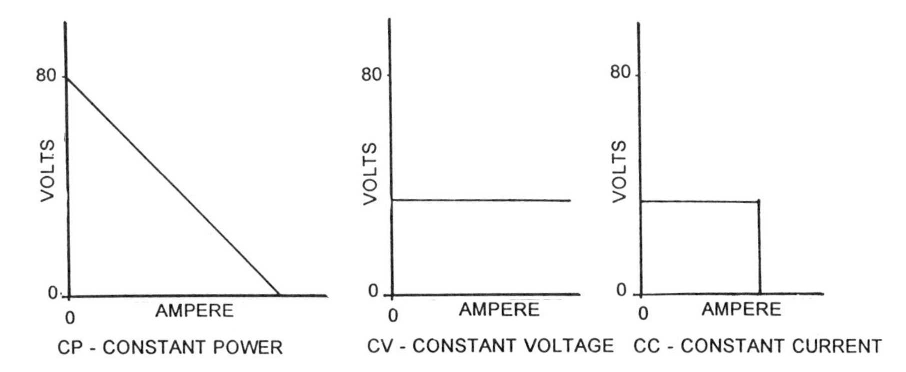

Figure 8.1 Three output curves—constant power (CP), constant voltage (CV), and constant current (CC).

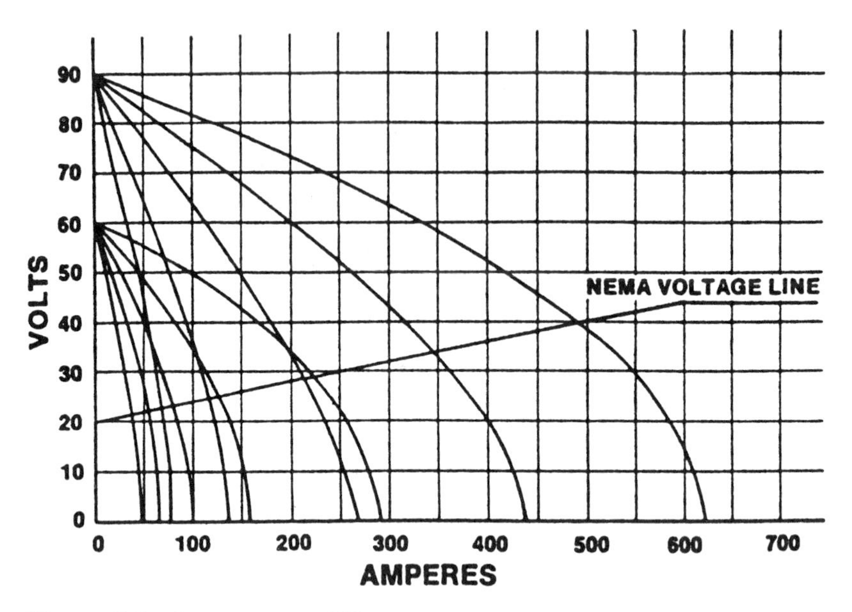

Figure 8.2 Constant-power (CP) curve.

decreases, the welding current increases. The human welder can vary the current in the arc, the "welding heat," by lengthening or shortening the arc.

For mechanized welding, a voltage-sensing controlled wire feeder is used. By using a reference or preset arc voltage, the wire feeder can be speeded up or slowed down to maintain the proper arc voltage. The current may vary because of the increased or decreased wire feed rate, which requires more or less welding current. This system is becoming less popular for mechanized welding except when large-diameter electrode wire is used for submerged arc welding.

Constant-voltage (CV) welding machine

A constant-voltage (CV) power source, sometimes called a constant-potential (CP) or modified constant-voltage power source, is a welding machine that provides a nominally constant voltage to the arc regardless of the current through the arc. Characteristic output curves are shown by Fig. 8.3. This type of machine is used for welding processes using a continuously fed electrode wire, where the weld metal is transferred across the arc. This includes gas metal arc (GMAW-MIG), flux-cored (FCAW), and small-wire submerged arc (SAW) welding. The machine produces direct current only. In continuous wire welding, the burnoff rate of a specific size and type of electrode wire in a specific atmosphere is proportional to the welding current. Higher welding current increases the amount

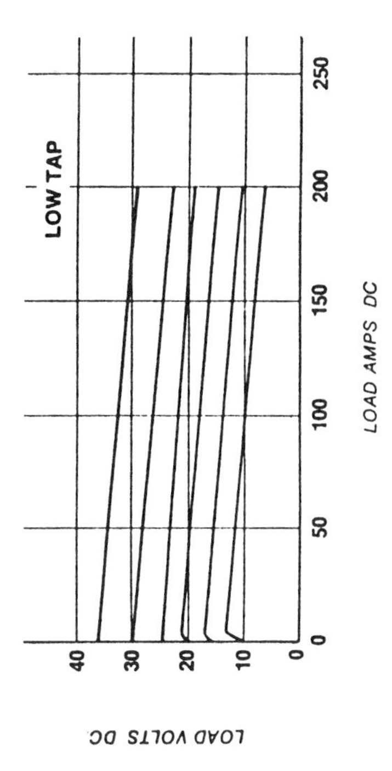

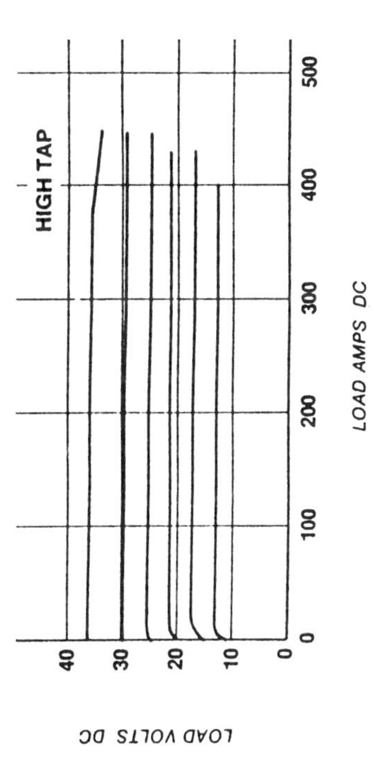

Figure 8.3 Constant-voltage (CV) curve.

of wire melted. The wire is fed into the arc by means of a wire feed motor with a constant but adjustable speed. The speed of the feed motor can be adjusted to increase or decrease the rate of electrode wire feed. The welding machine automatically provides the correct output depending on that rate of speed. The voltage produced is adjusted by moving the curve upwards or downwards, which is accomplished by the voltage control adjustment in the power source. The arc voltage and arc length depend on the setting of the welding machine. With an absolutely flat curve, the change in current does not change the arc voltage. However, the curve of most constant-voltage welding machines has a slight downward slope. The slope of the curve can be adjusted to match the welding process, type of work, electrode wire size, etc.

Constant-current (CC) welding machine

The constant-current welding machine is primarily used for gas tungsten arc welding (GTAW; TIG) and plasma arc welding (PAW). It can be operated manually or mechanized. Its static output curves are shown in Fig. 8.4. The vertical portion of the curve is adjusted by moving the welding current to the right or left. This is done by adjusting the current control in the power source. The arc voltage is changed by changing the arc length. With a true constant-current machine, the current does not change when the arc voltage changes. This allows

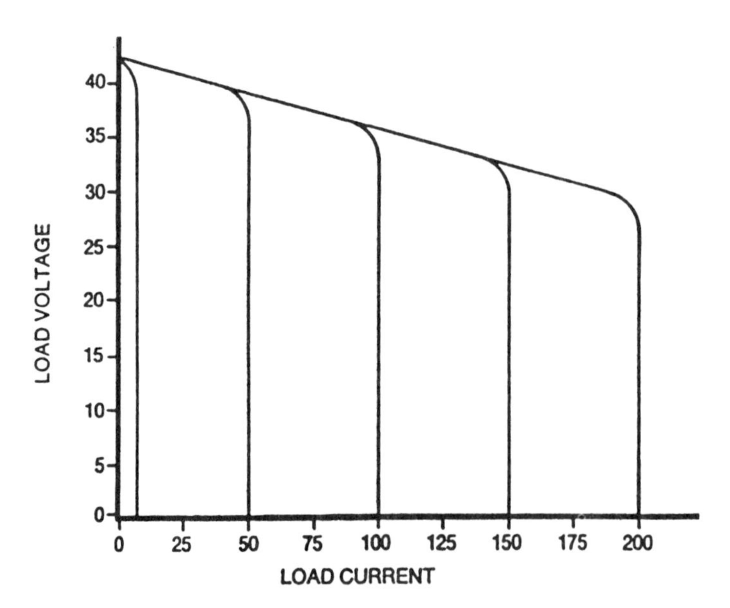

Figure 8.4 Constant-current (CC) curve.

for the independent adjustment of current and voltage desirable for TIG and plasma welding.

The static output curve of a welding machine depends on the machine's design. A static welding power source has no moving parts except a cooling fan. The most common type is the transformer welding machine. It produces alternating welding current of power-line frequency. The transformer takes power directly from the power company's utility line and, by means of magnetics, turns ratios, inductors, etc., provides the desired volt–ampere characteristics, normally the drooping type required for arc welding. Alternating current is used for the submerged arc welding process and for gas tungsten arc welding but is not used for gas metal arc or flux-cored arc welding.

Direct-current power is the most popular for mechanized arc welding and is produced by a transformer/rectifier power source. A rectifier connected to the transformer changes the alternating current to direct current. Various methods are used to smooth out the current. Industrial rectifier welding machines normally use three-phase input power, which is desirable because it is easier to smooth and eliminates utility line imbalance.

The output power of the transformer/rectifier power source is controlled by mechanical, electrical, or electronic means. This adjustment on most modern industrial machines is made electronically or by special electric circuits. It is possible to provide continuous-current adjustments from the minimum to the maximum output with a single control. The electronic method for controlling machine output is the most common. A block diagram of this type of machine is shown in Fig. 8.5. Using solid-state control, the output of the power source is adjusted to meet the needs of different applications for different welding processes.

One of the newer types of power sources is the inverter, shown schematically in Fig. 8.6. It involves special circuitry that provides a smaller and lighter weight power source that is extremely controllable. In an inverter power source, the power from the utility line is rectified into direct current. It is then converted

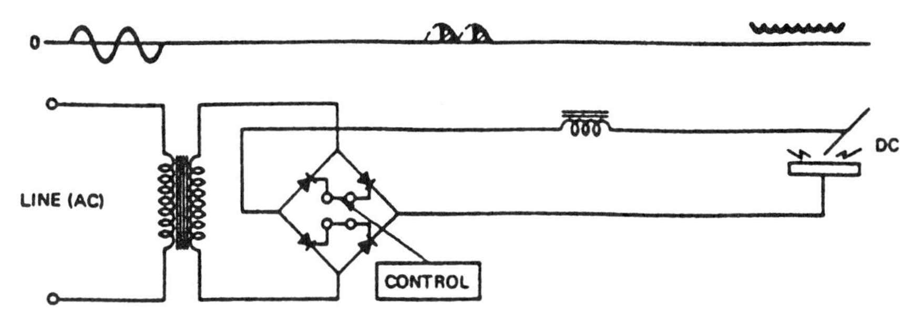

Figure 8.5 Block diagram of a static welding machine.

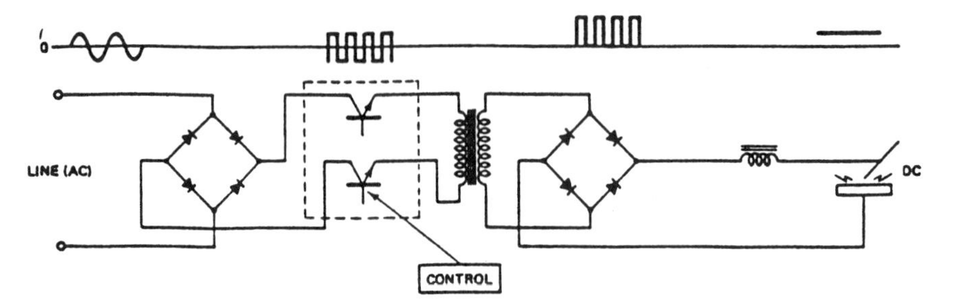

Figure 8.6 Block diagram of an inverter welding machine.

into alternating current of a very high frequency, usually above 20 kHz (kilohertz). A very light weight high-frequency transformer is used to change the current to meet the requirements of the welding job, and the current is then rectified to direct current. The inverter is adjusted by electronic circuits to provide the characteristics needed for the specific arc welding application. The inverter type of power source can provide either constant voltage or constant current and can be controlled by signals from a computer via a robot or other automated device. Such power sources are very popular for automated applications. Many have built-in programs that can be selected to match the welding procedure. Others allow the user to choose pulse settings and provide power for specific welding processes.

Special features such as pulsing or different output waveforms can be added to the power source. They can also be used to produce square-wave alternating current with variable frequencies or direct current of different amplitudes and cycle times.

The variable-output DC type with matched wire feeders is able to produce synergic power that offers extremely attractive characteristics for many welding applications. It provides a very stable arc with absolute starting and an arc that produces little or no spatter. These power sources are computer-controlled and will accept feedback signals from external sensors. Synergic machines must be selected for the correct output characteristics for the welding process and procedure to be used. The wire feeders and control systems must be compatible with the power source.

The *duty cycle* of a welding machine is defined as the ratio of arc time to total time. The National Electrical Manufacturers Association (NEMA) Standards Publication EW 1 [1] specified a 10-min cycle. Thus for a 60% duty cycle machine, the welding time could be applied continuously for 6 min and off for 4 min. Most industrial power sources are rated at 60% duty cycle or 100% duty cycle. Machines for mechanized or automatic welding must have a 100% duty cycle rating. To determine the duty cycle of a welding machine, use the formula

$$\% \text{ Duty cycle} = \frac{(\text{rated current})^2}{(\text{load current})^2} \times \text{rated duty cycle}$$

For example, if a machine is rated at 60% duty cycle at 300 A and is required to operate at 350 A continuous, the formula gives

$$\% \text{ Duty cycle} = \frac{(300)^2}{(350)^2} \times 60 = 44\%$$

Figure 8.7 illustrates some curves obtained with this computation. Rather than work out the formula, this chart can be used. For example, a question might arise as to whether a 400 A, 60% duty cycle machine could be used for a fully automatic requirement of 300 A for a 10-min welding job. In Fig. 8.7, the line for the 400 A, 60% duty cycle machine passes through the 100% duty cycle vertical slightly above the 300 A horizontal. This shows that the machine can be used at slightly over 300 A at 100% duty cycle.

Conversely, there may be a need to draw more than one rated curve from a welding machine for a short period. This figure can be used to compare various machines. All machines should be related to the same duty cycle for comparison.

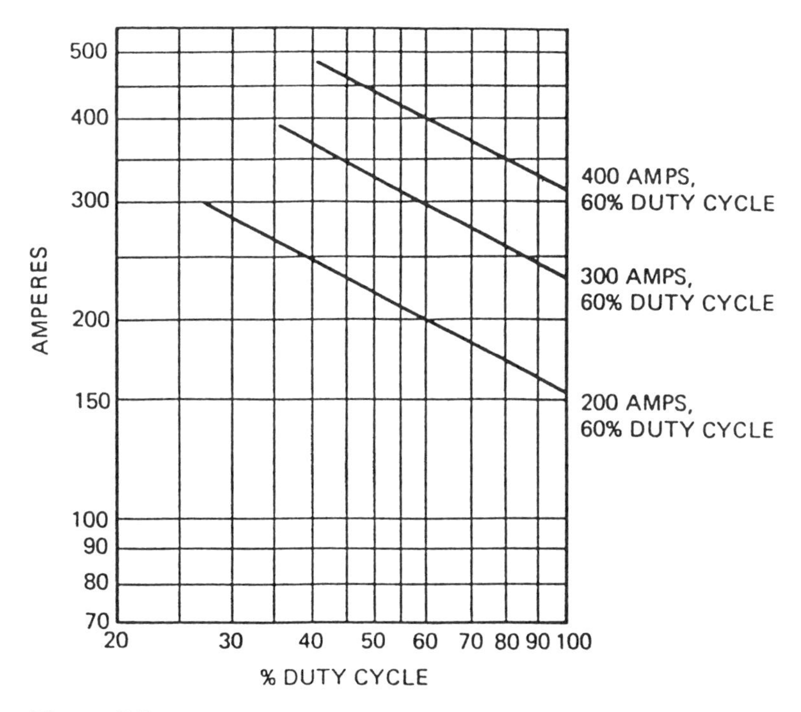

Figure 8.7 Duty cycle vs. welding current load.

Specifying an arc welding power source

The selection of a welding machine is based on

1. The process or processes to be used
2. The amount of current required
3. The power available at the job site
4. Economic factors and convenience

The information given earlier concerning each of the arc welding processes indicates the type of machine to be used. The size of the machine depends on the welding procedure, which includes the current, voltage, and duty cycle. The appropriate welding current and duty cycle are determined by analyzing the welding job, the weld joints, weld size, etc., and consulting the welding procedure tables. The incoming power available dictates the input voltage and frequency. Finally, the job situation, economic considerations, and personal preferences narrow the field to the final selection.

To properly specify a welding machine the following data should be provided:

1. Manufacturer's type designation or catalog number
2. Manufacturer's identification or model number
3. Voltage incoming power (normally 460 or 230 V)
4. Frequency of incoming power (normally 60 Hz or 50 Hz)
5. Number of phases of incoming power (normally three)

It is probably helpful to also indicate the rated load current (amperes), the rated load voltage, and the duty cycle according to NEMA. It is also wise to indicate what welding process is to be employed.

8.2 Electrode Wire Feeders

The wire feeding system must be matched to the welding process and the type of power source being used. There are two basic types of wire feeders. The type used for the consumable electrode wire process is known as an *electrode wire feeder*. The electrode is part of the welding circuit, and the melted metal from the electrode crosses the arc to become the weld deposit. The second type is known as a *cold wire feeder* and is used for gas tungsten arc welding and other welding processes. The electrode is not part of the circuit, and the filler wire fed into the arc area melts from the heat of the arc and becomes the weld metal.

The wire feeder or feed head must feed the electrode or filler wire into the arc at a specific feed rate and must be able to feed electrode wires of different sizes and types. The feeder must have a variable-speed drive motor and be adjustable over a fairly wide range of speeds. The feeder must feed the electrode wire uniformly and smoothly and grip the wire firmly so that there is no slippage and the wire is fed into the arc at the proper rate to comply with the welding procedure. To cover the entire range of feed rates required, different gear ratios may be used. The wire feeder must be specified to provide the range of wire feed rates required by the welding process, type and size of electrode wire, and specific feed rate range of the welding procedure. Feed rate is measured in inches electrode wire per minute, in millimeters per minute, or, in some cases, meters per minute.

Cold wire feeders are filler metal feeders that feed the wire into the arc to be melted in the heat of the arc. They are used for gas tungsten arc welding, for plasma arc welding, and sometimes for high-energy beam welding processes. Normally, the filler metal does not carry current as it is fed. The feed rate of the cold wire feeder must be very accurately controlled, and the wire feed motor must have continuous speed regulation over a wide range. In general, cold wire feeders feed wire at a much lower rate than electrode wire feeders. Table 8.1 shows the speed range for steel wire.

The electrode wire feeder pushes the electrode wire through the torch and contact tube to the arc. In some cases the electrode wire is fed through a flexible conduit so the feed head mechanism is a distance from the arc. Feeding wire through a conduit of any type increases the drag on the wire and requires more power.

There are two basic types of electrode wire feeders. The constant-power or drooping curve power source requires a voltage-sensing wire feed system in which the feed rate may be changing continuously. The constant-voltage (CV) system requires a constant feed rate during the welding operation. The control circuits for the wire feeder are different for each type.

The motors for wire feeders range from a small permanent magnet type for feeding small-diameter wire to a heavy duty universal type for feeding large-diameter wire. Other types such as the print type (pancake) motor or stepping

Table 8.1 Feed Rate for Cold Wire Feeder

Process	Current range (A)	Wire size (gauge)	Feed rate (in./min)
GTAW (TIG)	100–300	0.035	25–200
	100–300	0.045	13–100
	100–300	1/16	6–50
	10–100	0.035	6–50
	10–100	0.045	3–25
PAW (plasma)	50–100	0.035	25–200
	50–100	0.045	13–100
	10–50	0.035	6–50
	10–50	0.045	3–25

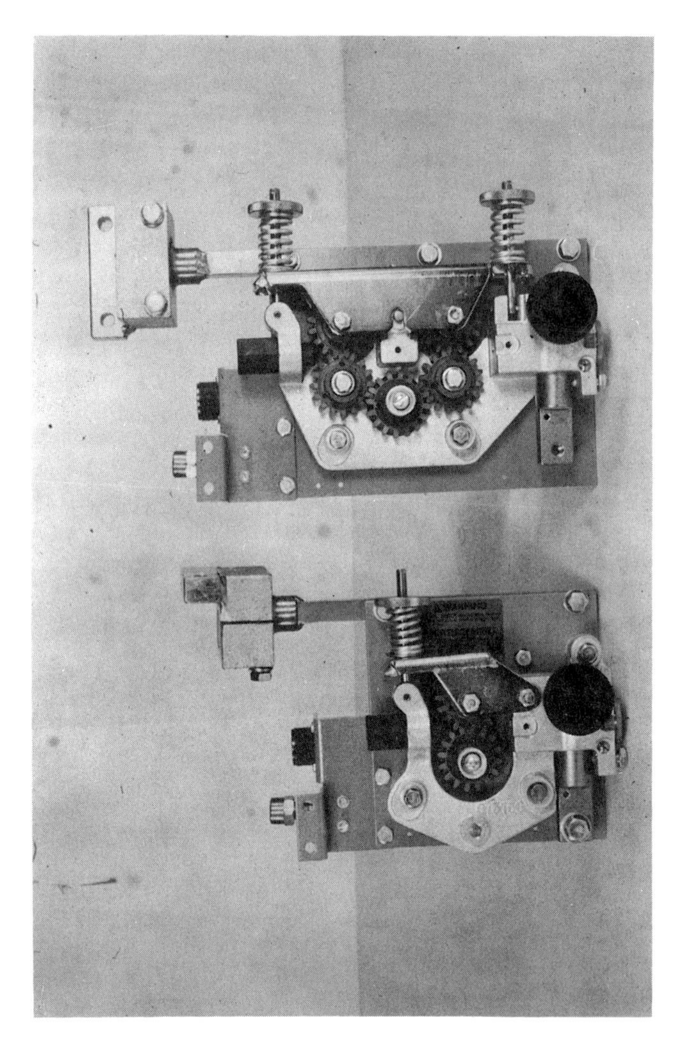

Figure 8.8 Two- and four-roll drive systems.

motor may also be used. The type of motor employed depends primarily on the power required to feed the wire.

The power required depends on many factors. The large-diameter electrode wire normally used for submerged arc welding requires fairly high power, especially if it is fed through a conduit and if the current pickup system is spring-loaded. The wire-dispensing system and the distance through which the wire is pulled also affect the power required. The third factor is the response time, or how fast the speed must be changed. This is very important when feeding aluminum wire. For small-wire applications on a spool gun, the amount of power needed is extremely low.

Transmitting the feed motor power to the electrode wire involves transforming rotary motion to linear motion. The most common means of doing this is to use pinch rolls that grip the wire on opposite sides and exert pressure against it. Grooves and knurls on the surface of the rolls move the wire linearly. Two-roll drives are commonly used, although four-roll drives are almost equally popular. Figure 8.8 shows both a two-wire drive system and a four-wire drive system. There is an advantage to the four-wire drive system since there is less pressure on the electrode wire. This is important when flux-cored wires are being fed, for the sheath of the wire may collapse if the pressure is too great. The design of the drive rolls is extremely important, and different types of rolls are specified for different types of electrode wires. The chart in Fig. 8.9 shows the types of drive rolls commonly used for feeding wires with the two-roll and four-roll designs. The design of the drive rolls is important and is shown by the chart. Normally,

ELECTRODE WIRE DIAMETER		ELECTRODE WIRE TYPES					
in.	mm	Hard Wire	Hard Wire	Hard and Tubular Wire	Soft Wire	Hard and Tubular Wire	Tubular Wire
0.024		X	X	–	X	–	–
0.030	0.75	X	X	–	X	–	–
0.035	0.9	X	X	–	X	–	
0.045	1.1	X	X	–	–	–	
3/64 (0.047)	1.2	–	–	–	X	–	
0.052	1.3	X	X	–	X	–	X
1/16 (0.063)	1.6	–	–	X	X	X	X
5/64 (0.078)	2.0	–	–	X	X	X	X
3/32 (0.094)	2.4	–	–	X	X	X	X
7/64 (0.109)	2.8	–	–	X	X	X	X
1/8 (0.125)	3.2	–	–	X	X	X	X
5/32 (0.156)	4.0	–	–	–	–	X	X
3/16 (0.188)	4.8	–	–	–	–	X	X
7/32 (0.219)	5.6	–	–	–	–	X	X
1/4 (0.250)	6.4	–	–	–	–	X	X
FEED ROLLS SELECTION		Flat – Smooth	Flat – Knurled	Smooth Vee	Smooth Vee	Knurled Vee	Cog
		Smooth Vee	Smooth Vee	Smooth Vee	Smooth Vee	Smooth Vee	Cog

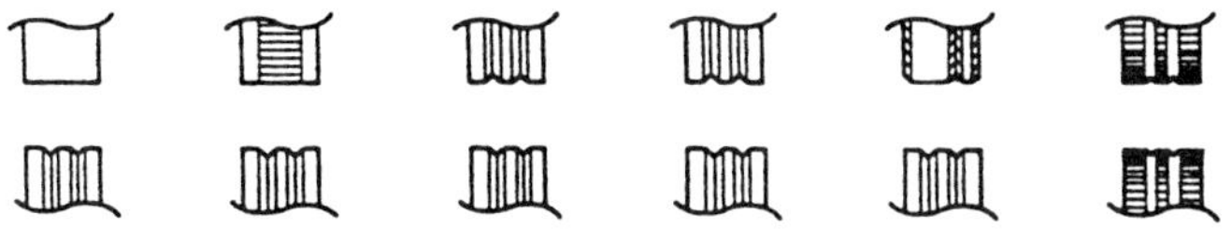

Figure 8.9 Drive roll selection chart.

the drive rolls have sufficient power to feed the wire up to the stall point of the drive feed motor. If too much pressure is exerted, more power will be required, which will deform the wire and possibly stall the motor. Pressure adjustment on wire feed drive systems is extremely important and in all cases should follow the wire feeder manufacturer's instructions.

The vee groove in the wire drive roll is important. U-shaped grooves are not recommended because the electrode wire diameter can vary (±0.001 in.). If the wire is too large, it will not fit into the groove and may require too much force to drive the wire. If the wire is too small, it will slip within the groove, and it will not be fed accurately. In addition, the surface of the roll wears, causing the roll's dimensions to change.

Wire feed systems come in different sizes to match the different sizes and types of electrode wires. The manufacturer's literature usually indicates the wire sizes that can be fed with a specific wire feed system. Figures 8.10 and 8.11 are charts showing wire feed speed versus welding current for various sizes of steel and nonferrous metal electrode wires. This is useful as a guide for selecting the wire feed system. However, it is important to refer to the welding procedure to be employed to determine the exact feed rate that should be used.

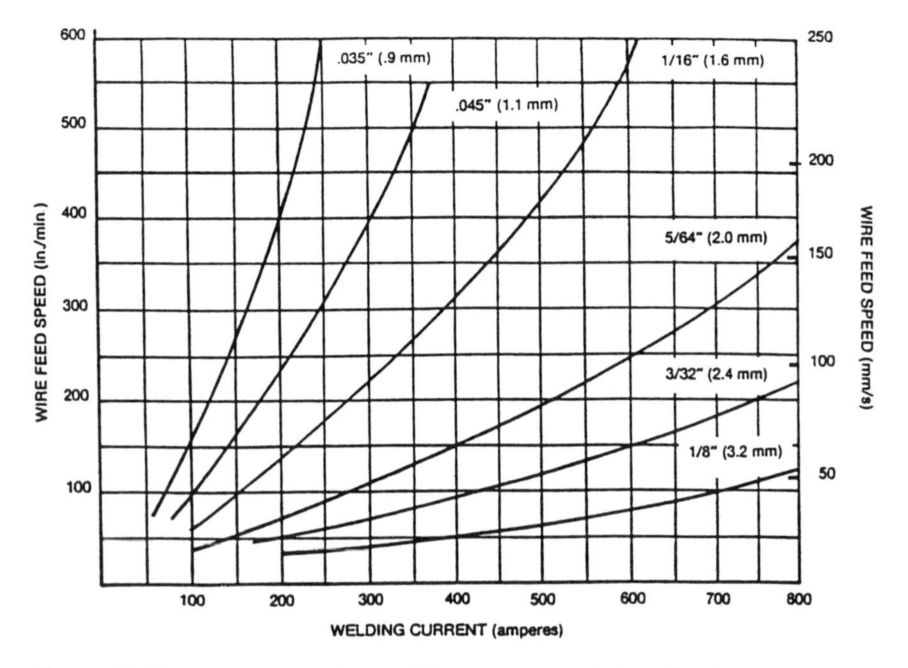

Figure 8.10 Wire feed speed vs. welding current—steel electrode wire.

The control circuits for the electrode drive system must match the power source type. For a constant-voltage power source, the wire feed control system must be adjustable so that it can be increased for higher welding currents or decreased for lower current. Relate it to the type of power source and wire feeder.

The other control circuit is for voltage-sensing, variable-speed wire feeders used with the constant conventional power source. The control circuit senses arc voltage and is continually changing speed to maintain a preset arc voltage. The voltage can be adjusted for a shorter or longer arc. The relationship between power source type and wire feeder type is illustrated by Table 8.2.

Push-pull system two-wire drive motors are used to feed small aluminum electrode wire. This is a trouble-free system when small-diameter "soft" wire is being fed through a long conduit. Motor speed balance must be maintained.

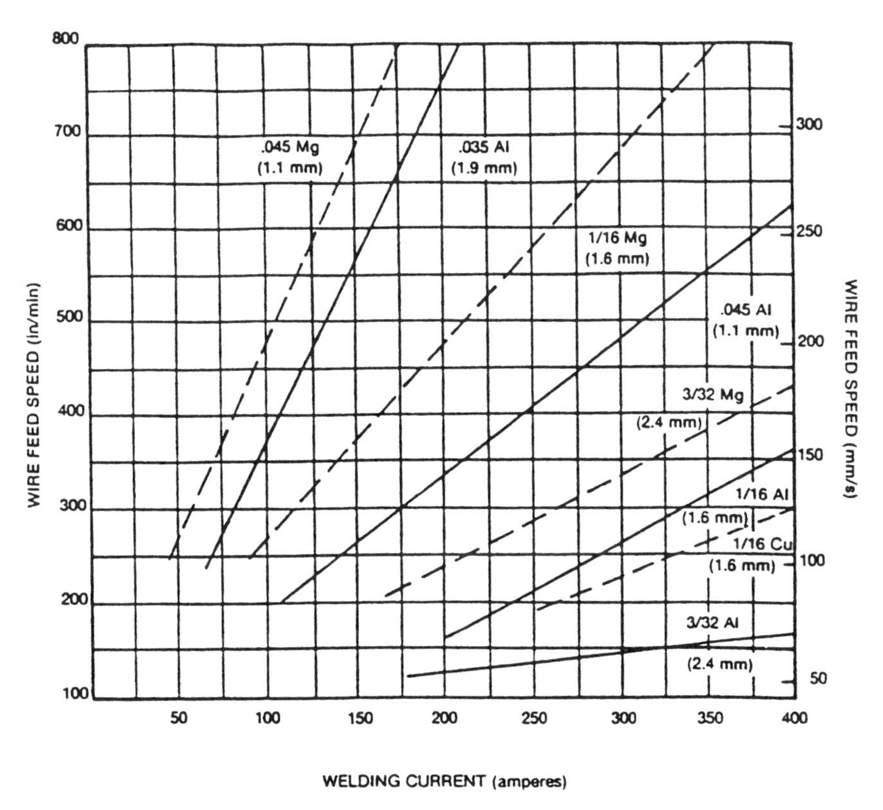

Figure 8.11 Wire feed speed vs. welding current—nonferrous electrode wire.

Table 8.2 Power Source to Wire Feeder Relationship

Power source type	Electrode wire feeder type	
	Voltage sensing	Constant speed
CV, direct current	Difficult to adjust; seldom used; self-regulating within limits	Best for gas metal arc welding; best for flux-cored arc welding; best for submerged arc when using small-diameter electrode wire; self-regulating
CC, direct current	Best for submerged arc when using large-diameter electrode wire; used for GMAW on aluminum; self-regulating	Difficult to control; not used for small-wire GMAW; not self-regulating
CC, alternating current	Used for submerged arc (medium and large electrode diameters)	Difficult to control; not used for GMAW; not self-regulating

8.3 Electrode Wire Dispensing Systems

The electrode wire feeding (dispensing) equipment should accommodate the packaging of the electrode wire. Welding electrode wire, solid or tubular, is packaged in various ways to suit the needs of the production operation. These include small and medium-size spools, small coils, reels, drums or payoff packs, and very large coils. Medium-size spools normally require a spool adapter that is designed to fit the inside diameter of the spool and engage a hole in the spool. This will provide a braking action so the wire will not unwind from the spool. Small coils also require an adapter or spider arrangement to center and retain the coil and allow uniform unwinding. Figure 8.12 shows typical adapters and spiders. Spiders may also have braking systems and are designed to fit the needs of the smaller coils. Coils are supplied with cardboard inner liners and come in weights of 50–60 lb (23–27 kg).

Large reels require special dispensing equipment. These reels contain 250–1000 lb (113–453 kg) of electrode wire. The reels are made of wood with holes for an axle. The larger reels can be unwound with the axis horizontal, in which case an axle must be inserted in the reel and the reel is carried on a dereeling device such as the one shown in Fig. 8.13. There are two general types of dereelers with a horizontal axis. The dereeler shown in this figure allows the wire feeder to pull the wire and rotate the reel. It may incorporate a brake to stop the reel to avoid overrunning and unwinding and the overlapping of loops

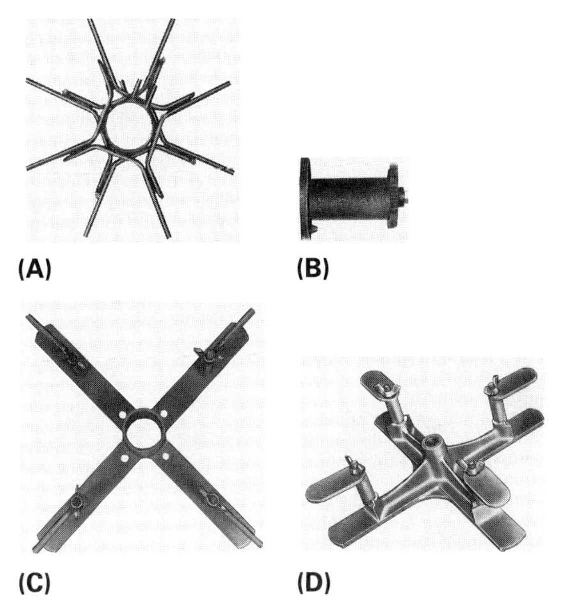

Figure 8.12 Typical adapters and spiders. (A,C) Slip-on adapters; (B) spool kit; (D) wire reel kit.

that may cause tangles. Wire feed motors are designed to use relatively small spools or coils that have relatively low inertia loads as they start to unwind. The big reels with a large amount of electrode wire are heavy, and the inertia loads for starting the reel rotating are high. This may cause poor arc starting and premature failure of the wire feed motor. This problem can be overcome with the use of motorized dereelers such as the one shown in Fig. 8.14. When the wire feed motor exerts a pull on the supply reel, the dereeler motor starts to rotate the reel. It has a variable-speed motor that is adjusted to match the speed of the wire feed motor. Extra load on the wire feed motor also occurs if the electrode supply is remote and the electrode wire is fed through a long conduit. This extra resistance load on the wire feed motor can be overcome by using a motorized dispensing system.

Electrode wire reels can also be used with the axis vertical, in which case special rotary dispensers are used, as shown in Fig. 8.15. These rotary dispensers are usually used for 1⁄16 in. (1.6 mm) diameter and smaller electrode wire. The problem with this type of dispenser is that it introduces a twist in the electrode wire, and this may cause the arc to wander as the electrode wire feeds from the tip of the welding torch.

Another method of supplying welding electrode wire is by means of

Figure 8.13 Dereeling device for larger reels.

drums or payoff packs. These are drums of heavy cardboard construction that will contain 250 lb (113 kg), 500 lb (227 kg), or more than 700 lb (318 kg) of electrode wire. Wire is placed in the drums to ensure a snarl-free payoff of the wire even if the drum is not rotating. A special dereeling system sits on top of the drum and uses a rotating pickup arm and a pulley system to supply the wire to the wire feeder. This does not involve the inertia of the entire supply of wire. The electrode wire rotates or twists through one revolution per loop as it is unwound.

Figure 8.14 Motorized dereeler for large reels.

Another method of dispensing wire from a drum or payoff pack is by means of a rotating table that revolves the drum at the correct speed as the wire is fed. This eliminates the twist or flip of the wire as it leaves the welding torch. A dereeler of this type is shown in Fig. 8.16.

Welding electrode wires are also supplied in extremely large coils weighing up to 1000 lb (453 kg). Coils of this type are usually strapped to a pallet for shipping. Once the pallet is in place, the straps are removed. A special dereeler is attached to the coil and dispenses the wire to the wire feed by means of a rotating arm. This type of system is shown in Fig. 8.17. The same twisting of the wire still occurs but can be overcome with a rotary wire straightener.

It is important to match the dispensing method to the automatic welding

Figure 8.15 Rotary dispensers for large reels.

system. The decision should be based on the cost savings that accrue from purchasing large amounts of electrode wire and from decreasing downtime versus the cost of the dispensing equipment.

Dispensing systems for different electrode wire packaging forms are normally available from the welding distributor.

8.4 Welding Torches

A welding torch is used in an automatic welding system to direct the welding electrode into the arc, to conduct the welding power to the electrode, and to provide shielding of the arc area. There are many types of welding torches, and the choice depends on the welding process, the welding process variation, welding current, electrode size, and shielding medium.

Welding torches can be categorized according to the way in which they are cooled. They may be water-cooled with circulating cooling water or air-cooled with ambient air. A torch can be used for a consumable electrode welding process such as gas metal arc or flux-cored arc welding, and shielding gas may or may not be employed. Some torches are designed for submerged arc welding, which

Figure 8.16 Dispenser for wire from drums.

uses granular flux for shielding. Some torches are designed for a nonconsumable electrode welding process such as gas tungsten arc welding or plasma arc welding. The GTAW and PAW torches are vastly different, as the plasma torch has a constricting gas orifice. A torch can be described according to whether it is a straight torch or has a bend in its barrel. A torch with a bend is often used

Figure 8.17 Rotary dispenser for large coils.

for robotic arc welding applications to provide access for the weld. Another category for differentiation is whether the torch provides concentric delivery or side delivery of the shielding gas. In addition, all welding torches are specified by their welding current carrying capacity. Six types of torches are illustrated in Fig. 8.18.

In some types of automatic welding equipment the torch is incorporated into the welding head. Orbital heads for automatic pipe and tube welding have built-in torches. In many cases, they use a standard welding gas nozzle.

The major function of the torch is to deliver the welding current to the electrode. For a consumable electrode process this means transferring the current to the electrode as the electrode moves through the torch. The transfer of current

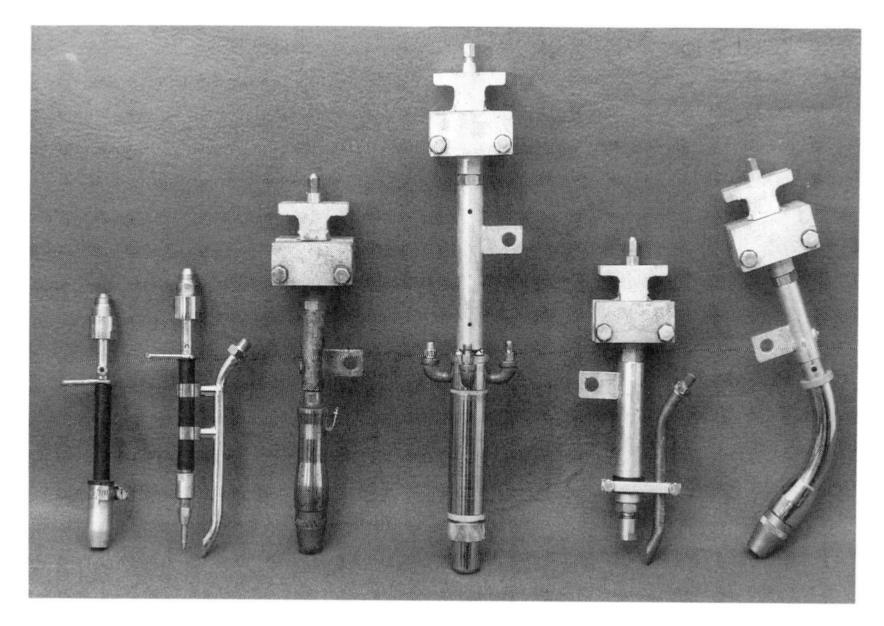

Figure 8.18 Some types of torches for mechanized welding.

is done by means of a contact tip or contact jaws. Efficient current transfer is important to avoid overheating of the contact tip, which is made of copper or copper alloy. Pure copper is very soft and will wear rapidly. Copper alloys can be quite hard and will provide much longer wear but are more expensive. Special tips with alloy inserts are also available. A major maintenance problem with robotic welding is the wear of contact tube tips.

For feeding aluminum, extra-long contact tubes are desirable. In some of these tubes a slight bend is incorporated to make sure that there is sliding contact between the aluminum electrode wire and the contact tip. Submerged arc applications may have spring-loaded contacts of copper alloy to make sure that the wire-to-contact tip electric sliding connection is efficient. For heavy-duty submerged arc welding, spring-loaded contact jaws are employed. These are loaded against the wire to efficiently transmit the current to the electrode wire.

A second major function of the torch is to deliver the shielding gas, if one is used, to the arc area. Gas metal arc welding uses a shielding gas that may be an inert gas or carbon dioxide or a mixture of inert gas, normally argon, with CO_2 or oxygen. CO_2 is often used for welding steel. CO_2 absorbs heat as it expands from the compressed gas cylinder. This tends to keep the arc torch cool. A torch can be given a higher current rating when CO_2 shielding gas is used. Side-delivery nozzles (Fig. 8.19) are often used with CO_2 shielding. CO_2 gas

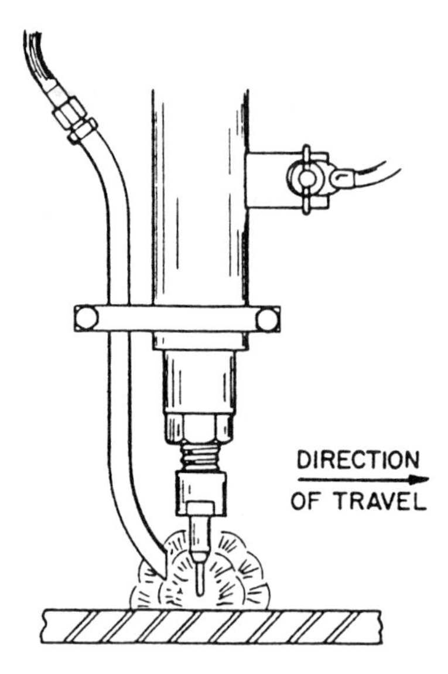

Figure 8.19 Side-delivery automatic torch.

is much less expensive than inert gases and is used at a higher flow rate. Side-delivery systems normally do not pick up spatter as quickly and are favored for those automatic systems that can use them. When inert gas is used, the concentric-delivery torch is required. For high-current applications, water cooling is necessary.

The current-carrying capacity of an electric torch is determined by its temperature rise during operation. NEMA Standards Publication EW 3 [2] provides these data for semiautomatic guns. The temperature rises allowed for hand-held torches are given. Torches are rated by their manufacturers for specific applications. For automatic applications it is best to use a torch rated at a higher current level than will be used. If this practice is followed, the torch will require less maintenance, which will reduce downtime.

8.5 Auxiliary Welding Equipment

There are a number of auxiliary devices employed in automatic or mechanized welding systems that greatly improve the operation of the system.

Wire straighteners are often required for automatic systems. A wire straightener is used to remove the inherent cast and helix of the spooled or coiled electrode wire and make it straight. Two three-roll wire straighteners arranged in two planes at a 90° angle to each other will remove the majority of the cast and helix from the electrode wire. Figure 8.20 shows an adjustable three-roll wire

Figure 8.20 Rotary wire straightener.

straightener commonly used in automatic systems. Wire straighteners of this type are usually placed ahead of the torch so that the electrode wire extending from the end of the torch pick-up tip will come out straight. This is to prevent the arc from wandering as the wire bends after leaving the contact tip.

Rotary wire straighteners are also sometimes used. They must match the electrode size and type and require a motor to provide rotational motion. They are most often used on GTAW and PAW systems when precision wire placement is necessary.

Smoke exhaust devices are sometimes used in mechanized welding systems. The exhaust system draws the air from the immediate area of the arc, passes it through a filter, and exhausts it to the outside atmosphere. A closed-circuit system is sometimes used, and the cleaned air is recirculated. These systems are more often used for flux-cored arc welding applications, because this process generates more smoke than gas metal arc welding. Exhaust systems are required for welding on hazardous materials or coated materials such as galvanized steel. The filter system must capture the particulate matter completely if the air is to be recirculated. Typical smoke exhaust systems are illustrated in Chapter 17.

Water coolers or circulators are used with high-current gas metal arc, flux-cored arc, and gas tungsten arc applications. All plasma torches require water cooling, but the submerged arc normally does not. Water-cooling devices have a specific heat removal capacity based on the size of the radiator or heat exchanger and the capacity of the pump. They are rated in heat removal in British thermal units per hour (Btu/hr) at a given flow rate and pressure. A typical water

Figure 8.21 Cooling water circulator.

circulator is shown in Fig. 8.21. It is important to specify the size of the heat exchanger needed for a particular welding application. This can be calculated by determining the heat generated in the torch and matching the capacity to this requirement. In general, the circulator should prevent the temperature of the water from exceeding 100°F. Heat exchangers are constructed of stainless steel or coated steel to avoid rusting.

References

1. *Electric Arc Welding Power Sources,* Specification EW 1, National Electrical Manufacturers Association, Washington, DC.
2. *Semiautomatic Wire Feed Systems for Arc Welding,* Specification EW 3, National Electrical Manufacturers Association, Washington, DC.

9

Arc and Work Motion Devices

9.1 Arc Motion Devices

Automatic arc welding requires motion, relative movement between the welding arc or torch and the part being welded. The first automatic welding machine, built in 1920, used linear and rotary motion for manufacturing automotive rear axles (see Fig. 1.3). Combinations of arc motion or workpiece motion have been used successfully for many years, particularly during the early 1940s, to help produce ordnance materiel. The production of combat tanks was greatly enhanced by the use of the submerged arc welding process with arc motion devices placed over work-positioning devices. All of the major joints of an Army tank hull were made in this manner. The U.S. shipbuilding program also used submerged arc automatic welding. In this case, the mechanization was slightly different in that welding heads were mounted on small carriages that could be guided along a long seam for ship plate assemblies. Arc motion devices and work motion devices are more or less universal devices made to specific sizes and specifications. Figure 9.1 is a diagram showing the axis of motion of a simple motion system.

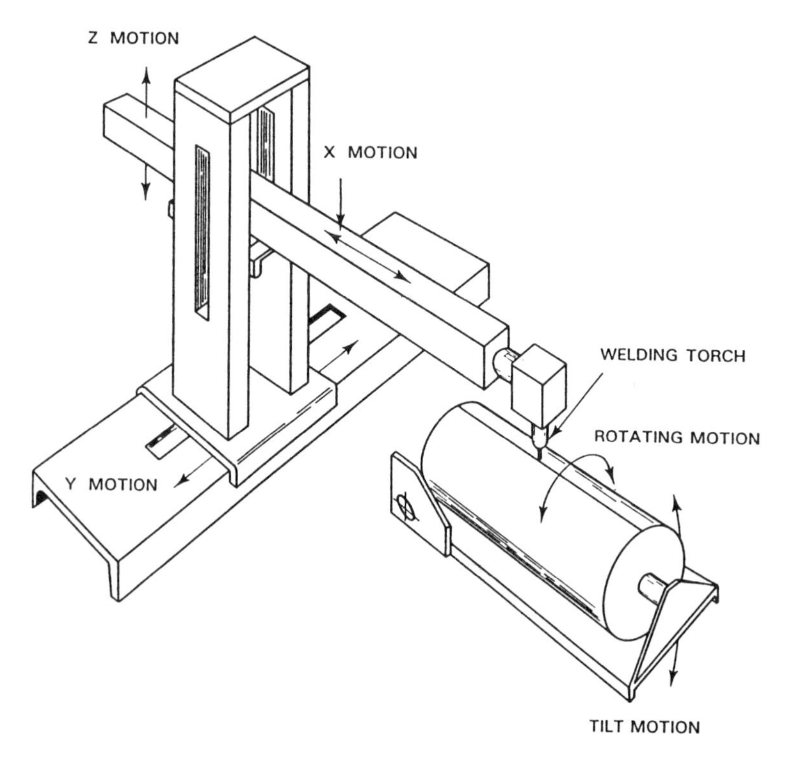

Figure 9.1 Axis of motion of simple motion system.

The use of mechanized equipment greatly improves the productivity of welding. The machine moves the arc, or torch, and welding head along a joint, and the welding operator performs a supervisory role, such as adjusting the torch to guide the arc or changing one or more welding parameters to overcome deviations. Since the welding operator is removed from the immediate vicinity of the arc, higher currents and higher travel speeds can be used than in manual welding. The fatigue factor is greatly reduced, and the duty cycle or operator factor is increased.

There are five standardized arc motion devices that can be purchased for automatic welding:

1. Tractors for flat-position welding
2. Carriages for all-position welding
3. Side beam carriages
4. Manipulators, boom-and-mast assemblies
5. Gantry or straddle carriages

The arc motion devices carry the welding head and torch. It can be used for all continuous wire processes, GMAW, FCAW, and SAW, and can also be used for GTAW and PAW. The device should be specified for the process to be used because of the difference in travel speed requirements.

Welding tractor

Welding tractors were the earliest systems used to provide arc motion. They were converted from oxy gas flame cutting tractors by changing the brackets to hold a welding head and a welding torch. Figure 9.2 shows a typical welding tractor for submerged arc welding being used in a shipyard. The tractor rides on a special track that is laid on the material to be welded. In other cases, the track can be made part of a fixture with in and out (X) or up and down (Z) directional adjustments so that the automatic head will make longitudinal welds in the Y direction. The tractor must be heavy enough to hold the welding head, the control panel, the electrode, and the flux supply. It should have sufficient power to pull the welding cables. Most travel carriages have provisions for adjusting the welding torch so that it follows the weld seam. Mechanical seam-following devices are sometimes employed.

A more complex welding tractor with two heads is shown in Fig. 9.3. This is known as a stiffener welder because it is commonly used in shipyards to weld stiffening angles to deck plates. Usually two arcs are employed to make welds

Figure 9.2 Welding tractor for mechanized welding.

Figure 9.3 Welding tractor with two heads.

simultaneously in the same joint. The speed of travel must accommodate the welding procedure.

All-position welding carriages

The wide use of gas metal arc and small-diameter flux-cored arc welding electrode wires brought about the requirement for all-position mechanized welding. Special carriages are available for making vertical and horizontal welds and even for overhead welds. These carriages use a special track that must be fixed to the part being welded or be part of a fixture that holds the workpiece. Figure 9.4

Figure 9.4 All-position carriage for vertical welding.

shows an all-position carriage operating in the vertical welding position. Such a carriage must be sufficiently rugged to carry the welding head, gun, and control panel and travel smoothly in the welding travel speed range.

All of the carriages just mentioned require close supervision by a welding operator to ensure the production of acceptable quality welds.

The third type of arc motion device in common use is known as a side beam carriage. A typical side beam carriage is shown in Fig. 9.5. It rides on a track that is usually attached to a wall or is self-supporting. Side beam carriages come in different configurations for particular applications. The unit shown is for gas metal arc or flux-cored arc welding but can be used for submerged arc welding. Side beam carriages are used in connection with a work-holding fixture and are

Figure 9.5 Side beam carriage for gas metal arc welding.

widely used for making longitudinal seams on tanks, for making structural shapes, and for almost any kind of weldment requiring long seams. In some cases, a number of similar parts are placed in the fixture and one pass of the side beam carriage welds all parts.

A more precise side beam carriage is shown in Fig. 9.6. This is more precise because it is normally used for gas tungsten arc or plasma arc welding. In these cases, the arc-to-work distance must be very closely controlled, and the travel speed must be regulated very closely. Special sensors are sometimes used to provide Z direction adjustment to control the work-to-torch distance for quality work. The track and travel system is more accurate than those for GMAW.

Manipulators

The manipulator is the most versatile type of arc motion device and consists of a vertical mast and a horizontal boom that carries the welding torch and head. Manipulators are sometimes called boom-and-mast positioners. Figure 9.7 shows a typical boom-and-mast manipulator being used to weld a large tank. In this case the welding operator is riding on the manipulator and is in close proximity to the weld. The boom and mast come in various sizes and configurations. In some cases, the mast will rotate, and sometimes the entire boom-and-mast assembly is placed on a heavy-duty travel carriage for Y motion or to do work at more than one workstation. The welding equipment is normally mounted on this carriage.

The work envelope (Fig. 9.8) specifies the machine size. This relates to

Figure 9.6 Side beam carriage for GTAW precision.

Figure 9.7 Welding manipulator.

the length of the boom and the height of the column. Manipulators are specified by the height under the arc and the length of boom travel. They are known as 8 by 8 units, 10 by 10 units, etc. Smooth travel speed of the arc must be provided. The travel speed range must include the speed range of the welding procedure to be used. The return speed should be higher to save time. The boom elevation control is mechanized for ease of operation. Usually the manipulator is mounted on a carriage, and the carriage speed should be specified to fit the welding procedure and provide a rapid carriage return. All travel motions must be smooth and constant.

Precision manipulators are available for more precise welding requirements. An example is shown by Fig. 9.9. In this case the welding head is mounted at the end of the boom, and the boom traverses within the mast assembly. This

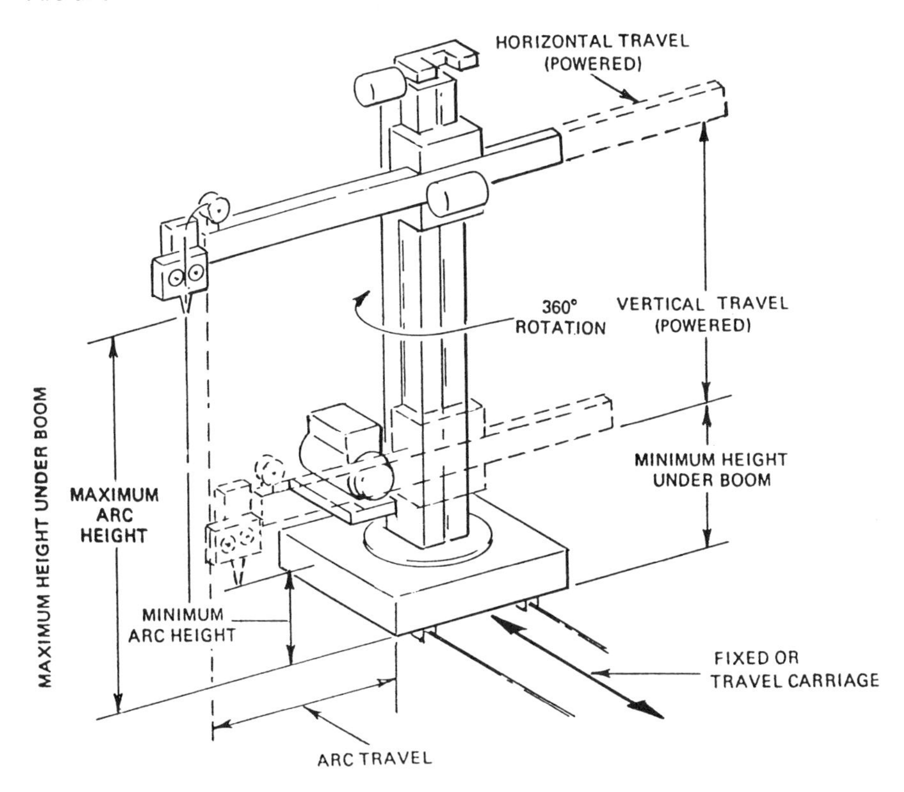

Figure 9.8 Welding manipulator work envelope.

type of manipulator is usually used for smaller assemblies but provides a more precise travel speed. It must also have more rigidity, as the gas tungsten arc welding arc length must be constant.

Gantry arc welding carriage

A gantry arc welding carriage is an arc motion device that provides motion in both the X and Y directions. The gantry consists of a horizontal beam supported at each end by powered carriages. The gantry straddles the work, and the carriages run on two parallel rails secured to the floor of the factory. This provides the longitudinal (Y) motion and can be very long. The length of the gantry bridge determines the width of parts that can be welded in the X direction. One or more welding arc torches are mounted on carriages, with vertical adjustment, that move along the gantry beam. The travel speed of the gantry carriages and the floor rail carriages must be smooth and match the welding speed of the welding procedure. Rapid travel speed should be available on all carriages for return motion. The one

Figure 9.9 Precision manipulator.

Figure 9.10 Gantry arc motion system.

or two welding heads on the gantry bridge will have power travel for the *X* direction. Adjustment must be possible in the vertical (*Z*) direction. A gantry system straddling a weldment is shown in Fig. 9.10.

9.2 Work Motion Devices

Another way to provide relative motion between the work and the arc is to move the work under the arc. There are three basic types of work motion devices:

1. Universal or tilt-top positioners
2. Turning rolls
3. Head and tail stock positioners

These types are standardized and are available from various manufacturers in various sizes with different features and specifications.

The universal or tilt-top positioner is the most widely used for work positioning and is shown in Fig. 9.11. Positioners of this type must be very strong

Figure 9.11 Universal tilt-top positioner.

and well built, must be anchored firmly to the floor, and must be sized for the weldment to be carried. In mounting a weldment on a universal positioner, it is very important to know the location of the weldment's center of gravity. This center of weight should not be too far from the faceplate or too far from the center of rotation of the faceplate. Data for off-center weight allowance are provided by the catalog of the positioner manufacturer. Off-center loading is explained by Fig. 9.12. The rotational speed and tilt speed must be specified because welding can be done while one or the other of these motions is in progress. Rotational or tilt motion must be very smooth and constant without backlash. Universal positioners should not be overloaded, and strict attention should be paid to the location of the center of gravity of the load. Universal positioners are available in many sizes from many manufacturers. Some have special features that are helpful. They were used originally for manual shielded metal arc welding and later for submerged arc welding.

There are several problems with universal positioners. They are relatively expensive, the weldment must be firmly attached to the positioner, and the time required for loading and unloading should be considered in cost calculations justifying positioners. In addition, the side of the weldment facing the positioner table top is usually not accessible for welding. Most positioners provide work-stations high above the factory floor, and therefore ladders or scaffolding and safety equipment are required.

Turning rolls are another type of work motion device. These are commonly

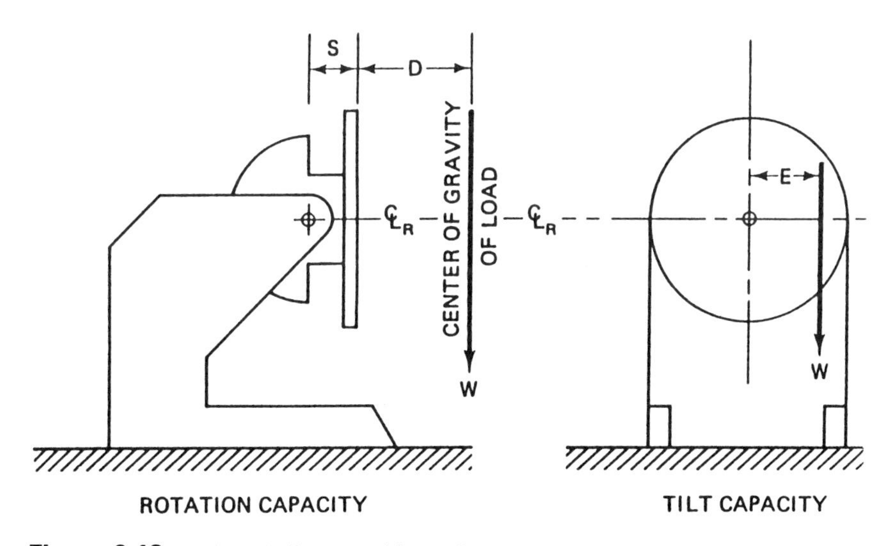

Figure 9.12 Universal tilt-top positioner diagram.

used for cylindrical parts such as storage tanks. Figure 9.13 illustrates the use of turning rolls for submerged arc welding of a large, long cylindrical weldment. Rectangular or odd-shaped parts can also be rotated on turning rolls with the use of special split fixtures cut to fit the particular part being rotated but round on the outside. There are two types of rolls: idler rolls, which are merely rolls on shafts, and power or driving rolls, which are rotated by a motor. Usually, a set of one idler and one power roll is used. Rolls can be steel-tired or have a tire of rubber or insulating material. If rubber or insulating tires are used, the part cannot be highly preheated because that would destroy the tires.

Spacing of the rolls should be adjustable to accommodate weldments of various diameters. The capacity of the rolls should be adequate to handle the weight of the weldment. Different manufacturers offer turning rolls with various features. Turning rolls must be carefully specified to cover the size and weight of the weldment, the welding process, and welding travel speed. One problem with turning rolls is the possibility that the weldment will move longitudinally if the rolls are not properly aligned.

The final standardized device for moving and positioning the work is known as a head and tail stock positioner. The head stock is very similar to

Figure 9.13 Turning rolls.

a universal tilt positioner except that it does not tilt. It has power for rotation. The tail stock is a similar device that does not tilt and does not have power for rotation. The head stock and tail stock are both used for long weldments. The head and tail stock positioners perform the same function as turning rolls. Holding devices or fixtures are provided on both the head and tail stocks so that the part can be loaded readily and rotated for optimum welding procedures. They come in various sizes and must be specified for off-center loads, which will affect the power for rotating the positioner. Figure 9.14 shows the application of a head and tail stock positioner.

Figure 9.14 Head and tail stock positioner.

9.3 Combination Arc and Work Motion Systems

Combinations of standardized products that provide arc motion or work motion can be used to provide automatic welding for many weldments. In some cases, one is used for adjustment of position while the other provides the travel motion for making the weld. In other cases, both units provide motion to make the weld and require coordinated motion control. An example of a manipulator and universal positioner being used together is shown in Fig. 9.15. Each unit must be

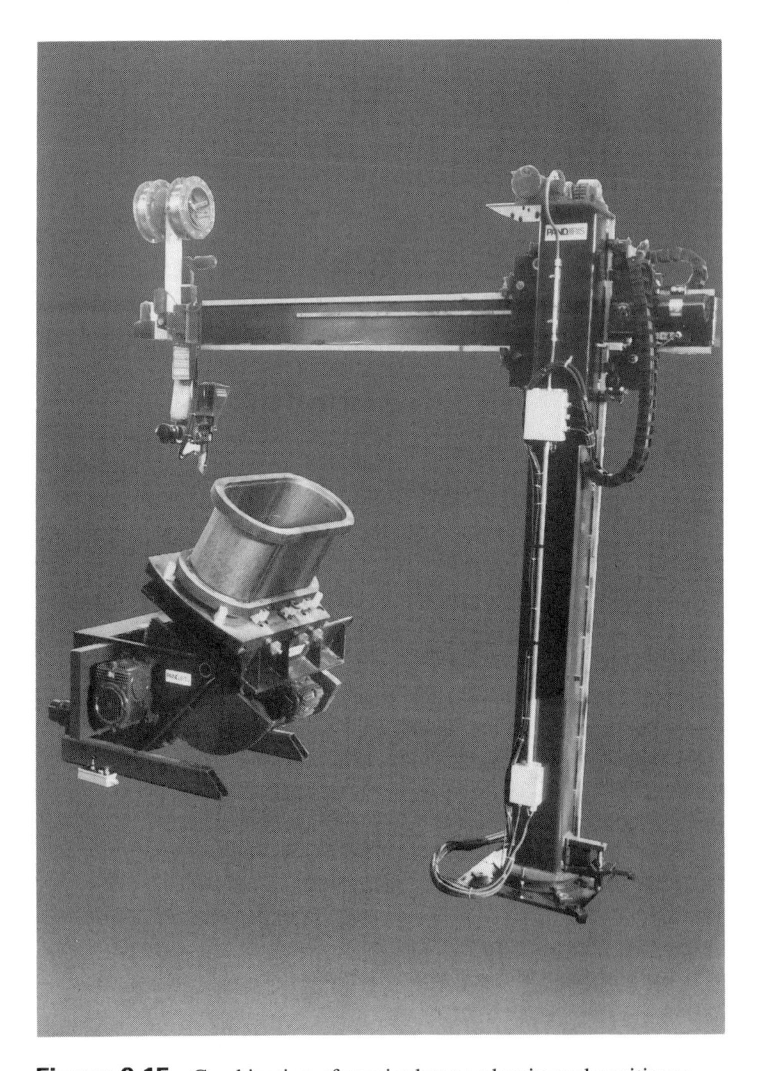

Figure 9.15 Combination of manipulator and universal positioner.

specified with respect to the work to be done, the weight of the weldment, and all relevant welding parameters.

A combination of a manipulator and turning rolls is shown by Fig. 9.16. This is used in making internal welds in a tank. The boom is sufficiently long to extend through the tank. Coordinated motion control is not needed. Another example of combined equipment is shown in Fig. 9.17. This shows a side beam carriage that has a boom mounted over a universal positioner. Vertical adjustment is provided by a vertical travel device mounted on the boom.

The selection and specification of the specific arc motion device and work motion device is a matter of analyzing the job to be done and choosing the combination that will accomplish the weld in the least time.

Arc motion and work motion devices must be properly constructed to perform their duties. It is not the intent here to provide specifications for every type of unit but to provide general specifications for the devices. They should be matched to the products to be manufactured. Specifications are necessary to ensure that the desired equipment is obtained.

The purchase specifications should state that the device must be designed to move an automatic head or work in a manner suitable for making welds. It

Figure 9.16 Combination of manipulator and turning rolls.

Figure 9.17 Combination of side beam carriage, boom, and universal positioner.

must be designed to provide a specific type of motion and rate of speed of motion, and it must be convenient, efficient, and safe. All parts subject to wear must be easily accessible for adjustment, replacement, and repair. The device must be designed to operate satisfactory in ambient temperatures of up to 125°F (52°C). The design must be approved by the buyer prior to manufacture.

The product will consist of desired components such as bases, frames, cars, columns and booms, crossheads or positioner tables, etc. All components must be of welded steel construction of sufficient strength to maintain original alignment and to minimize deflection. The component may require stress relief to ensure the maintenance of flat surfaces that do not distort at capacity load operation. Machine surfaces must be flat within an agreed-upon amount. Bases and frames must provide for floor mounting that will withstand overload conditions.

Motion devices such as rotational drives, tilt mechanisms, and linear motion systems must be capable of moving the part within the speed range specified. Smooth constant motion must be ensured, and gearing must be adjustable to avoid backlash. Accurate machine cut gears must be employed. Rapid traverse should be provided as specified and should not interfere with the normal set speed during welding. When required, electric brakes should be provided to lock equipment at a specific position. Rotational speed accuracy and linear speed accuracy must both be within ±1% of set speed. Adjustments, when required, should be provided for. Manual adjustments must also be provided where required. Safety guards must be provided where dangerous motion equipment is used.

Safety guards and control devices must be provided in accordance with OSHA requirements.

Optional equipment such as air-clamp devices, floor-lock devices, and magnetic clamps must be provided as specified.

The supplier is responsible for inspection and quality control and the performance of the device. The buyer reserves the right to perform any inspection to determine that the device meets its design specifications.

The supplier must provide a technical manual with complete information for installation and maintenance of the equipment, including parts list, wiring diagrams, dimensional diagrams of components, and other information required by the buyer to properly install, maintain, and operate the equipment.

9.4 Controllers, Accuracy, and Specifications

The controllers for arc motion and work motion devices have become increasingly complex in recent years. Originally, motor speed controllers were very conventional, but now they are controlled by computer. This makes it possible to make the arc motion device and the work motion device work together. An example is shown in Fig. 9.15, where a weld is being made on a more or less rectangular shape, and the speed and location of both units are controlled by the computer.

It is important to specify the accuracy of both the arc motion device and the work motion device, each of which must be specified for the size and weight of the weldment and the welding processes to be used. Gas tungsten arc and plasma arc welding require more precise travel speed control and accuracy of movement and location than other processes. As the units get larger, these tighter specifications are more expensive but necessary. In addition, sensors, which are discussed in Chapter 14, may be used to overcome variations in joint location. The tolerances for precision units should be only half the tolerance normally allowed for gas metal arc or submerged arc welding. The normal manufacturing tolerance is $1/32$ in. per foot (2.5 mm per meter) of reach runout for the assembly. The precision item would have a specification not to exceed a "tracking" tolerance of 0.015 in. per foot (1.2 mm per meter) of reach runout for the entire system. Gear backlash is controlled for precision units. Motion devices can usually be obtained with either of two accuracy levels—one with a tighter tolerance than the other. It is necessary to order the correct tolerance for the welding process to be used. Units with tighter tolerances are more expensive. Control systems can be ordered from the arc motion or work motion device manufacturers. In this way they can be tied together for coordinated motion. Controllers with sufficient capacity to control the welding parameters plus six axes of coordinated motion are also available.

Controllers with coordinated motion were designed specifically for robots; however, they can be used to make two units, such as an arc motion device and a work motion device, work together. Examples of combination units are shown in Figs. 9.15 and 9.16. Coordinated motion control systems have never become very popular with this type of equipment owing to the rapid spread in the use of robots.

All control cabinets must be in accordance with NEMA specifications. Remote controls or pendants must be used where specified. Electric circuits in pendants should not exceed 115 V AC. Pendant controls should consist of on/off controls, on/off pilot light, emergency stop, forward/reverse switch, and others as agreed upon. All wiring and interconnecting cables must comply with NEMA specifications. Insulated wires must be of the best quality, oil- and moisture-proof, with mechanically attached terminal lugs. Circuits must be protected with reset-type circuit breakers. All circuits must be identified and color coded to facilitate circuit tracing. A master power disconnect switch must be provided. Electric motors must be drip-proof, equipped with ball bearings, and in compliance with applicable NEMA specifications. Electronic variable-speed drives should be mounted in appropriate enclosures.

10

Standardized Automatic Arc Welding Equipment

10.1 Development of Standardized Automatic Welding Machines

Manufacturing productivity is continually under review because of the pressure to reduce costs. Of all the manufacturing processes, welding is the most often studied, because it is considered a highly labor intensive process. This is certainly true of manual welding. The quest is for additional improvements to further reduce costs. Mechanized welding increases the productivity of the welding operation by a factor of 2, 3, or even 4 over manual welding. This can be understood by reviewing the welder operator factor, or duty cycle, for the various methods of application (see Fig. 10.1). The operator factor is the ratio of arc time to the total paid time.

Manual welding, the popular shielded metal arc welding (stick welding), has the lowest operator factor. Semiautomatic welding, which is normally GMAW or FCAW, has a much higher operator factor, often double that of manual welding, ranging from 10% to 60%. The operator factor for mechanized welding is even

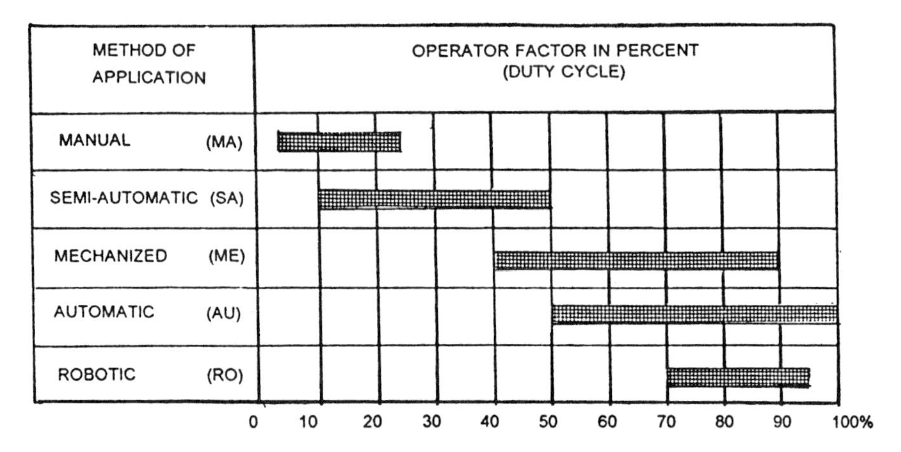

Figure 10.1 Welder operator factor—method of application.

higher, 40–90%; and that of automatic welding is the highest, 50–100%. The exact value in a particular case depends on the type of work and the particular plant. It is easy to see why mechanization and automation increase the productivity of the welding operation.

Through the years, the welding industry has used automatic welding machines for producing specific parts. Many special machines have become standardized because they are used to produce welded products that are manufactured in many different factories. They can be considered welding machine tools. Other types of standardized machines are used for producing specific welds on a repetitive basis.

Standardized arc welding machine tools have evolved through time. They are more expensive than equipment for manual or semiautomatic welding, but they are usually less expensive than robotic welding systems. They reduce the cost of welded products by operating at a higher welding speed with a higher duty cycle and sometimes higher welding current.

These standardized welding machines use the mechanized or automatic welding method of application. They usually consist of a single axis of arc or work motion. The work-holding device includes a work-holding fixture and sometimes a weld backup bar. These machines feature modular design and construction, which allows combinations to cover different sizes of products within the same family.

Four basic types of standardized welding machines have evolved.

1. Seamers allow linear motion of the arc, with the work held stationary.
2. Welding lathes use a stationary arc and rotate a round workpiece about a horizontal axis.

3. Weld-around machines supply rotary motion of the arc around a round workpiece whose axis is vertical.
4. Bore welders operate with the workpiece stationary with its axis horizontal and the arc rotating around it, producing an all-position weld.

Many of these machines use two arcs.

Standardized welding machines can be used for manufacturing products, for maintenance and repair, and for pipe welding. In some cases, a standardized machine will improve productivity to a greater extent than a robot, especially if two or more arcs are used. Additionally, they can make welds that a robot cannot make, such as a weld on the inside of a small bore.

The control system normally controls only the parameters of one arc and one axis of motion. More complicated machines offer programmable coordinated control of two or more motion axes and arcs simultaneously. As adaptive welding and improved sensors become available, they will be incorporated. The level of complexity of the weld determines the type of controller required and the cost of the equipment.

These machines are designed to handle a family of similar products, for example, tanks that require welding different thicknesses of material in different widths, lengths, or diameters. Standardized machines have the same problems as other automatic systems with respect to piecepart tolerances.

10.2 Types of Standardized Welding Machines

The following is a brief description of some examples of the four types of welding machine tools introduced in the previous section.

External seamers are used for making longitudinal welds to produce the shells of tanks. They employ a linear motion device such as a beam and carriage on a track positioned over the joint. The work is held in a fixture with a backup bar. Seamers will accommodate different thicknesses of metal rolled to different diameters. They are automatic machines or mechanized systems that require the supervision of a human operator. The welding machine operator loads and unloads the machine and may also align the joint and monitor the weld. A typical external weld seamer is shown in Fig. 10.2.

Seamers use a square groove weld joint with a small or zero root opening for the longitudinal seam. The welding process depends on the thickness of the material. External seamers must produce full-penetration welds with one pass of the welding head suitable for the service intended. They normally employ backup bars with a relief groove and backing gas supply. The gas metal arc welding process is used for thin steel and most nonferrous metals. Accurate fit-up and alignment is required. For thicker steel, the flux-cored arc welding process may be used. For heavier thicknesses with the square groove joint design, the sub-

Figure 10.2 External seam welding machine.

merged arc welding process with a flux recovery system will be specified. For an even heavier thickness, a special joint design or special welding procedure must be used. External seamers may be positioned to make welds in the downhill position, which will increase travel speed.

Internal seamers are similar to external seamers except that they make a weld on the internal side of the longitudinal seam. They are used when the design of the tank requires such a weld. A typical internal seamer is shown in Fig. 10.3. Internal seamers are usually used with an external seamer when the thickness of the material requires two passes of the welding head and they usually use a square groove weld. The two-pass weld must provide the strength required. Internal seamers are usually manually loaded and unloaded and must be monitored by a welding operator. They are often used to produce "code" tanks.

Tank head welders, or welding lathes, rotate the work on a horizontal axis

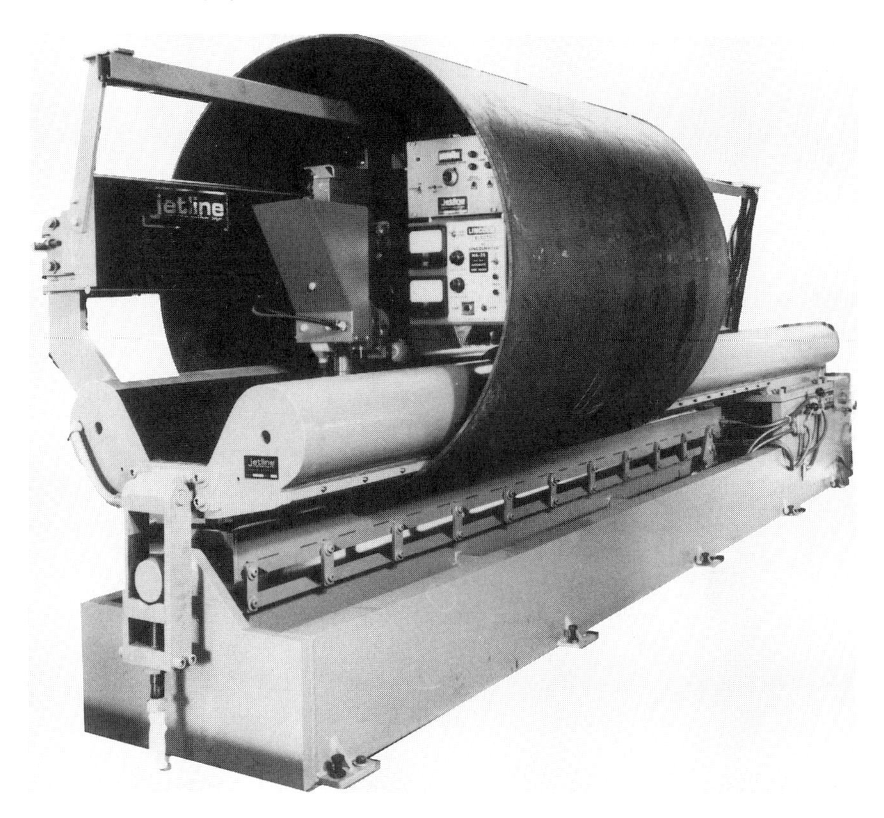

Figure 10.3 Internal seam welding machine.

and usually operate with two welding arcs for welding the tank heads to the tank shell. They are used to make circular welds when the weldment can be rotated and provide automated movement for a variety of parts that require circumferential welding. An assortment of such parts is shown in Fig. 10.4. Welding lathes are most popular for making all sorts of tanks, from small vacuum brake system tanks through large tanks holding gases and liquids. They are designed to weld tanks of different thicknesses, different lengths [up to 20 ft (6 m)], and different diameters, usually ranging from 12 to 48 in. (305–1220 mm). These machines are monitored by a welding operator, who does the loading and unloading. A typical tank head welder is shown in Fig. 10.5. Figure 10.6 shows one that is used for GTAW.

These machines use different weld joints for joining the heads to the tank cylinder. The joint may require a fillet weld, a flare bevel weld, or a square or bevel weld. If a groove weld is used, the welder will usually employ a backup bar, which may be left inside the tank. The weld joint detail, welding process, and

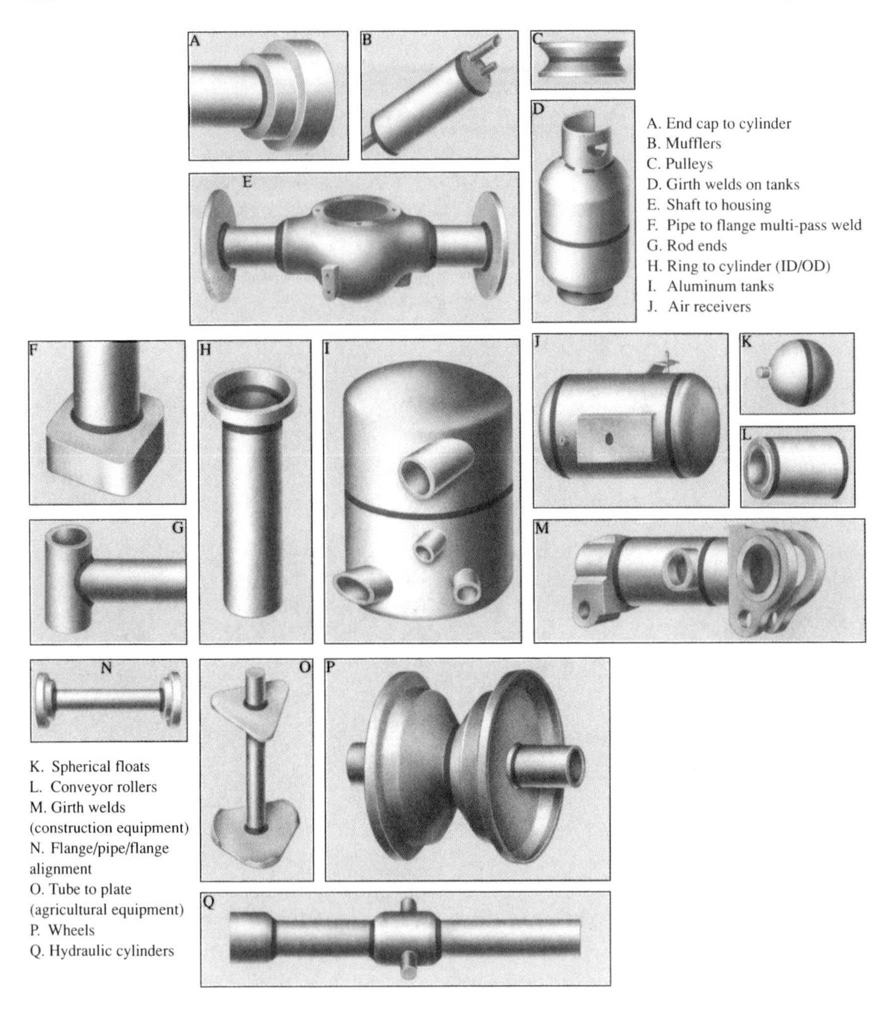

Figure 10.4 Assortment of weldments made on a weld lathe.

procedure must be carefully selected, because tanks are usually manufactured to strict codes or specifications.

Another standardized automatic welding machine is known as a weld-around machine or a spud welder. The torch rotates around a round work-piece whose axis is vertical. It makes a circular weld to join around part to another round part or to a flat or curved surface. Figure 10.7 shows a variety of parts that can be welded with a weld-around machine, which can also be used to weld spuds to a tank or to a large pipe. In some cases, two heads are used for faster production.

Figure 10.5 Tank head welding machine.

Figure 10.6 Lathe welder for gas tungsten arc welding.

Figure 10.7 Variety of parts welded with weld-around machine.

The spud welder or weld-around machine is usually mounted on a floor stand and resembles a drill press. An operator loads and unloads the machine and monitors the weld. A typical spud welder (Fig. 10.8) has a rotating welding head, torch, and electrode wire supply. The torch can be adjusted to the diameter of a round part. The angle of the torch can be adjusted for different fillet weld procedures. Normally a single-pass weld is produced. Some spud welders have cams for adjusting the torch angle if the spud is welded to a circular surface. Weld-around machines are used with the gas metal arc or flux-cored arc welding process.

There are several variations of the weld-around machine. One variation for small parts uses a welding rotating fixture on the table, as shown in Fig. 10.9. This allows longer "arc on" time. Another variation uses two torches, which reduces welding time because the rotation needed is 180° (with overlap) rather than 360°. An example is shown in Fig. 10.10. Other variations are made for GTAW and PAW and cutting.

A similar circular welding machine known as a nozzle welder is used for attaching nozzles to tanks or boilers. This equipment can make a full-penetration groove weld or a fillet weld on heavy material. The machine mounts on the nozzle

Figure 10.8 Spud welder.

and provides rotary motion for making the weld. It may use the submerged arc, flux-cored, or gas metal arc welding process. Complex control and motion are required for thick material and for following complex joints on small-diameter tanks. The machine must be closely supervised because the welds must meet strict code requirements. These machines usually make multipass welds. A typical nozzle welder is shown in Fig. 10.11. A nozzle welder is much more complex

Figure 10.9 Variation of weld-around machine.

Figure 10.10 Two-torch variation of weld-around machine.

Figure 10.11 Nozzle welder.

than a spud welder because it produces a full-penetration groove weld and requires multiple passes in basically the horizontal position. The root or initial pass is extremely critical and requires continuous monitoring. The placement of the beads on a large-diameter nozzle requires the exercise of operator judgment. Nozzle welders are taken to the work and can be adjusted to accommodate nozzles of different diameters and different wall thicknesses and the contour of the tank to which the nozzle is being welded.

Bore welders are also rotating arc machines that do the same job time after time. They are used to make welds on the inside of holes with different variations of inside diameters ranging from slightly less than 1 in. (25 mm) up to 9 in. (228 mm). A similar machine with special attachments is used to weld inside diameters from 12 in. (305 mm) to 28 in. (711 mm), and other accessories allow welding inside bores up to 10 ft (3 m) in diameter. Bore welders can be used with the axis of the hole horizontal or vertical. Figure 10.12 shows the equipment being used for horizontal axis welding with the welds made in all positions. Figure 10.13 shows the equipment being used for making large-diameter overlays with the axis of the hole vertical.

The bore welder uses a conventional wire feeder and a CV power source plus a rotating device that rotates the gun around the inside diameter of the bore

Figure 10.12 Welding inside surface—bore axis horizontal.

Figure 10.13 Welding inside surface—bore axis vertical.

and indexes at every revolution to provide a smooth interior weld surface. It is used for rewelding bores that have become oversize due to wear or have been mismachined. Figure 10.14 shows the surface of welds made in both positions.

The steel industry does not roll all sizes of wide-flange and special-size I-beams on hot roll mills. Beams of special sizes with particular characteristics such as different thickness of the top and bottom flanges are constructed in special welding mills known as beam welders. These mills weld three steel bars or strips together to form a special wide flange or I-beam. In a beam welder a fixture holds the bars in the correct position, and a carriage carrying two welding heads travels the entire length of the beam and makes two fillet welds simultaneously. The beam is then turned over, and the machine makes two more fillet welds to produce a beam of the desired design. The submerged arc welding process is normally used. In most machines the arc moves; however, in some cases the work is pulled through the welding station. These machines can be adjusted to make beams of

Figure 10.14 Inside weld surface—bore welds.

various sizes with different sizes of bars. Beam welders are closely monitored by the operator. A typical beam welder that travels the length of the beam is shown in Fig. 10.15. Beam straightness is maintained by making the second two fillet welds shortly after the first two and adjusting the fillet size.

Another automatic welding machine is known as a strip welder or splicer. This is a machine with a welding head that travels across the width of a strip or skelp quickly making the joint. Splicers are used in continuous mills prior to the forming operation. The tail end of one strip is welded to the front end of another strip, usually in a loop preceding the forming mill. For thin nonferrous material, gas tungsten arc welding is used. For thicker material, gas metal arc welding is used. These machines sometimes incorporate a shearing operation as well as holding clamps. A typical strip welder is shown in Fig. 10.16. The reason for the quick splice is to avoid stopping the continuous forming mill. They are placed in

Figure 10.15 Beam welding machine.

Figure 10.16 Strip welder.

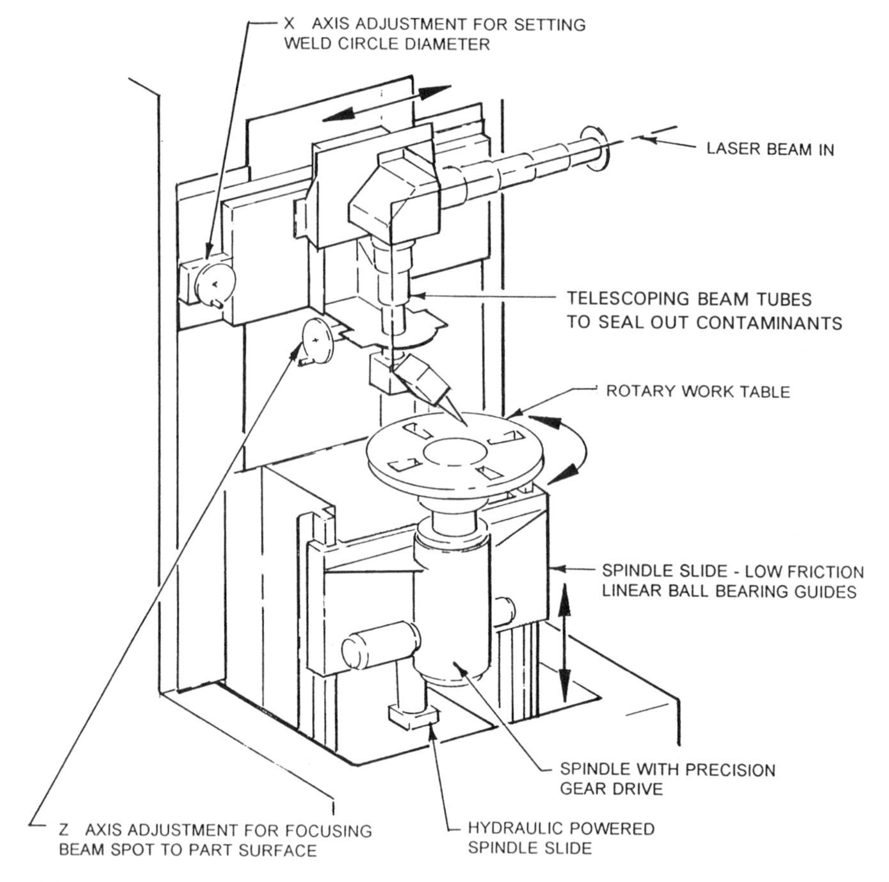

Figure 10.17 Laser welding cell.

the production line with a loop following them. The material in the loop is used up in the mill while the splice is being made. Strip splicers must be automatic because they are not supervised by a weld operator. The procedure must be designed for the particular material type and thickness and the time allowed for making the weld. In some cases, the welds are ground flush prior to entering the forming mill.

One of the newer standardized automatic welding machines is the laser welding cell (LWC), which is designed for the laser welding of round or almost round parts. It can also be used to perform other types of laser processing, including cutting, drilling, and cladding. It has a stationary laser beam head mounted over the table. The laser beam head can be adjusted horizontally to allow for welding different diameters, and it can be adjusted vertically for focusing the

(B)

laser beam. Figures 10.17A and B show a diagram and photograph, respectively, of a typical laser cell. Note the round product being produced at the lower right corner of the photo. Figure 10.18 shows a laser welding a round part. Either CO_2 or solid-state lasers can be used. The control panel will allow preprogramming of the welding procedure. Material handling can be automated or manual. This type of equipment will handle a large variety of parts in any quantity and is much less expensive than dedicated tools.

Other standardized welding machines are used for maintenance work. These include machines used for building up the track shoes of crawler tractors. Other machines are used for building up track idler rollers. This is routine scheduled maintenance work performed in mines and quarries. Worn parts are regularly rewelded to build them up to their original size or to give the part a hard surface. The machines for building up track shoes usually include several welding heads on a linear carriage with limit switches that stop and start the arcs as the carriage progresses the length of the shoe assembly. They use special flux-cored electrode wires to provide the wear-resistant surfaces. Figures 10.19 and 10.20 show machines for building up crawler tractor rollers and tractor track shoes, respectively. These machines are normally monitored by a welding operator.

Figure 10.18 Laser welding a round part.

Figure 10.19 Weld buildup—tractor rolls.

Figure 10.20 Weld buildup—track shoes.

Another overlay operation is shown in Fig. 10.21. The machine illustrated will provide a corrosion-resistant surface on the inside of a digester tank used in a paper mill. It rotates around the center of the circular vertical tank, making a continuous weld. At each revolution it indexes upward to provide a continuous surface. Various welding procedures are used.

Plasma transferred arc (PTA) welding is now being used to apply tough wear surfaces and corrosion barriers to critical high-wear areas on parts such as engine valves, engine valve seats, glass and plastic molds, and feed screws or flights. Overlays are deposited in the exact shape and thickness needed, from as thin as 0.05 in. thick to as thick as 3⁄16 in., in a single pass. Figure 10.22 shows a PTA system being used to build up a feed screw. The linear motion of the arc is coordinated with rotation of the workpiece to build up the edges of the flights. The plasma transferred arc process metallurgically bonds a metallic surfacing powder to any base metal. The resulting surface is bonded to the base metal with as low as 5% dilution per pass. Note the powder feeders on top of

Figure 10.21 Overlay inside a tank.

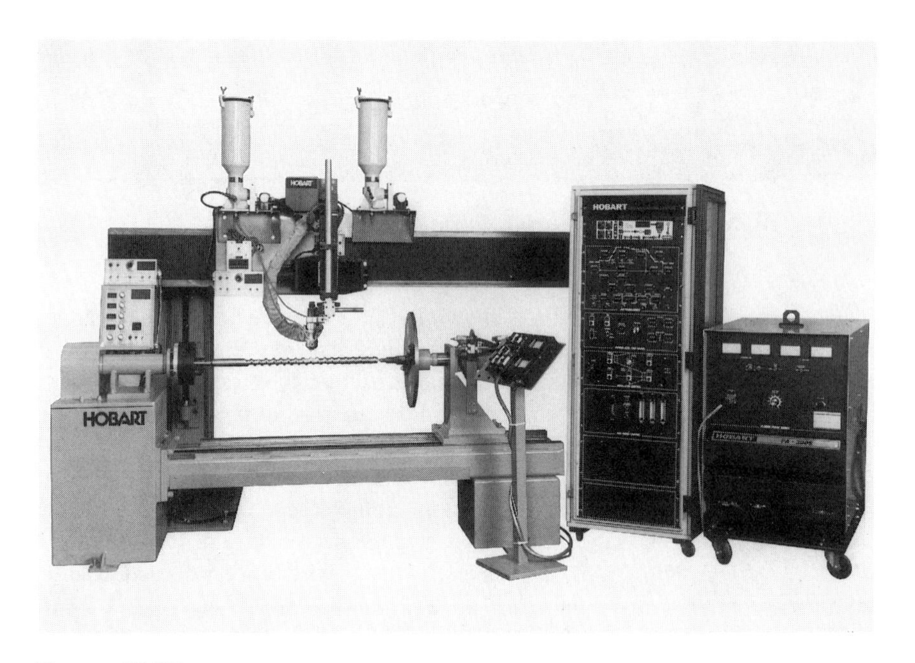

Figure 10.22 PTA overlay system.

the feedhead. Two different powders can be blended as required for the particular application. The PTA process is finding increased use for precision deposits of very hard alloys.

Standardized automatic welding equipment can be used to apply the dabber welding method for rebuilding knife-edge seals for jet engines. The work rotates about a horizontal axis as shown in Fig. 10.23. This method uses a dabbing motion similar to that of an individual doing manual welding with GTAW. Dabber welding is designed to automatically rebuild thin edges by depositing narrow weld beads one on top of another. A cross section of a dabber weld is shown in Fig. 10.24. The dabbing is synchronized with pulsing of thc weld current for most applications. The dabbing stroke length is sufficient to quickly pull back the end of the filler wire, from which a droplet has just been detached. This removes it from the heat of the arc and into the cool copper wire guide nozzle. At the same

Figure 10.23 Dabber welding method.

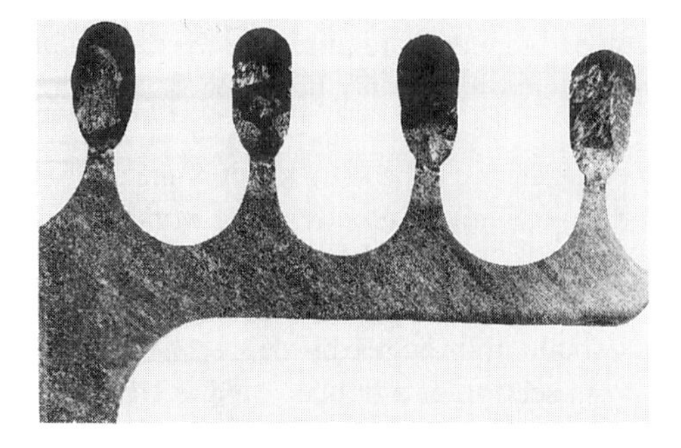

Figure 10.24 Cross section of dabber weld.

time the torch moves toward the work and the arc is shortened. The end of the filler wire reapproaches the arc at a speed sufficient to prevent premelting with the associated formation of large globs, and melts and deposits another small portion of metal. At this time the arc is lengthened. The amount of weld deposited at each stroke depends on the specific welding and dabbing parameters. The dabber method produces a uniform narrow weld bead with relatively small heat input, precise deposition of the filler metal, and repeatability. It is used primarily to weld titanium and nickel alloys in repair welding of jet engine seals. It is also used to repair turbine blade tips, valve and impeller wheels, etc. It is applicable to any part that requires the placement of a narrow bead on a thin edge. It is also used to salvage rotary saw blades, valve seats, milling cutters, large drill bits, and dies.

10.3 Automatic Welding of Pipes and Tubes

Approximately 10% of the steel produced is made into pipe and tubing. The majority of pipe is installed by welding. Butt welds on pipe or tubing require a high degree of skill when made manually or semiautomatically. Many automatic tube and pipe welding machines are now available. Automatic tube welding was pioneered by the dairy, brewery, and aircraft industries. Equipment was developed using the gas tungsten arc welding process for welding relatively small diameter tubing. Units rotate the arc around the tube and are designed as a family of automatic heads that operate from a single power source and controller. They make butt joint welds in various sizes of tubing. Most are for thin-wall tubing where filler metal is not added. A typical tube-to-tube welder and its power source are shown in Fig. 10.25. Heads of various sizes can fit on this weld tool to accommodate tubing of different sizes.

A different style of welding tool is used for heavier wall tubing. In this case, filler metal is added to the weld and multiple passes are made. Figure 10.26 shows

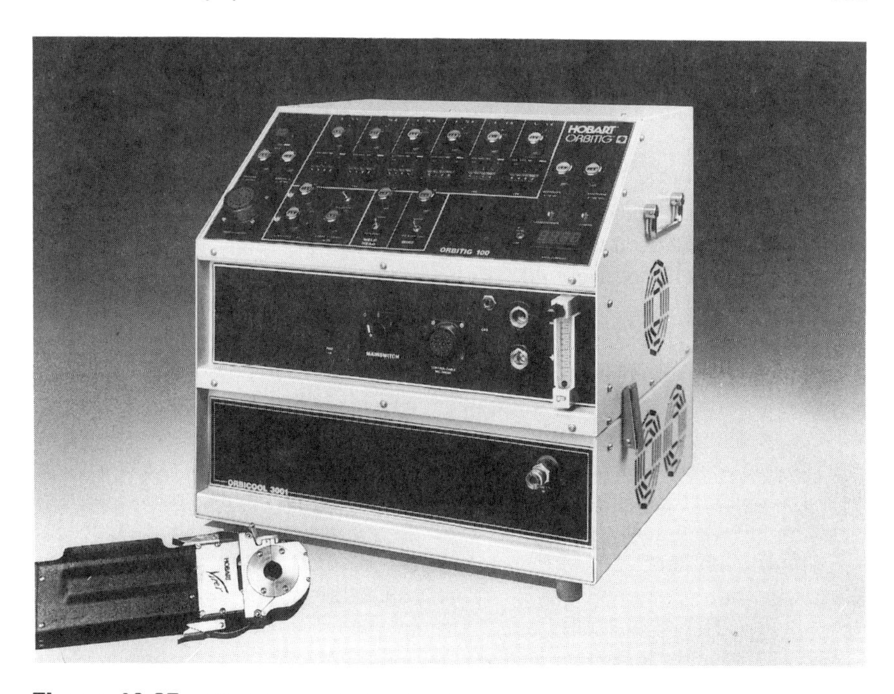

Figure 10.25 TIG tube-to-tube head and power source.

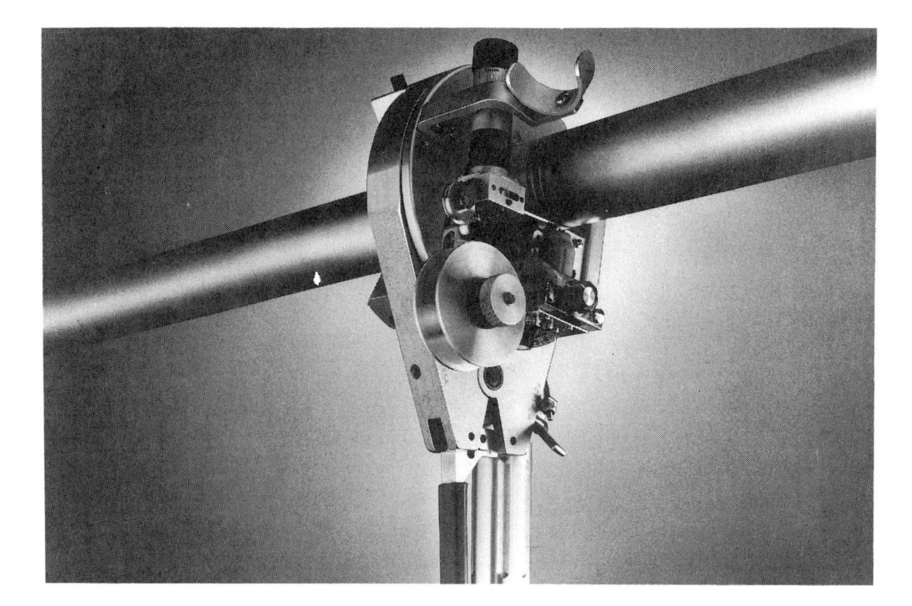

Figure 10.26 TIG tube-to-tube head with wire feed.

this type of head with the small spool that supplies filler wire for making the thicker joint. Once the operation is started it does not require operator assistance. The operator moves the head from joint to joint and aligns it at each weld.

A family of machines similar to the tube-to-tube head is used for making tube-to-header welds. These machines used the gas tungsten arc welding process and are programmed to make the weld at each joint where a tube is joined to the header. Weld joints of different designs can be made in all tube sizes. Some machines incorporate cold wire feed when heavy-wall tubing is welded. All of the machines use a welding programmer and are moved from joint to joint by an operator. A typical tube-to-header welding head with cold wire feed is shown in Fig. 10.27. These machines are very popular in the heat exchanger industry, in which literally thousands of code welds are made, every one of which must be perfect.

A similar standardized family of machines is used for making gas metal arc welds in pipe. These machines are used for industrial piping and for making welds on cross-country pipelines. The more sophisticated of these machines are computer-driven and can weld pipe in any position. Some employ internal mechanical line-up clamps that incorporate a backing bar. They are very rugged for field use. They are sized according to the pipe diameter and wall thickness. Some machines use more than one arc, and some have arcs on the inside of the pipe. They are operated by skilled welding technicians who perform continuous supervision and maintenance. A typical pipe welding machine used for large-

Figure 10.27 Automatic cross-country pipe welding machine.

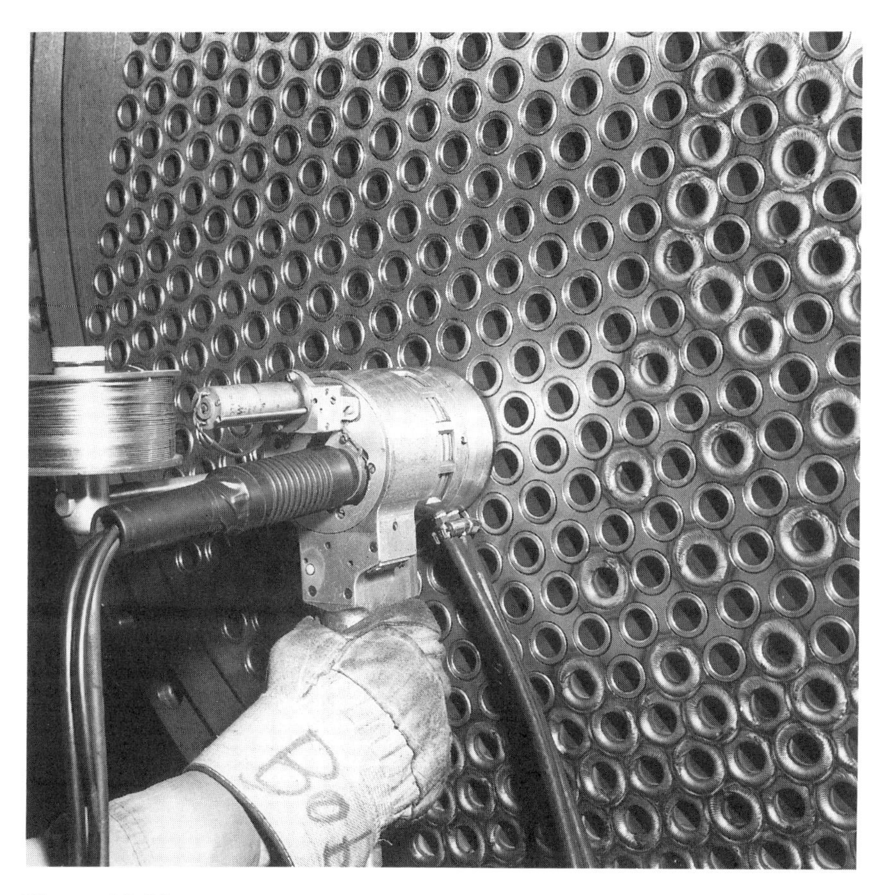

Figure 10.28 Tube-to-header welding machine.

diameter, thin-wall pipe is shown in Fig. 10.28. This is a cross-country pipe welder mounted on a side beam carried by a crawler tractor.

10.4 Specifying Standardized Automatic Welding Equipment

Many companies offer standardized automatic arc welding machines to produce many different products. The January issue each year of *Welding Design and Fabrication* offers the "Welding and Fabricating Buyers' Guide." This includes a section titled "Positioning, Manipulating, Assembly, and Handling Equipment: Complete Arc Welding Machines." This section has many subdivisions. The sections of specific interest here deal with welding lathes, gantry welders, welding beams, seamers, travel carriages, spud or boss welders, and tank assembly welders. The buyers' guide includes a list of over 60 companies and supplies the addresses and phone numbers of each. Most manufacturers of tube heads and automatic pipe welding machines advertise in *Welding Design and Fabrication* or in the *Welding Journal.* It is a good idea to contact each company to determine the type of automatic equipment they provide and request their catalog.

Obtain and study the catalogs of those companies that produce the type of equipment in which you are interested. Usually they illustrate their equipment and indicate the limits of capacity for thicknesses, sizes, diameters, etc. A catalog should also indicate the tolerances of each dimension and the types of weld joints that can be made. It normally indicates the recommended welding process. If the exact machine you desire is not available, many companies will customize their product to fit your needs. However, a customized automatic welding machine can be much more expensive than the standardized machine.

Manufacturers' catalogs provide information indicating how a particular standardized automatic machine is specified. Important information here relates to the tolerances of pieceparts produced in your plant. Automatic welding equipment will perform properly only if the pieceparts are made accurately. Standardized automatic equipment rarely comes equipped with seam trackers or adaptive controls.

If additional specifications are required, refer to Section 9.3. For a discussion of machinery acceptance, refer to Section 12.6.

11

Dedicated Automatic Arc Welding Machines

11.1 Hard Automation—Dedicated Customized Automatic Welding Machines

A dedicated automatic arc welding machine is a welding machine custom-designed to weld one specific part on a high-volume production basis. With quick-change tooling, a family of similar parts can be welded. Dedicated or customized machines, sometimes called "hard automation," are used whenever identical parts are manufactured in sufficient quantity or on a continuous basis. This type of machine is very popular in the automotive and appliance industries for manufacturing the same part day in and day out. The very first automatic arc welding machine, mentioned earlier in this book, was made to automatically weld differential housings for automobiles.

Dedicated fixtures have long been used in the welding industry, and in the past they were designed for manual shielded metal arc (stick) welding or semiautomatic (wire) welding. These fixtures were designed to hold the parts in proper alignment so that when they were manually or semiautomatically welded identical weldments would result. This type of equipment was popular to eliminate the setup to blueprint and tack welding operation, which is extremely time consuming. As continuous wire welding became more popular, and as automation became more desirable for cost reduction, arc motion systems were added to dedicated fixtures to convert them for use in automatic welding. This combination of arc motion and work-holding and work-motion device created the automatic dedicated arc welding machine. The major advantages of these machines is that the human welder is absent from the arc area, production rates are high, and high-quality welds are produced.

Cost surveys have shown that dedicated automated single-purpose welding machines greatly reduce the cost of manufacturing weldments. This is particularly true if two duplicate fixtures are used so that one fixture can be loaded and unloaded while the other is welding. This allows arc-on time to approach 90%. It is the most economical welding production method because it reduces labor expense by increasing labor productivity. If the customized equipment can operate on a full-time continuous basis it will quickly pay for its high initial cost.

Dedicated machines are usually manually loaded. Automatic loading is possible but rarely used. Unloading is more often done by a mechanical system. The automated welding machine can be made a part of a total production line and integrated into the manufacturing operation.

There are two basic types of automatic machines. One requires the supervision of an operator who must load and unload the machine and observe the welding operation. This *mechanized* method of application is considered a closed-loop system due to the presence of the operator. The second type is the *automatic* method of application—the operator loads and unloads the equipment but does not observe the welding operation. This is considered an open-loop system. The operator does not normally visually inspect the part during the welding operation. Some dedicated machines are able to produce a family of parts that are very similar. These machines can be quickly adjusted to allow for these variations and can make similar parts with only a short shutdown for adjustments.

The major disadvantage of customized dedicated machines is the need to redesign or modify the machine when the design of the product is changed. This is expensive and time-consuming.

11.2 Types of Dedicated Automatic Machines

Customized automatic welding machines are as varied as the weldments they produce, which range from tiny heart pacemakers to giant components of earth-moving equipment, from small appliances to large bulldozer blades. Dedicated arc welding machines can be classified according to the number of welding arcs employed, the number of axes of motion, and the number of workstations. For example, a dedicated machine with a single welding process may be

Single arc, single axis of motion, one workstation
Single arc, multiple axes of motion, one workstation
Multiple arcs, single axis of motion, one workstation

or

Multiple arcs, multiple axes of motion, one workstation.

It is also possible for a dedicated machine to carry out different welding processes at more than one workstation.

Machines that employ two or more arcs simultaneously will normally have higher welding productivity than a robotic cell, which usually has only one arc. Multiple workstations using turnaround positioners with two identical holding fixtures will greatly increase productivity.

To obtain a better understanding of the variety of dedicated automated welding machines, it is best to consider some typical applications.

Example 1

One early application of automatic and semiautomatic welding was the manufacture of domestic furnace heat exchangers (see Fig. 11.1). These are made in two halves pressed from thin gauge plain carbon steel or aluminized steel. Baffle plates and spacers are first tack welded to both the right- and left-hand press-formed pieces of the heat exchanger. Two welders using semiautomatic gas metal arc welding equipment do this work adjacent to the automatic welding machine. The right- and left-hand pieces are then manually placed in a fixture that clamps them together.

Only the edges of the two halves to be welded are exposed. However, each corner is different. It was decided to simplify the automatic machine to make only the straight welds on each side. After loading, the unit is fully automatic. The weld produced is an edge weld on all four sides of the rectangular sections. Because of the different corner designs, each side of the workpiece has a weld of a different length. In operation, the machine automatically selects the weld length for the first side, makes the weld automatically, indexes the workpiece 90°, selects the next weld length, and repeats the sequence. This progresses automatically until all four sides are edge welded. This amounts to a total of 70 in. of weld made at a speed of 42 in./min, which greatly exceeds the 15 in./min attained with semiautomatic welding. Semiautomatic welding is still used to finish the corner joints adjacent to the openings, as shown in Fig. 11.2. The semiautomatic welding operator removes the partially welded heat exchanger, inserts two new pieces, and then finishes welding the corners

Figure 11.1 Domestic furnace heat exchanger being automatically welded.

of each heat exchanger semiautomatically. This is done while the automatic welding machine is making the straight welds. This automatic machine increased production by 60–65% over semiautomatic welding.

Example 2

Figure 11.3 shows a single-arc, rotary motion, two-station machine that is used to weld automobile torque converters. It is a relatively simple machine because

Figure 11.2 Semiautomatic welding on corners of heat exchanger.

the weld is a circle; however, the torch is retractable to allow for loading and unloading. The impeller cover is a thick sheet metal stamping, and the hub is a steel tube. The second station allows loading and unloading while the first station is welding. The parts are self-jigging because of the hole in the impeller and the collar on the hub. Figure 11.4 shows the hub being inserted with the welding torch retracted, and Fig. 11.5 shows the welding operation with the head in place. Only one station welds at a time, and a single power source is used. In normal operation a curtain automatically comes between the operator and the arc, so a welding helmet is not required. The centerline of the hub is at a 25° angle, which provides a fillet weld that has the proper contour for maximum strength.

Example 3

An automatic machine that produces heavy duty truck wheels is shown in Fig. 11.6. This machine has two welding stations, two power sources, and two wire feeders but only one control system. It has the ability to weld wheels of

Figure 11.3 Welding an automobile torque converter.

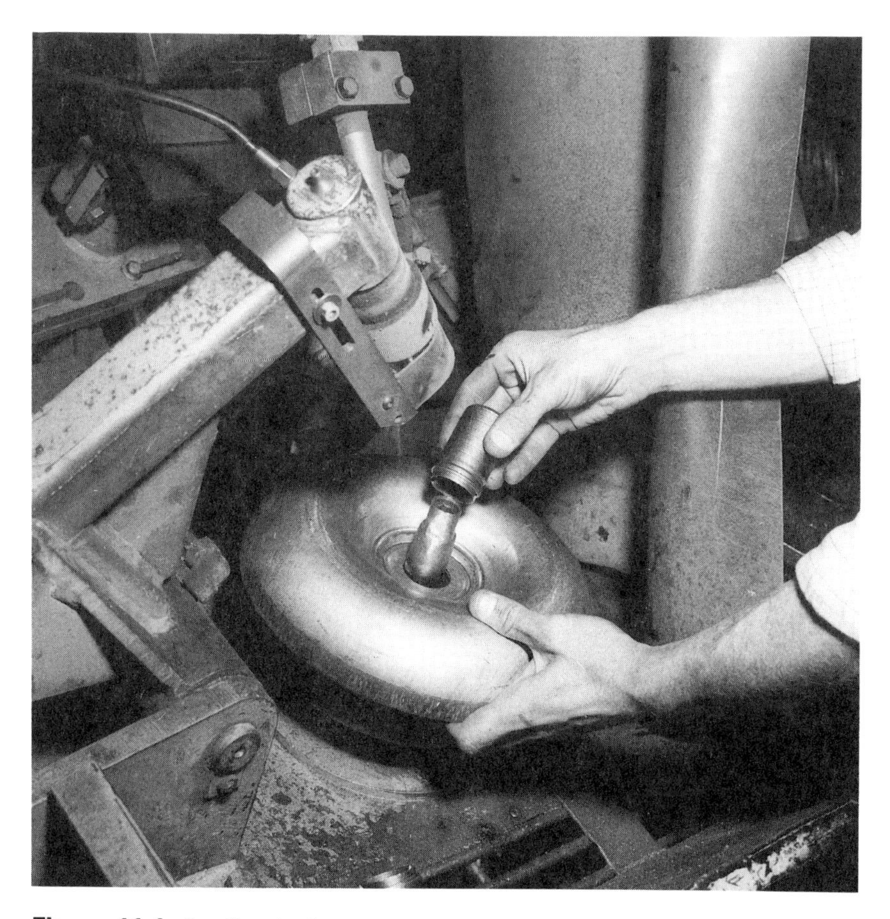

Figure 11.4 Loading the fixture.

various sizes, which are shown in the foreground in the photo. Loading time is negligible because the parts are self-jigging. This unit produces 84 wheels per hour. It is interesting to note the use of large spools of electrode wire. This greatly reduces the downtime for electrode wire changing.

Example 4

This automatic machine, also used in the automotive industry, is more complex, using two arcs at two workstations. The product is a clutch or brake pedal for a recreational vehicle. It is a three-piece assembly consisting of a hub, a curved

Figure 11.5 Making the weld.

steel arm, and a pedal. The welding operation is shown in Fig. 11.7. The operator handles both welding stations, unloading and loading one while the other is welding. At the left-hand station, the hub and the arm are placed in the fixture. The hub is clamped at the inside diameter and the arm is attached to the same fixture; the fixture rotates through approximately 120°. The operator initiates the arcs at the left-hand station and then goes to the right-hand station and inserts the partially welded arm and hub assembly and also the pedal. Each arc at this station makes a 2-in. longitudinal weld on each side, joining the arm to the pedal. Cycle time is such that the two arcs are almost continuously in operation. All four welds are made with solid wire and CO_2 gas shielding.

Figure 11.6 Automatically welding truck wheels.

(A)

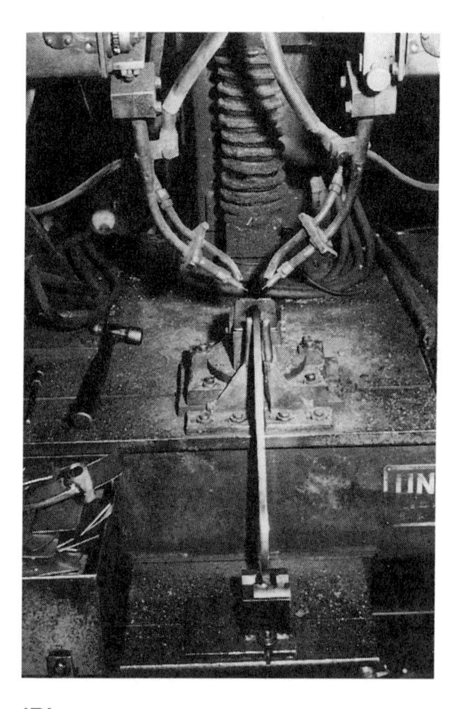

(B)

Figure 11.7 Welding automotive pedals (steps A–D).

(C)

(D)

Example 5

Heavy-duty truck axles for large highway trucks can be produced on various types of automated welding equipment depending on the design of the axle and the quantity required. Figure 11.8 shows the welding of a truck axle using four arcs to make fillet welds joining the hubs to the axle housing. Two fillet welds made with flux-cored electrodes are required on each side of each hub as the assembly is rotated. The four arcs require four power sources and four wire feeders. A different truck axle design made on a different type of equipment is shown in Fig. 11.9. In this case, two robots are used with a special holding fixture.

Example 6

This example shows how two different welding processes can be used to produce a specific part. A dual-process automatic welding machine is used to weld laminations for an electric motor rotor. The stamped round laminations are stacked together and welded. The first station, shown in Fig. 11.10, has a four-position index table that selects the correct number of 0.025 in. thick laminations and then clamps and holds the lamination assembly in the proper position. The machine moves the assembly vertically in front of four gas

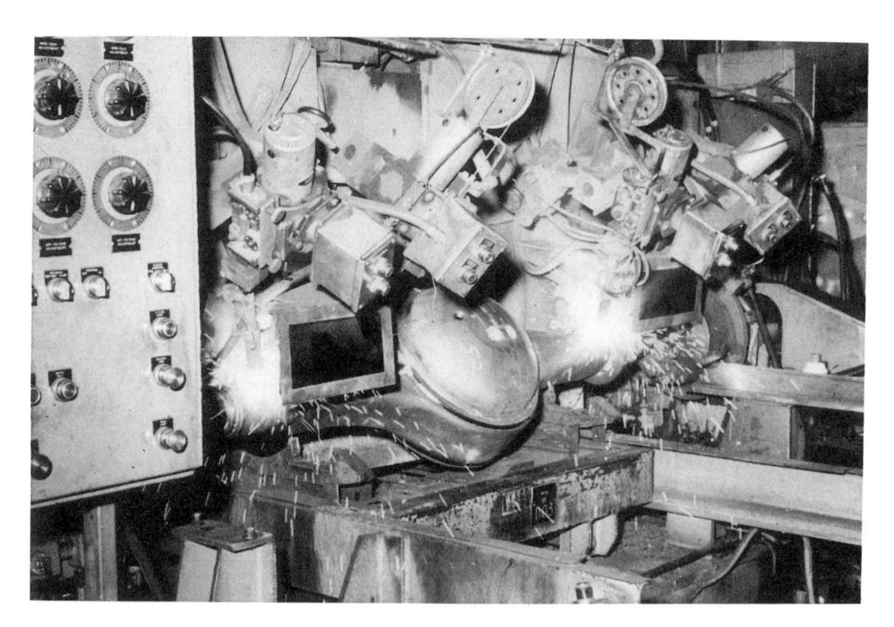

Figure 11.8 Welding a truck axle—"hard automation."

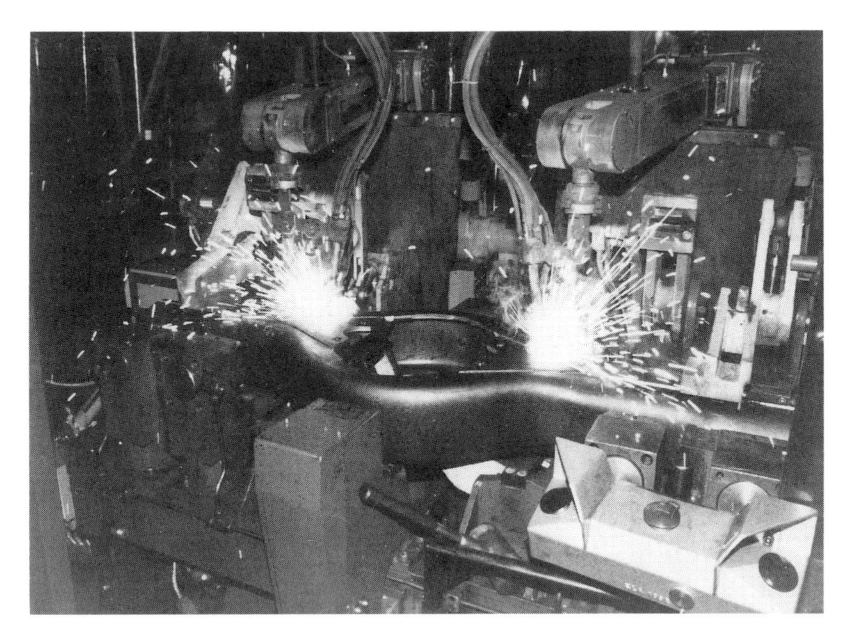

Figure 11.9 Welding a truck axle using a robot.

tungsten arc welding torches, which simultaneously make four bead welds across the stacked assembly to hold it together. The second workstation, shown in Fig. 11.11, presses a cap ring around the lamination assembly. A gas metal arc plug weld is made through prepunched holes. The mandrel that holds the stator turns after each plug weld is made to bring the next hole in line with the torch. The total operation, which combines the two welding processes, produces 200 five-inch stators per hour.

Example 7

An example of a single-arc machine with a single axis of rotary motion inclined at a 45° angle is shown in Fig. 11.12. The 45° incline is optimum for materials handling and for the welding operation. This dedicated machine is designed to weld a base cup to a tube that is part of an automotive MacPherson strut assembly. The strut assemblies are of different lengths, and the fixture and quick-change tooling were designed to accommodate different models. A weld is made 360° around the tube, and the torch is retractable. Weld parameters vary from model to model, leading to the use of a dual-schedule weld controller. A typical part is

Figure 11.10 Gas tungsten arc welding of electric motor laminations.

shown standing on the welding machine. The welding machine also has an exhaust collection system, material loading and unloading is done manually, and the machine cycle time is approximately 10 sec.

Example 8

The customized machine shown in Fig. 11.13 uses a single arc and a single rotary axis on an inclined angle. A special feature of this machine is the two-position fixture base. This feature allows two inside pipe flange fillets to be

Figure 11.11 Gas tungsten arc welding of an end cap ring to laminations.

welded sequentially. The welding torch assembly retracts to allow the shifting of the fixture base. It also retracts to allow for loading and unloading. The welding current to the rotating fixture is supplied by means of a rotary brush assembly. The total machine cycle time to complete the tube welds sequentially is 19 sec.

Example 9

Figure 11.14 shows a dedicated machine that uses a single torch but employs five axes of rotary motion. It is used to weld a crossover pipe to a tubular automotive

Figure 11.12 Inclined axis welding of an automobile part.

exhaust manifold assembly. The part in question is shown in front of the machine in the photograph. The cam-controlled rotary torch motion is the major feature of this machine. Each torch motion is controlled by a separate cam, with the cams mounted on a common shaft. This five-axis rotary motion provides a 360° circumferential weld even when the design of the part prevents full rotation. The 360° circumferential weld is made with only 270° of spindle rotation. Safety light curtains protect the operator from the motion of the machine. Loading and unloading is done manually, and the machine cycle time is 14 sec.

Example 10

A dedicated machine used to weld bicycle frames is illustrated in Fig. 11.15. It employs two torches and nine axes of cam-controlled rotary motion. Each welding torch has four independent axes of motion. The ninth axis is applied to offset simultaneous girth welds. Quick-change tooling and cams are available to

Figure 11.13 Two-position shifting machine.

accommodate a complete family of bicycle frames. This machine is manually loaded and unloaded. It employs two different welding schedules and has a cycle time of 9 sec.

Example 11

The dedicated welding machine shown in Fig. 11.16 is used to weld two separate models of an automotive catalytic converter assembly (shown in the foreground). It employs two arcs and nine axes of cam-controlled rotary motion. A two-piece rotating cradle is used to locate and clamp the converter body. Two

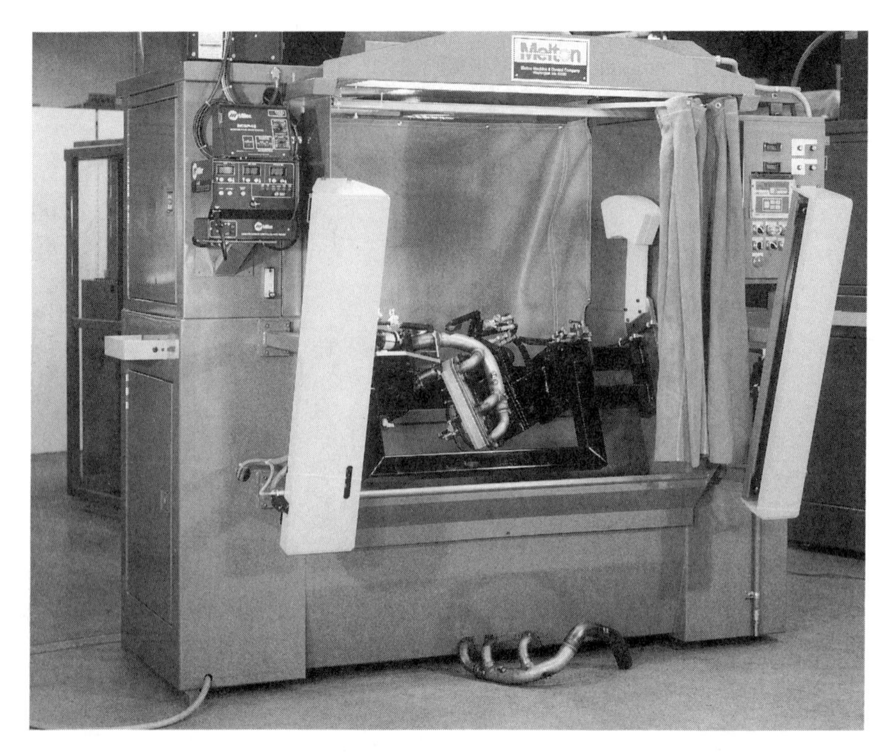

Figure 11.14 Exhaust manifold assembly.

different end pieces are located at the ends of the body section. Each torch is cam-controlled and welds across the body seam flanges and around the circumference of the pipes attached to each end. A single rotation cycle is used to weld both the large oval end and the smaller round end at the same time. It can be changed quickly to weld a different catalytic converter assembly. Two separate power sources and different weld schedules are provided by two separate welding controllers. The production rate for one converter model is 100 parts/hr, and for the other, 180 parts/hr.

Customized dedicated automatic arc welding machines are designed and built to meet specific requirements. It is wise to have written specifications covering the special machine that must be accepted and agreed to by the machine supplier. Section 15.5 in Chapter 15 discusses the items that should be covered in the specifications. See Section 12.6 for information concerning the acceptance criteria that can be used.

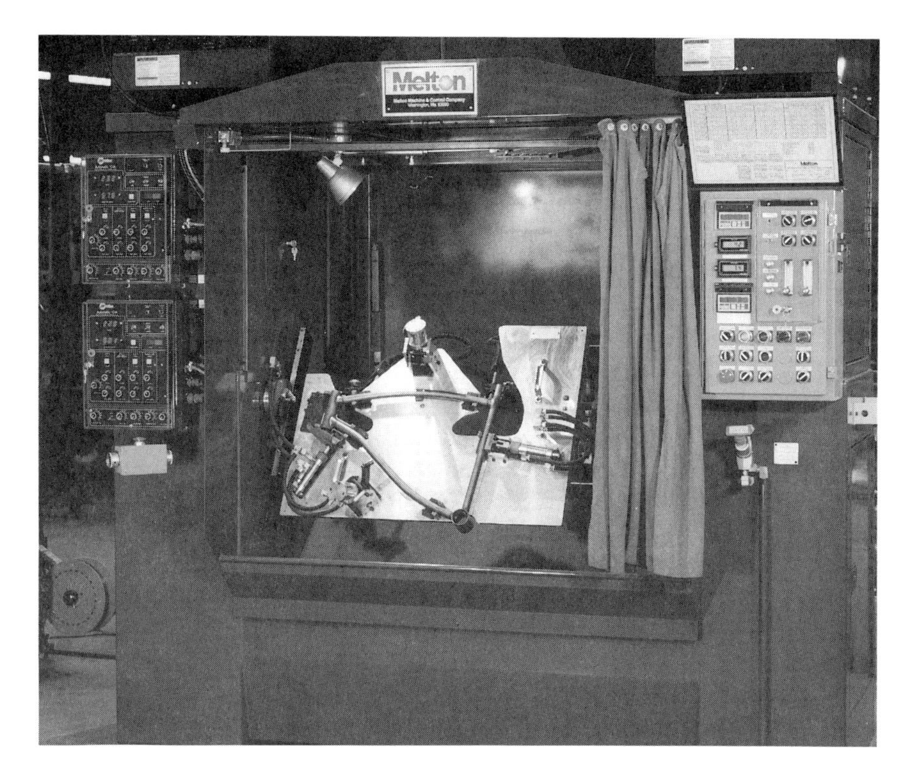

Figure 11.15 Welding a bicycle frame.

Figure 11.16 Welding a catalytic converter.

11.3 Temporary Portable Automated Tooling for Welding

Automated welding is usually thought of as being used in the mass production industries. However, it can also be used profitably in the construction industry and in the manufacture of one-of-a-kind large weldments. Automatic equipment is used to replace manual welding or oxyfuel gas cutting. It is used when it is impractical to provide a dedicated fixture due to high cost or the need to manufacture only a small number of parts.

Portable tooling can be taken to the welding operation of large parts where it will greatly reduce material handling costs. This type of equipment was briefly described in the discussion of arc motion devices. Some trade names are Skate, Weld Tooling, Bug-O, Pack Rat, Klimber, and Kat. Equipment of this type will greatly improve welding or flame cutting quality and increase production by the ingenious use of variable-speed motorized carriages that provide linear motion for the welding gun or torch. The carriages attach to a specially designed track that allows all-position travel. Matched modules of tracks and brackets can be assembled to provide complex travel mechanisms to meet the needs of the specific job.

Another type of portable device that includes a variable-speed motorized rotator allows the rotation of medium-size to large pipe set on a pair of idler rollers. An example is shown in Fig. 11.17. The drive chain can be lengthened or shortened to accommodate the pipe diameter. A variation uses powered rollers. Rotating the pipe allows mechanized welding or flame cutting, which gives better quality and higher productivity than manual operations.

Figure 11.18 shows an assortment of matched modules consisting of straight and curved tracks with appropriate carriages or drive units plus adjustable

Figure 11.17 Motorized rotator.

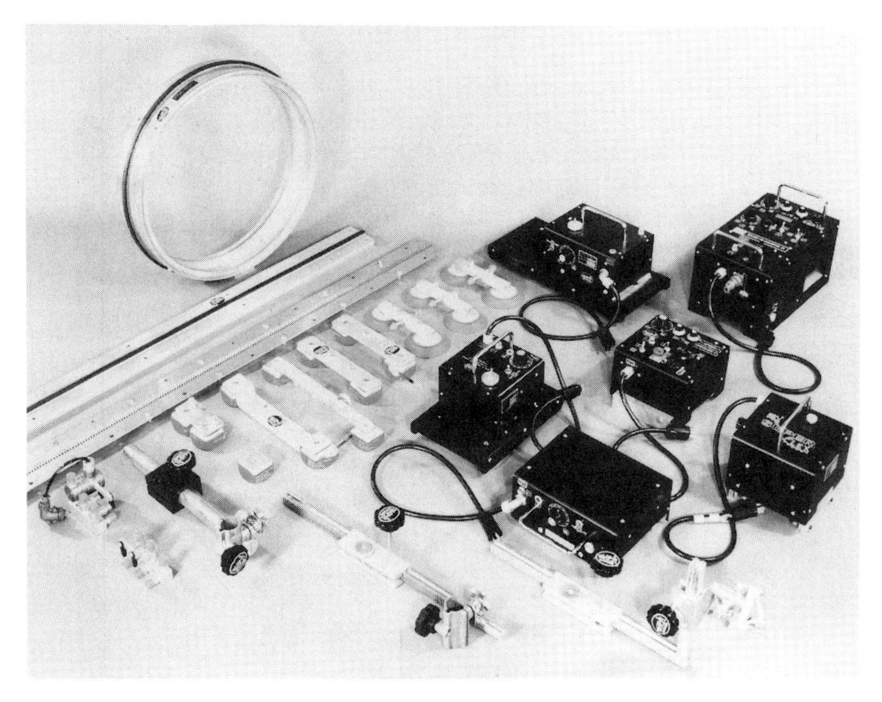

Figure 11.18 Assortment of tractors.

brackets and torch-clamping devices. The tracks or rails are made of extruded aluminum with machine-cut steel gear racks. Interlocks are provided on each rail to allow fast, accurate assembly. The carriages lock onto the rails, which can be mounted in any position to provide all-position travel, as shown by Fig. 11.19 for vertical welding and in Fig. 11.20 for horizontal welding. The rails come in straight lengths and in curved sections with a minimum radius of approximately 18 in. (460 mm). Flexible rails are available with a minimum radius of approximately 10 ft (3 m). Very flexible rails are available that mount on a workpiece with compound curves with a radius of 24 in. (609 mm). The tracks or rails are normally attached to the weldment with magnetic clamps. These are affixed to the rail and hold it at the proper distance from the workpiece. Vacuum attachment brackets with an appropriate pump are also available. Tracks can be mounted together to provide two-way travel of the carriage. The drive carriages have variable speed and come in various speed ranges. A pinion on the carriage engages with the gear rack on the tracks to provide motion. Different track designs and carriages are available from different manufacturers, and a particular carriage may not fit the track provided by another manufacturer.

The advantage of temporary and portable tooling are manyfold. Production

Figure 11.19 Vertical welding on a ship.

Figure 11.20 Horizontal welding on a storage tank.

is increased by using mechanized travel instead of manual travel. This also improves quality by providing precise speed and path control. It reduces distortion by providing uniform heat input. It improves the welder's environment by keeping the operator further from the arc; this also reduces operator fatigue. It reduces material handling by taking the lightweight portable equipment to the work, and it works in all positions. Portable tooling is widely used in shipbuilding and building construction. It can be used for flame cutting and for carbon air arc gouging as well as welding. Various adjusters and accessory devices are available. These can provide for oscillation, for the welding head to float and provide guidance by riding on a flange or other locating surface. This type of portable equipment utilizes semiautomatic guns. The carriage carries the welding gun, or if flame cutting is being done, it carries a standard gas cutting torch.

12

Robotic Arc Welding

12.1 Flexible Automation of Arc Welding

Years ago, when dedicated automatic welding systems were widely used, manufacturers objected to their inability to adapt to design changes and requested systems that would provide flexibility. They readily accepted automatic welding as the most economical way to produce high-volume production parts. However, they demanded automatic welding for low-volume job shop operation. They wanted to be able to switch from one product to another without the penalties of having to rework the automatic machine and incur changeover costs. They wanted the ability to respond quickly to modifications in a product. They wanted an automatic machine that, instead of being dedicated to the production of a single high-volume product, had the capability to manufacture many different products.

To understand this requirement, it is necessary to understand the variety in the production needs of metalworking manufacturing operation. Production requirements can be placed into three categories: job shop production for low volume, batch production for medium volume, and mass production for high volume. Table 12.1 compares these three categories.

Job shop production is used for low-volume jobs and can have a lot size as small as one, which is the least volume possible. Job shop production for large complex parts has a lot size of 1–10 parts; for small simple parts, the lot size is 1–300 parts. Batch production is used for medium-volume jobs, with the lot size varying from 10–300 for large complex parts to 300–15,000 for small simple parts. Mass production is high-volume production, with lot sizes of over 300 for large complex parts and over 15,000 for small simple parts.

Mass production is usually thought of as continuous production by dedicated production equipment, that is, dedicated automatic welding equipment. The need today, however, is for flexible manufacturing systems (FMSs) that can be used on low-volume or medium-volume projects and still maintain the economy of high-volume production. In the past, job shop production rarely used dedicated equipment to produce single units. Batch production could justify the use of simple fixturing and standardized machines but was still relatively labor-intensive. Mass production involving high volume or continuous production can justify customized welding equipment. Batch-type metalworking manufacturing accounts for about 40% of the total manufacturing employment. Mass production manufacturing accounts for less than 25% of metalworking parts manufacture. Seventy-five percent of such parts are manufactured in lots consisting of fewer than 50 pieces. This is important when you realize that manufacturing normally contributes approximately 30% of the gross national product of the United States and other industrialized countries.

The high-volume industries, specifically automotive and appliance manu-

Table 12.1 Comparison of Job Shop, Batch, and Mass Production

	Type of production		
	Job shop (low volume)	Batch (medium volume)	Mass (high volume)
Lot size			
Large complex parts	1–10	10–300	Over 300
Small simple parts	1–300	300–15,000	Over 15,000
Weld setup	Manual	Fixture, manual loading	Fixture, automatic loading
Welded	Manual or semi-automatic weld	Standardized welding machine	Dedicated welding machine
Estimated percentage of U.S. production	10–20%	60–80%	20–30%

facturing, can justify the expense of dedicated automatic welding equipment. However, high-volume mass production represents only a small part of the total manufactured products in this country. Hence the search for flexible automation. Prior to the introduction of arc welding robots, flexible welding automatic systems had come into use. Figure 12.1 shows a variety of parts that can be manufactured with flexible welding equipment. Welding these parts typically uses only one axis of motion, and the weld can be circular or short and longitudinal. These types of products can be produced with a flexible welding system that can be computer-controlled.

A flexible automatic welding station for welding small simple parts is shown in Fig. 12.2. A workstation of this type costs less than half as much as a robotic welding cell. The welding procedure for each workpiece is programmed and stored in computer memory. Simple holding and locating fixtures are provided for each weldment. When a particular part is to be manufactured, the operator places the fixtures on the work table and calls up the program from memory. Changeover from one product part to another is very quick. Once the equipment is set up, the operator loads the fixture, presses the start button, and the machine makes the weld without operator assistance. The operator then unloads the finished weldment. A machine of this type can be programmed for

Figure 12.1 Typical parts produced by flexible welding equipment.

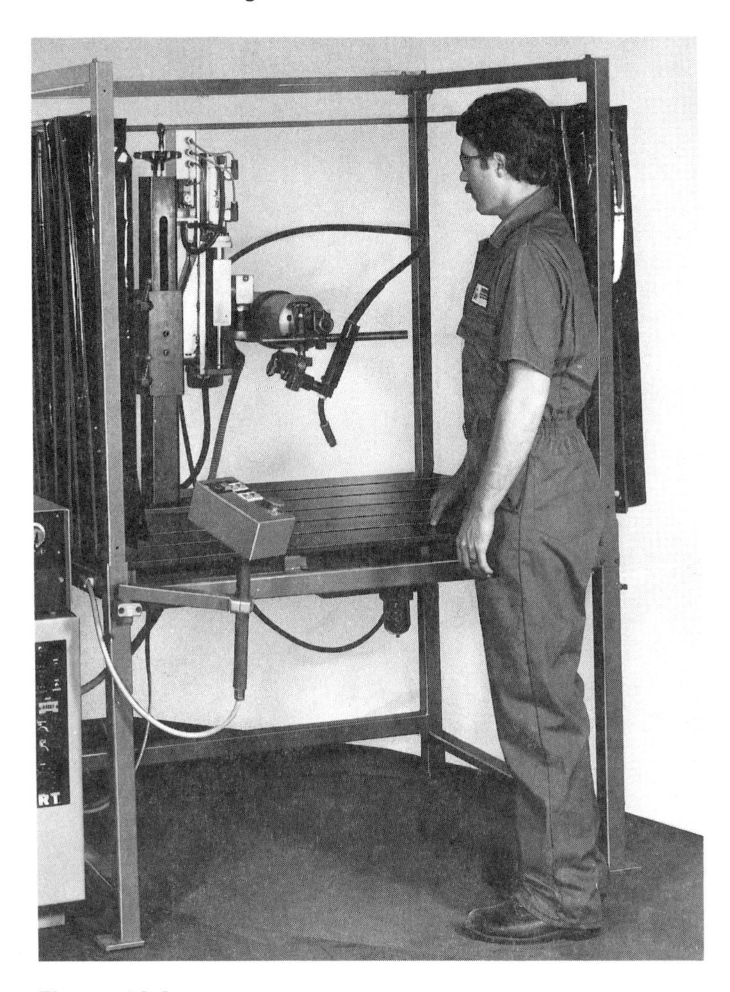

Figure 12.2 Flexible welding workstation.

linear arc motion using one or two torches. It can provide circular welds with the axis of rotation vertical or, when using headstock and tailstock rotary motion, with the axis horizontal. Figure 12.3 shows the headstock and tailstock with a bicycle fork being welded about the horizontal axis. Two torches can be used with linear motion. This type of flexible automated welding can be used for many short-run applications on simple parts. The equipment is quickly changed from one type of motion to another and is applicable to small lots or small batch lots.

Figure 12.3 Welding a bicycle fork.

Modules for flexible welding systems are available from many companies. Bases or work tables with slots are desirable for mounting the "bolt-together" modules. Separate stations should be provided for stationary torches (movable workpiece) and movable torches (stationary workpiece). Torch holders normally allow mounting the torches with flexible brackets to provide almost any work angle. When the torches move, they are mounted on a short linear drive motion device over a stationary workpiece. Rotary workpiece motion is usually accomplished with small tabletop rotators or small universal positioners. Motors and gear boxes are included with the motion devices, and all devices that provide motion are plugged into a simple computer controller. Integration is the key to having a practical flexible system. The teaching procedure must be simple, and the memory system should allow quick retrieval. Systems of this type are best limited to a maximum of 25 lb for the weldment. A system of modular components is shown by Fig. 12.4; this includes the base and, from the left, a rotating head, a torch holder with cross slides, a tailstock, and the control panel. Different components can be mounted on the base for different simple weldments for GMAW. Modular components for gas tungsten arc welding of smaller parts are diagrammed in Fig. 12.5.

Rotary indexing tables (Fig. 12.6) can also be used. These allow for parts loading and unloading, with the welding done at an adjacent workstation. The workstations themselves can also employ rotation for making the weld.

The search for flexible manufacturing methods for large weldments continued. When the industrial robot became available, it seemed to be the answer.

Figure 12.4 Modular components for GMAW.

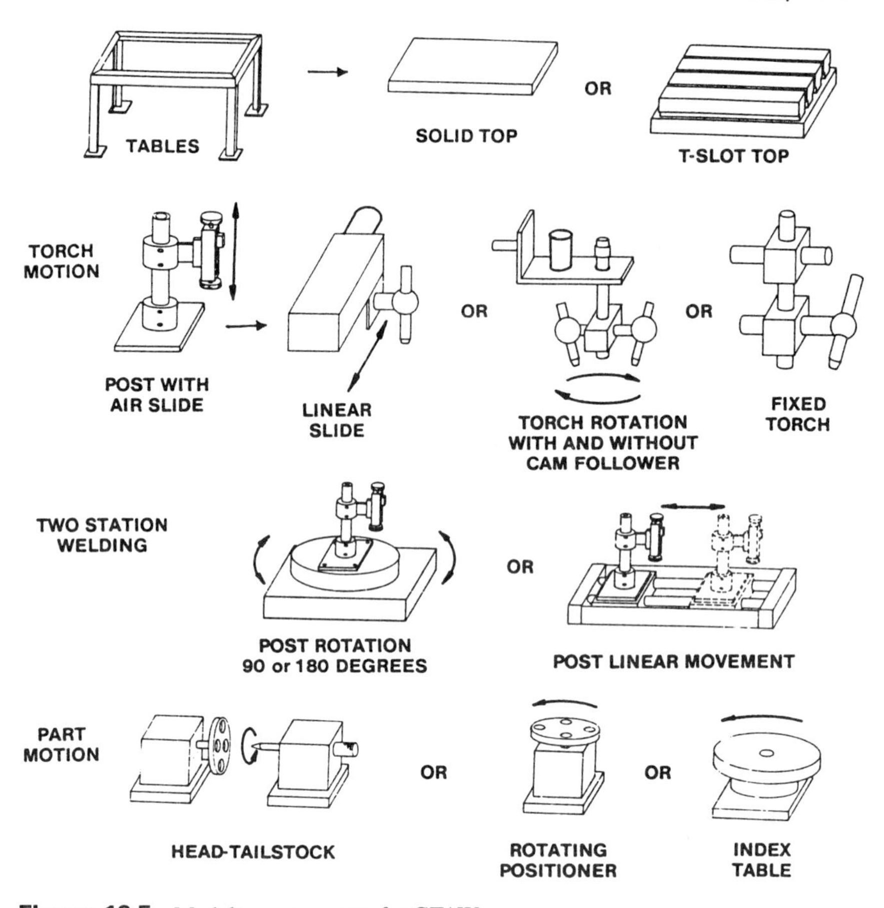

Figure 12.5 Modular components for GTAW.

12.2 Background and Types of Welding Robots

Robots came to the world's attention in the movie *Star Wars,* where little R2D2 and sidekick C3PO stole the public's heart when they marched across the screen. Industrial robots are different. The first successful industrial robot was patented in 1970 by George C. Devol. He married a manipulator to the then emerging technology of numerical control and called the result "program-controlled equipment." Joseph F. Engelberger, a co-worker of Devol at Unimation, promoted the concept and gave it the name ROBOT. This was based on the Czech word *robota,* which connotes forced labor that was depicted as a kind of automation in Karel Capek's 1920 play *RUR*. It could be interpreted as meaning slave labor.

Figure 12.6 Rotary table in use.

Originally, "robot" was defined as an automated device or machine that performs mechanical functions of a human being but lacks emotions and sensitivity. Today, at least in the United States, the robot is defined as a reprogrammable, multifunctional manipulator designed to move materials, parts, tools, or specialized devices by means of variable programmed motions for the performance of a variety of tasks. This is the definition of the Robot Industry Association. In Japan, the Japanese Industrial Robot Association (JIRA) provides a description of six categories of robots.

1. Manual manipulator, i.e., worked by an operator.
2. Fixed-sequence robot, i.e., provides repetitively successive steps according to a predetermined sequence and cannot be easily changed.
3. Variable-sequence robot—same as the fixed-sequence robot, except that the information can be easily changed.
4. Playback robot, a manipulator that, from memory, carries out operations originally executed under human control.
5. Numerical control (NC) robot, a manipulator that can perform a given task according to the sequence, conditions, and positions commanded via numerical data using punched tapes or cards or digital switches.

6. Intelligent robot, a manipulator incorporating sensory perception to detect changes in the work environment with decision-making capabilities.

The differences in these definitions are in part responsible for the larger number of robots in use, at least nominally, in Japan than in the United States.

In Europe, an industrial robot is defined, according to ISO/TR 8373, as

> An automatically controlled, reprogrammable multipurpose manipulative machine with several degrees of freedom which may be either fixed in place or mobile for use in industrial automation applications.

All industrial robots operate in much the same manner. A sequential program is entered into a memory device and is played back during operation. Playback is used to control power to the motion actuators and provide the three-dimensional travel. Feedback devices indicate to memory the progress of the directed actions. The memory unit evaluates the feedback signals and moves to the next step in the program. Point-to-point nonservo robots usually use simple memories to establish the sequence of operations. As this sequence is played back, power is directed and held on until a feedback signal indicates that the specific position has been reached. Once the directed action is accomplished, memory is indexed to the next step, and the process is repeated until all steps in the sequence have been completed. Servo-controlled continuous-path robots are indexed on a time basis, usually 60 times per second. These units index point to point so quickly that the motion is smooth. Servo-controlled continuous-path robots are usually taught by real-time lead-through, and the initial program can be edited to introduce necessary changes. A computer-based control system has the ability to perform mathematical computations of incoming data that are necessary for the modern robot. Electric robots have servo motors on each axis. Robot controllers are classified as servo-controlled or non-servo-controlled, and servo-controlled robots can be separated into point-to-point servo or continuous-path servo.

The original robots were point-to-point machines and were well suited for pick-and-place operations. Industry quickly adapted them for loading and unloading machine tools, die casting machines, and similar operations. They immediately became popular in the automotive industry, which adapted them for spot welding. Robots were adapted for auto body production, and soon the automobile body was completely resistance welded by robots, as shown in Fig. 12.7. Today almost every automobile body produced is spot welded by robots.

It was soon realized that a robot should provide positional accuracy, smooth motion, and a continuous path. By the late 1970s more powerful controllers were developed that provided coordinated motion of a programmable path for

Figure 12.7 Automotive body being welded by robots.

arc welding. Cincinnati Milacron used their minicomputer-controlled hydraulic jointed arm robot for arc welding (Fig. 12.8). At about the same time, Unimate adapted their hydraulic cylindrical coordinate robot for arc welding. It was soon determined that hydraulic robots were not the best suited for arc welding because of drift problems. ESAB in Sweden developed the first all-electric robot with coordinated motion in 1974. This was well suited for arc welding. Yaskawa in Japan followed with its Motoman, an electrically powered microprocessor-controlled unit, in 1976. Other robots were developed in the United States, Germany, Japan, and the United Kingdom, which caused arc welding robot production to grow rapidly. In the early 1980s, they became very popular in the automotive industry. In the mid-1980s other industries started using welding robots, and today their use is widespread and growing.

All arc welding robots consist of at least six major components (Fig. 12.9). The part we normally refer to as the robot is the manipulator, mechanical unit that provides the arc motion. The manipulator, however, is helpless without the brain,

Figure 12.8 Milicron early robot—arc welding.

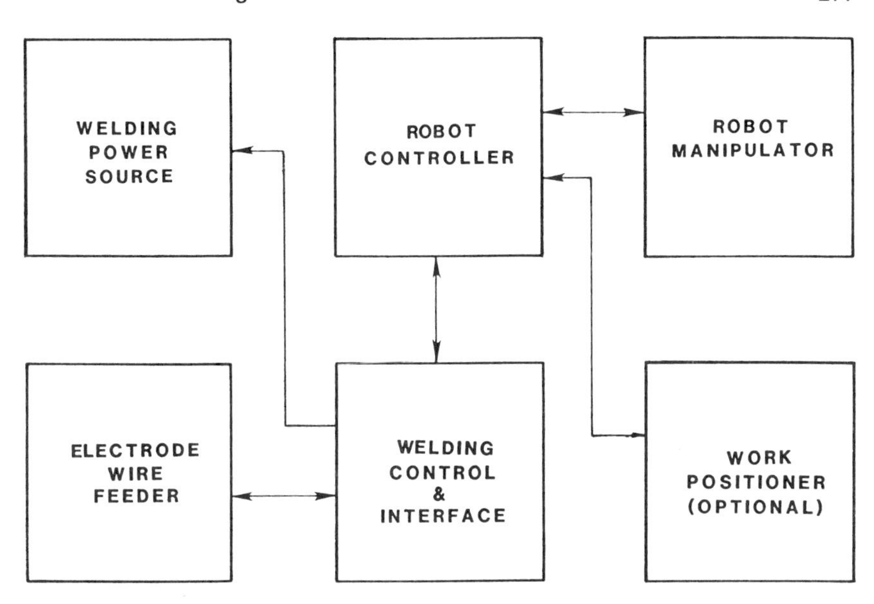

Figure 12.9 Robot arc welding system.

and the brain of the robotic system is the robot controller, which has grown more complex through the years. The other items necessary for the robotic arc welding system are the welding power source, electrode wire feeder, positioners or work motion devices, and control interfaces. The robot manipulator is a series of mechanical linkages and joints capable of moving in all directions to provide arc motion. Most of the early robots had hydraulic actuators or motors. Welding robots are now almost completely electrical. Refer to ANSI/NEMA/AWS D16.2, Standard for Components of Robotic and Automatic Welding Installations, for more details.

The mechanical manipulator robot can be categorized by its design. The more common types of manipulators are the anthropomorphic coordinate (revolute), Cartesian coordinate, spherical coordinate, cylindrical coordinate, and SCARA types. Each type has specific advantages and features, and all can be used to move a welding torch through a series of programmed motions.

In addition to their basic motions, most robots have two- or three-axis wrist motion. Each type of robot also has a particular work envelope within which it can make welds. The size and shape of the work envelope determine the size of the weldment that can be robotically arc welded. Originally, robot designers were attempting to match the work envelope of a human being.

To select the best type of robot for a particular weldment, it is necessary to have an understanding of the different types of robots, the motions they produce, and their work envelope. This information is summarized in Fig. 12.10.

The motions of the anthropomorphic robot (revolute or jointed arm robot) are all rotational with no sliding motion. The work envelope is irregularly shaped in the vertical plane and about two-thirds of a circle in the horizontal plane. This type of robot swings about its base to sweep the arm in a circle. It can bend the upper arm forward and backward at the shoulder and raise and lower the lower arm at the elbow. The electric-powered jointed arm design has become the most popular robot design for arc welding.

The Cartesian coordinate robot is based on the three-plane drawing system used for blueprints, which is often called the rectangular coordinate system. This robot moves within a box-shaped volume defined in the *X, Y,* and *Z* directions, where *X* movements are longitudinal and *Y* movements transverse or "in or out," in the horizontal plane, and *Z* movements are up and down in the vertical plane. This robot has sliding motion in all three directions—longitudinal, transverse, and vertical—and it can be fitted with wrist motions. Figure 12.11 shows a large Cartesian robot. It is claimed that the Cartesian design can be used to build the

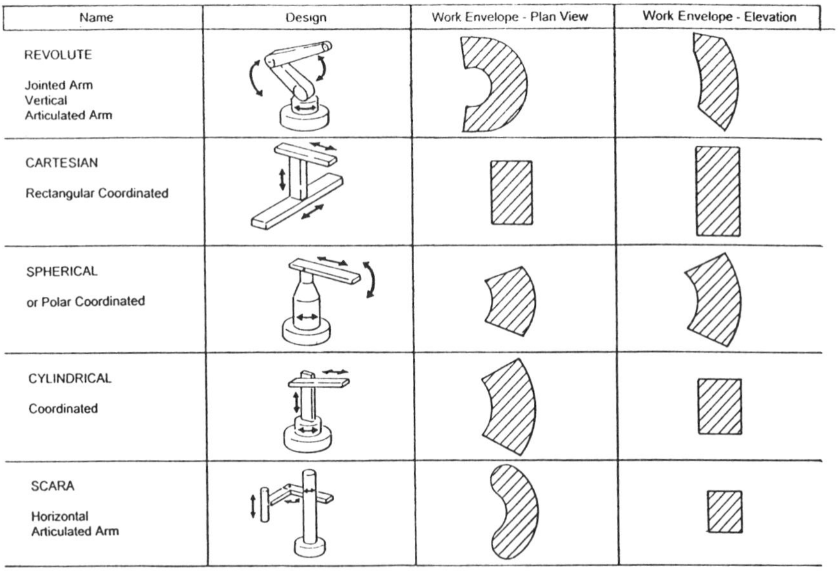

Name	Design	Work Envelope - Plan View	Work Envelope - Elevation
REVOLUTE Jointed Arm Vertical Articulated Arm			
CARTESIAN Rectangular Coordinated			
SPHERICAL or Polar Coordinated			
CYLINDRICAL Coordinated			
SCARA Horizontal Articulated Arm			

Figure 12.10 Types of robots.

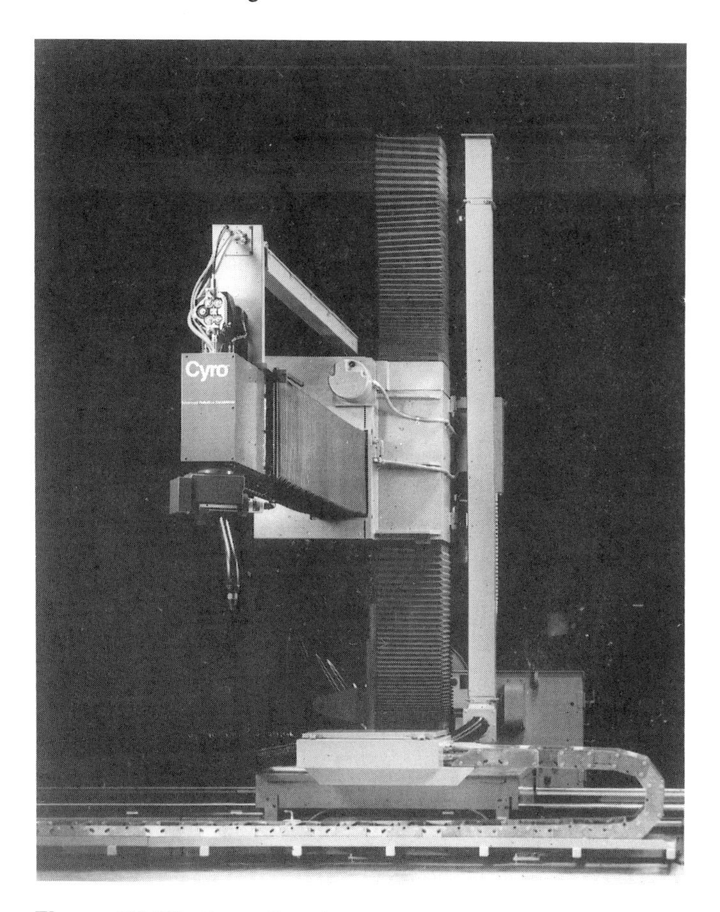

Figure 12.11 Large Cartesian robot.

largest robot, and it can also be used for very small robots. It is also claimed that the Cartesian design will produce the most accurate movements.

The spherical coordinate robot, also known as a polar coordinate robot, has one sliding motion and two rotational motions, around the vertical post and around a shoulder joint. The mechanism holding the arc swings about a vertical axis and rocks up and down about a horizontal axis. The arm slides to extend and retract. The work envelope is spherical but, as the elevation view shows, rotational motion is accomplished by the shoulder rotation. Figure 12.12 shows a popular hydraulic spherical coordinate robot. It was originally used for unloading machine tools but was adapted for arc welding. However, as hydraulically operated robots fell into disfavor for arc welding, it became less popular.

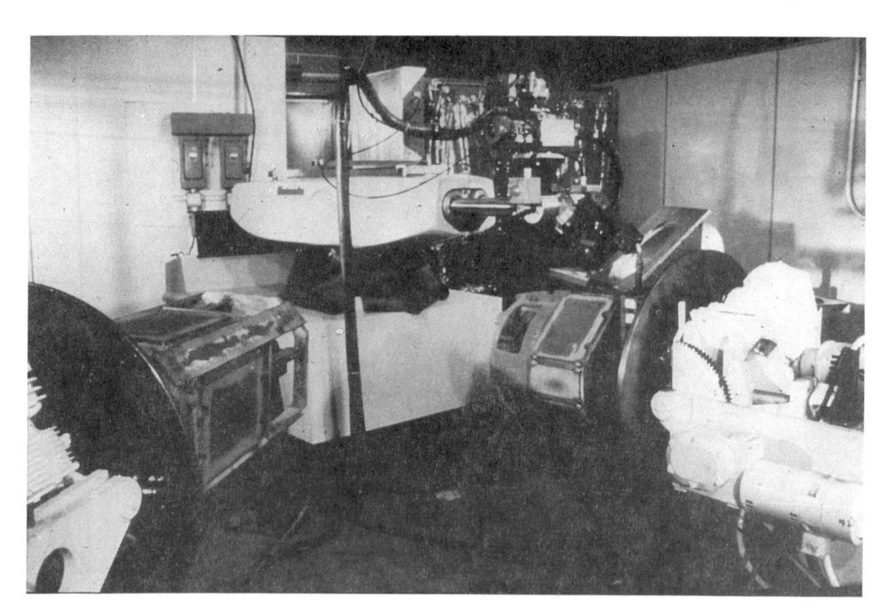

Figure 12.12 Spherical coordinate robot.

The cylindrical coordinate robot is similar to the Cartesian coordinate robot in that it uses sliding motion for two directions, the vertical and one extension, but it also has one rotational or swing motion. The work envelope is cylindrical in the plan view and rectangular in the elevation. The arm holding the welding torch moves up and down the mast and swings about the mast with less than a full circle. The torch can be extended and retracted. This design has not become popular for arc welding robots.

The fifth robot is the SCARA (Selection Compliance Assembly Robot Arm), also known as a horizontal articulated arm robot. Some SCARA robots rotate about all three axes, and some have sliding motion along one axis in combination with rotation about another, as diagrammed in Fig. 12.10. SCARA robots have four axes of motion (two of them parallel) but have limited vertical travel. They are used primarily for welding in a single plane. Their work envelope is a flat rectangular box. The SCARA robot is not popular for welding. Figure 12.13 shows a SCARA robot used for arc welding.

Several hybrid robots have been designed for arc welding. An early Japanese robot manufactured by Shin Meiwa mounted a welding torch that was angled and could rotate 360° on a vertically adjusting boom and mast. The workpiece or holding fixture was mounted on a vertical rotating table that could

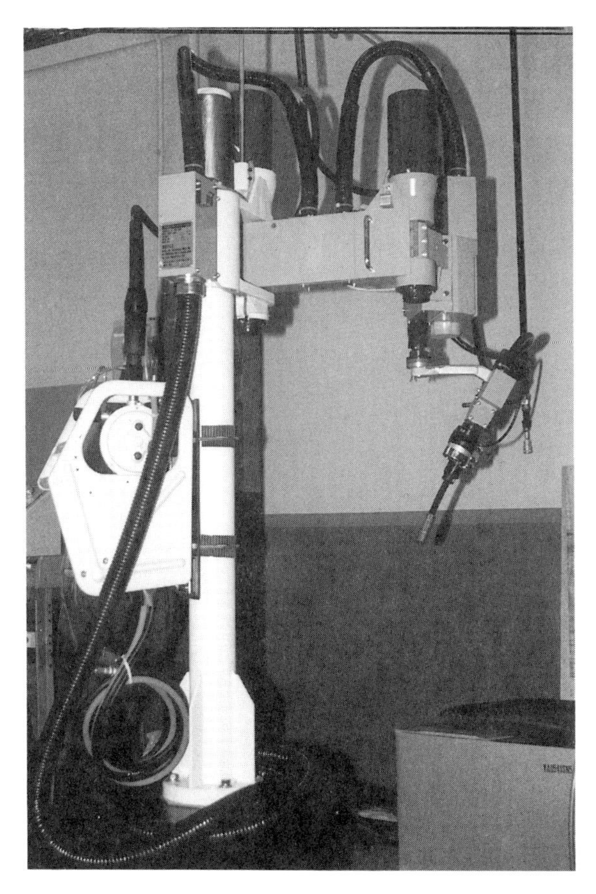

Figure 12.13 SCARA robot.

move in the *X* and *Y* directions. It had five degrees of motion. Its control system was relatively simple, and it was used for welding small assemblies. This unit is illustrated in Fig. 12.14. Other types of hybrid robots were developed but never became popular.

One disadvantage of the Cartesian unit is the large amount of floor space required. Movement along the *Y* axis is provided by a long boom moving in and out from the vertical column. To overcome this space problem, the *Y*-axis boom was replaced by a jointed arm member that can provide the same basic work envelope as the Cartesian robot. It is popular for certain applications. A large unit of this type is shown in Fig. 12.15.

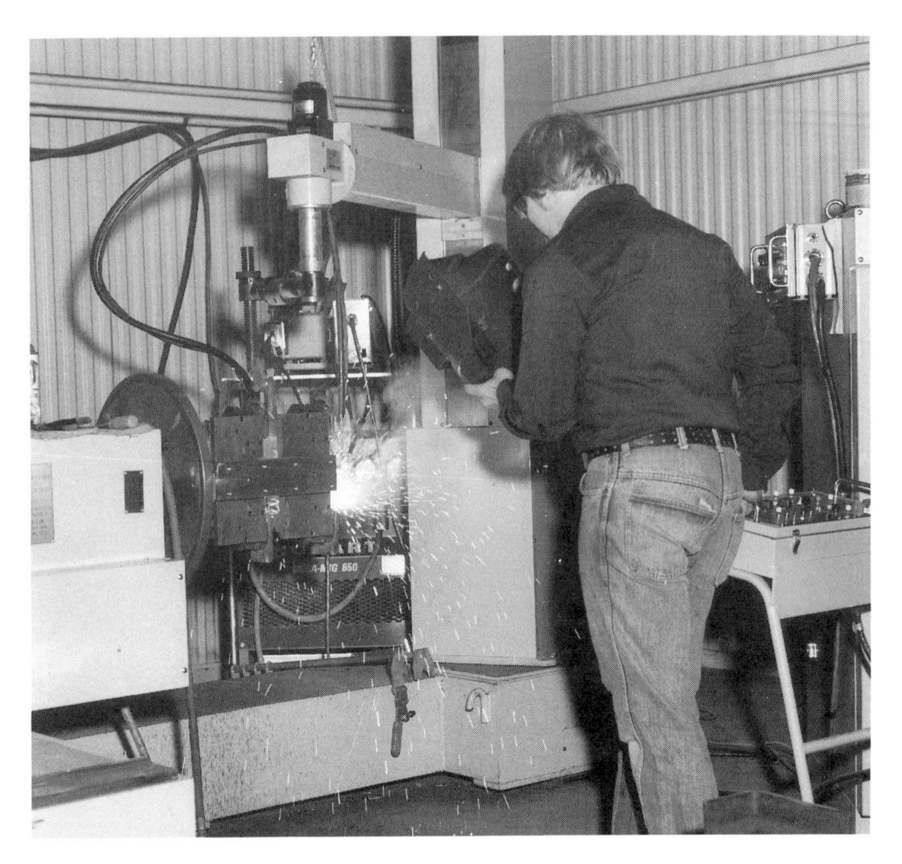

Figure 12.14 Hybrid Shim Meiwa robot.

The gantry robot (Fig. 12.16) has Cartesian coordinate motion. The gantry is a way of mounting a robot manipulator to provide a larger work envelope. A jointed arm or a two- or three-axis wrist can be attached to the gantry carriage to provide maximum movement within the work envelope. The robot can be mounted in the inverted position, which allows access to certain types of weldments. An elongated work area can be obtained by mounting the robot on a moving floor carriage or on a wall. The work envelope of the gantry robot is a large rectangular box. The gantry mounting system is popular for shipyards and for welding extremely large weldments.

A coordinated work-holding positioner can add additional axes of motion, usually rotation and/or tilt. A jointed arm robot with a three-axis wrist working with a two-axis manipulator would have eight degrees of freedom.

Figure 12.15 Large Cartesian robot with jointed arm.

The work envelopes of robots of the same type are basically the same. Variations in dimensions are due to the different lengths of the robotic arms. Most robot manufacturers offer models of various sizes and weight-handling capacities. The dimensions of the work envelope vary. In selecting a robot it is necessary to determine its work envelope and reach and the number of axes of torch motion. This will allow you to determine whether the robot can weld the weldment in

Figure 12.16 Gantry robot.

question. This is a difficult determination to carry out without making tests; however, computer design programs are available that can help you decide whether a given robot can accommodate the weldment. An actual test should be made before a robot is accepted.

12.3 Features of a Welding Robot

During the short time that industrial welding robots have been in use, the jointed arm or revolute type (Fig. 12.17) has become by far the most popular. For welding it has almost entirely replaced the other types except for the Cartesian, which is used for very large and very small robots. The reason for the popularity of the jointed arm type is that it allows the welding torch to be manipulated in almost the same fashion as a human being would manipulate it. The torch angle and travel angle can be changed to make good quality welds in all positions. Jointed arm robots also allow the arc to weld in areas that are difficult to reach. Even so,

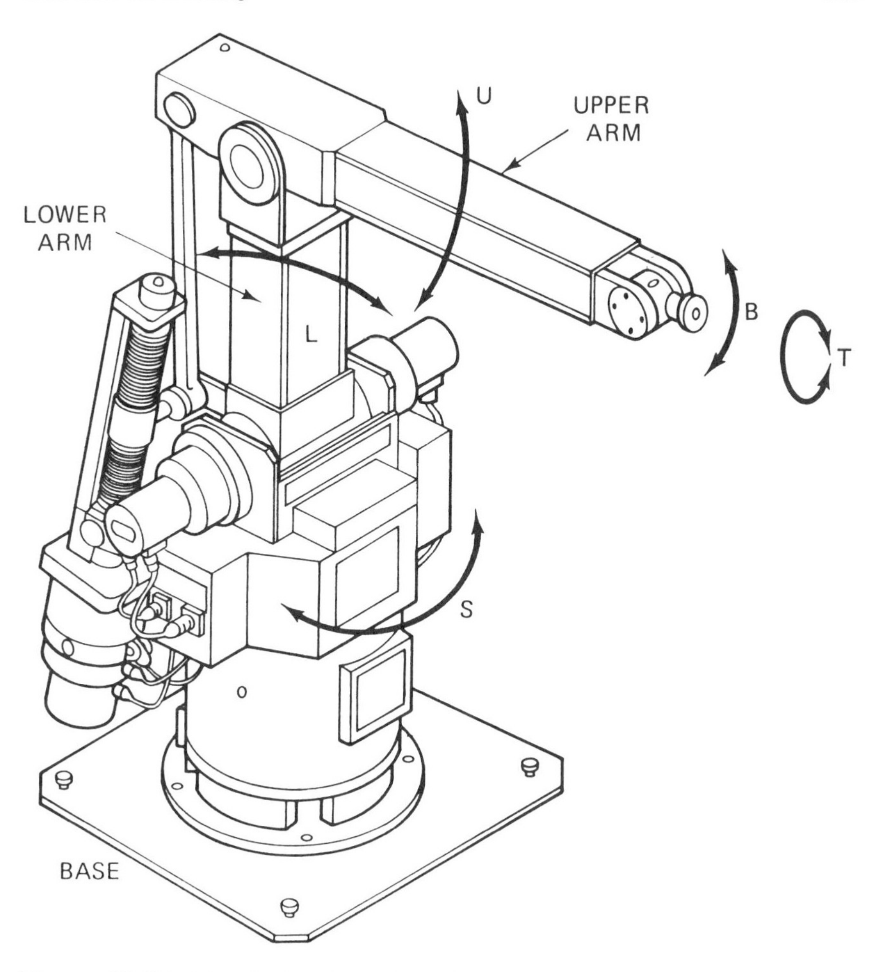

Figure 12.17 Jointed robot: S, rotation axis; L, lower arm axis; U, upper arm axis; T, wrist twist (roll) axis; B, wrist band (pitch) axis.

a robot cannot provide the same manipulative motions as a human being, although it can come extremely close. In addition, jointed arm robots are the most compact and provide the largest work envelope relative to their size. The work envelope of a typical jointed arm robot is illustrated in Fig. 12.18.

Additional range of movement is provided for any robot by a wrist device at the working end of the upper arm. The wrist is very similar to the human wrist and allows three more motions—pitch, roll, and yaw. (These are also flying or

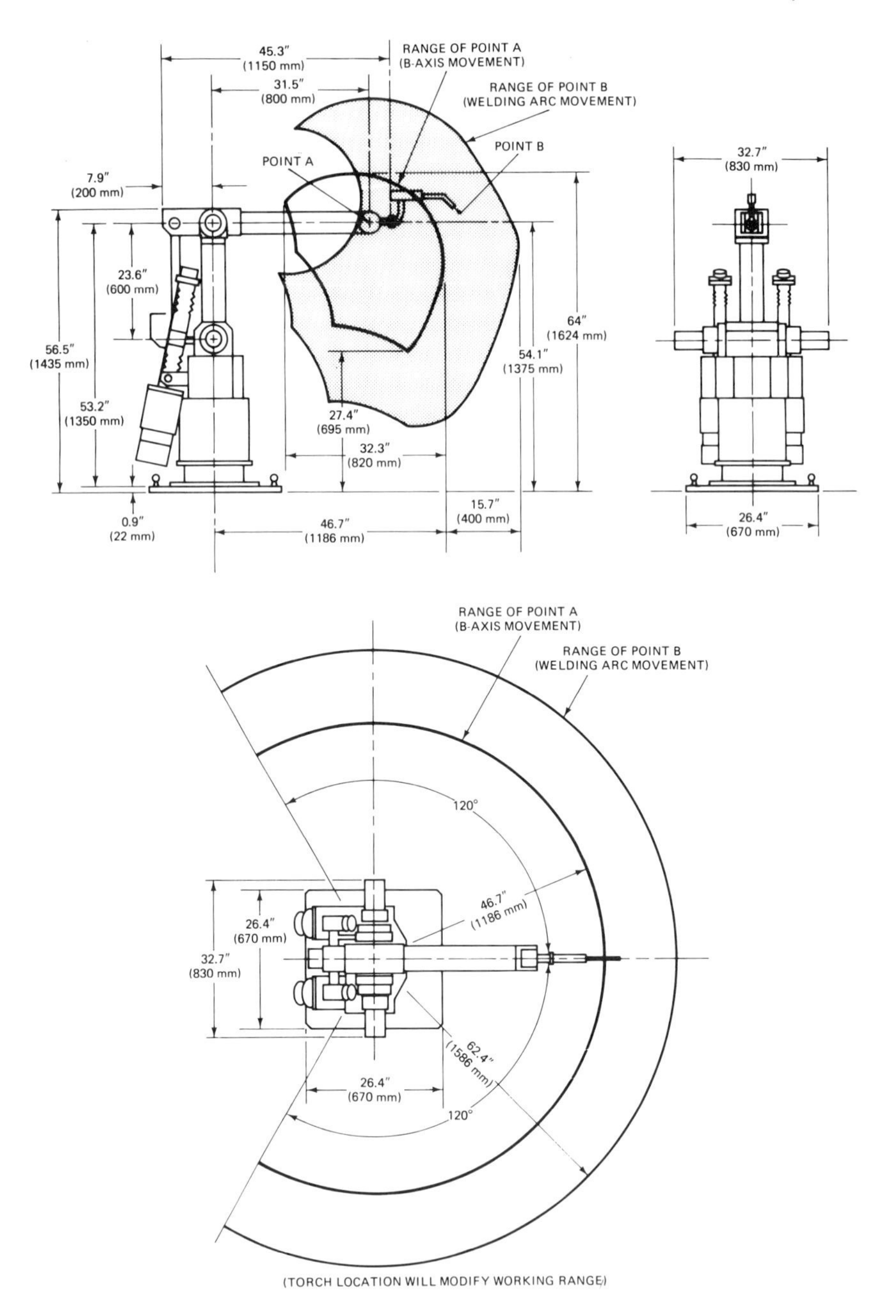

Figure 12.18 Work envelope—jointed arc.

boating terms for bend, twist, and tilt.) Figure 12.19 illustrates these wrist motions. The three-axis wrist is most popular, but two-axis wrists are also used. The wrist is used for mounting the end-effector or welding torch. It is extremely important to mount the gun in the proper manner. For continuous-electrode wire processes, a straight-line torch with an 18° bend is usually used. It is attached to the wrist by a special bracket as shown by Fig. 12.20. In this application the arc is in line with the axis of rotation of the wrist. Extralong torches can sometimes be employed to weld in hard-to-reach areas. The holder is normally 2.0 in. (50 mm) in diameter. Both water-cooled and air-cooled torches can be used. The gas tungsten arc gun and plasma arc torch are mounted differently. The straight-line GTAW torch is mounted in line with the axis of rotation of the wrist, as shown in Fig. 12.21.

A very useful accessory device is a breakaway mounting bracket (safety joint) for attaching the torch holder to the robot wrist. Crashes invariably happen, usually because of a fixture clamp, a change in the design of the weldment, or a mistake in the program. In any case, the robot halts immediately when the torch

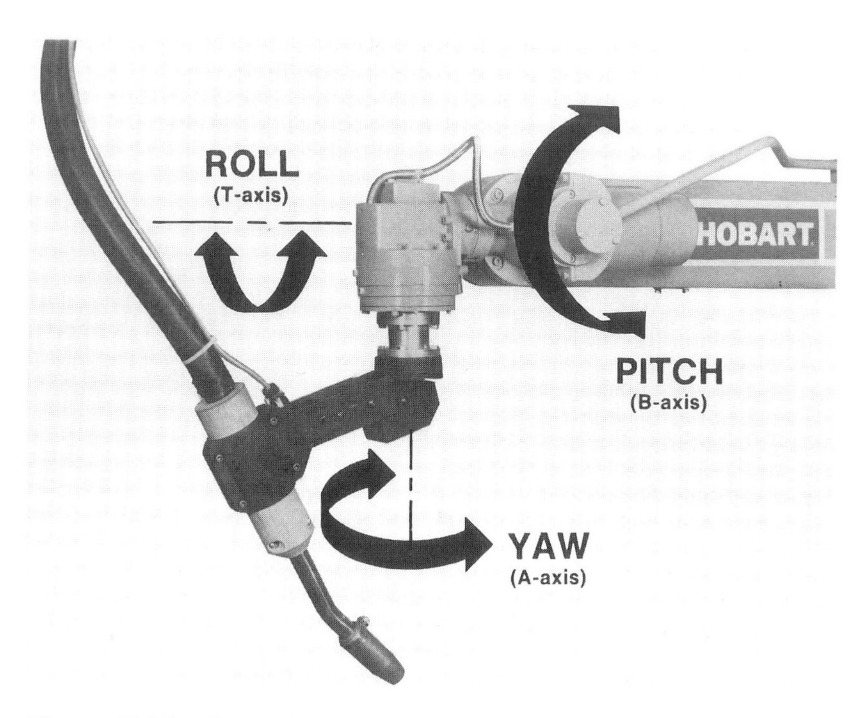

Figure 12.19 Wrist motions.

Figure 12.20 Method of welding gun attachment (GMAW).

impacts some unexpected obstacle, as shown in Fig. 12.22. This minimizes damage to the robot, torch, fixtures, and positioners. Pivot joints provide multi-directional deflection of the torch upon impact. The parts are keyed, and this ensures accurate realignment of the torch. Every robot should be supplied with a breakaway torch bracket.

Torch-changing devices are available that allow the robot to use one or more different torches for a particular job. The torches are completely wired and have electrode wire in the conduit. The robot is programmed to go to the bracket holding the appropriate torch, change torches, return to the welding operation, and continue to weld with a different procedure or process. This allows a work cell to make welds with different processes as required for the part being manufactured. Torches can also be changed to provide special arc accessibility.

Another accessory is the torch nozzle cleaner. Nozzle cleaning can be included in the welding program. The torch will be moved to the cleaning station, where the automatic nozzle cleaning device reams the inside diameter of the torch

Figure 12.21 Method of welding gun attachment (GTAW).

nozzle and an air blast blows out accumulated spatter. This station will also spray antispatter compound on the nozzle. Another type taps and vibrates the torch to remove spatter if it has not adhered solidly. Figures 12.23A and B show a nozzle-cleaning station with two operations.

An automatic electrode wire cutter is shown in Fig. 12.24. The cutter is used to trim electrode wire sticking out beyond the contact tube. It can be programmed to cut the electrode wire prior to starting the next weld.

Special heavy-duty, long-life electrode contact tips are available. These are made of hardened copper alloys or include special hardened inserts that eliminate the wear on the inside diameter of the current pickup tip. These special long-life

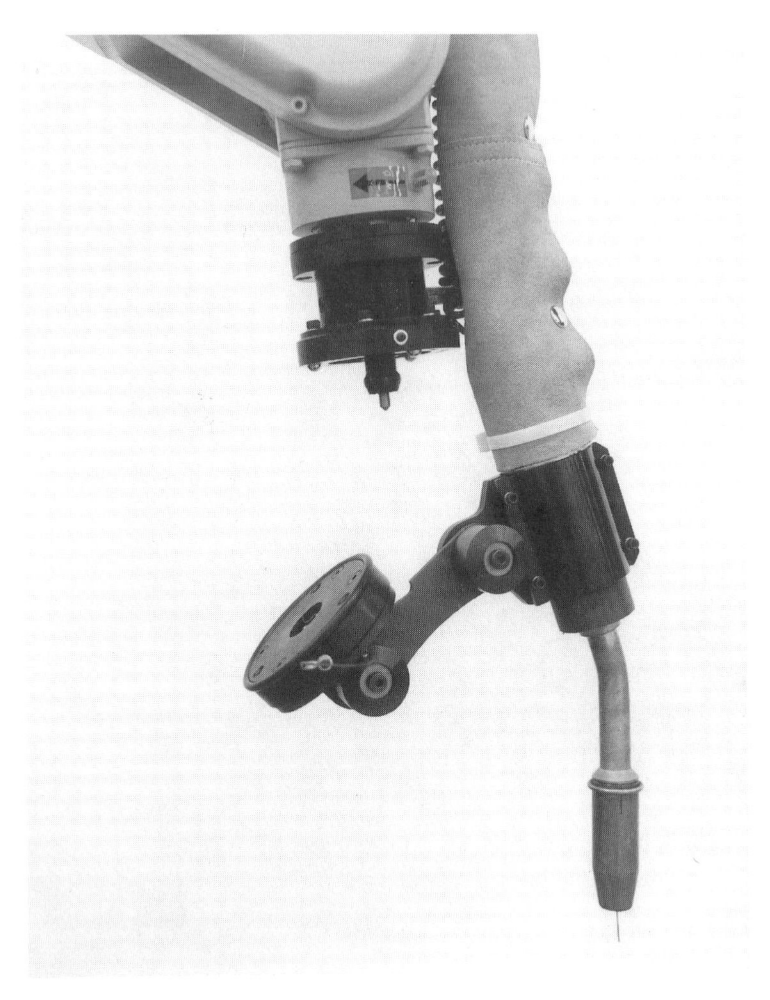

Figure 12.22 Torch breakaway bracket.

tips are important when sensors are involved or when it is inconvenient to shut down the robot for frequent tip changes.

The jointed arm robot can be given more welding capability by adding additional degrees of freedom. For large weldments, the most popular method is to mount the robot on a moving device of some type. The gantry mentioned previously is widely used for this purpose, with the jointed arm robot or robot arm mounted in the inverted position. A typical overhead mounting is shown in Fig. 12.25. Robots can also be placed on a moving platform mounted on the floor or on a side beam carriage mounted on the factory wall.

(A)

(B)

Figure 12.23 Nozzle cleaners.

Figure 12.24 Electrode wire cutter.

12.4 Robotic Part-Holding Positioners

Robotic part-holding positioners are used for at least three reasons. First, for safety—the operator remains outside the robot's work barrier to load and unload the fixture. Second, to increase the productivity of the robot—the robot arc-on time continues while the other fixture is being loaded or unloaded. Third, to improve the versatility and extend the range of the robot.

Specialized multiaxis coordinated positioners are used to improve the versatility and to extend the range of robotic arc welding systems. The usable portion of a robot's work envelope can be limited because the welding torch mounting method may not allow the torch to reach a joint properly. The lead and lag angles of the torch, which are controlled by the wrist movement, reduce the distances that the torch tip can be extended, which reduces the robot's working range. These special positioners reduce these limitations by making the workpiece more accessible to the robotic welding torch. They expand the range of the robot and provide additional axes of motion to the system.

A variety of part-holding positioners are used with robots. They must be more accurate, manufactured to tighter tolerances, and with less backlash, etc., than positioners for manual or semiautomatic welding. Refer to Chapter 9 for information concerning the capacity and specifications for positioners.

Two types of systems are used to drive the robotic positioners. The simplest

Figure 12.25 Robot mounted in inverted position.

motion system causes the positioner to move from one specific location to another. At each specified location it is locked into position. The robot provides the necessary travel motion to make the weld. An indexing axis can be used for shuttle devices, turnover devices, or turnaround devices. These can be pneumatically or hydraulically powered since coordination of motion is not required. Motion is usually initiated by an independent control station, not part of the robotic computer-controlled system. Motions that are made, however, must be recorded in the computer memory. This is necessary so that different welding programs can be carried out when the motion device is locked into different positions.

The other type of motion is true servo motion with coordinated or simultaneous movements controlled by the robot controller. The robot manipulator and the work-holding positioner must be controlled by the same controller. Movements are synchronized so that the arc directed by the robot and the work on the positioner arrive at a predetermined spot at exactly the same time. All motions must start and stop simultaneously. This is known as coordinated axis motion.

The most popular positioner is the turnaround positioner, which is a pneumatically powered 180° indexing positioner (see Fig. 12.26). It is controlled by the operator, not programmed by the robot. Its location, however, must be transmitted to the controller memory. This dual station unit is for small or medium-sized parts. Welding takes place at one station inside the work cell while the other end is being loaded or unloaded. It is used for safety and also to increase robot productivity. Different work-holding fixtures for making different parts can be placed at either end of the 180° indexing positioner. The robot uses a different program for each end of the positioner. For safety reasons, usually two palm buttons are required to cause the turnaround to index or rotate 180°.

A variation of the turnaround two-station indexing positioner has rotary motion tables at both ends, usually horizontal (see Fig. 12.27). The tables are servo-driven so that coordinated movements are obtained. This adds two more

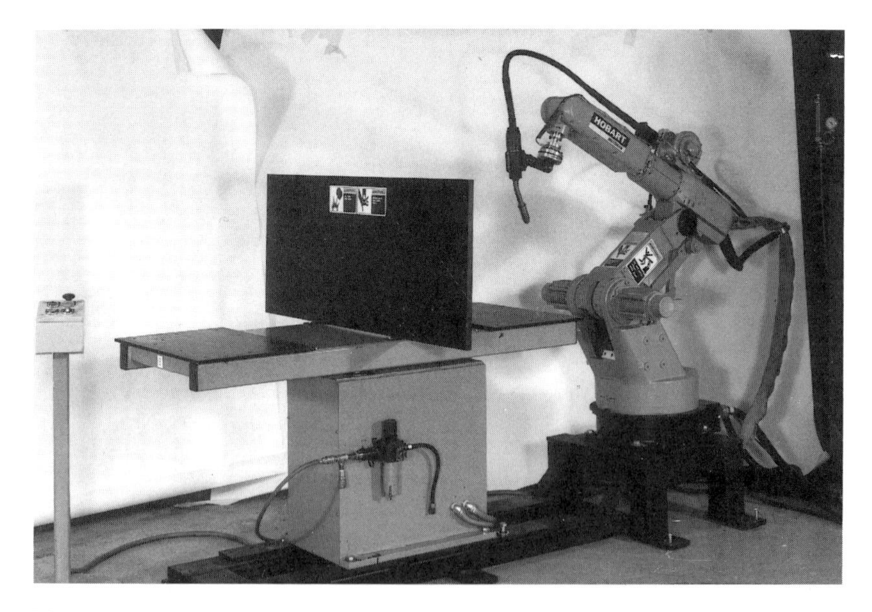

Figure 12.26 Turnaround positioner—pneumatic indexing.

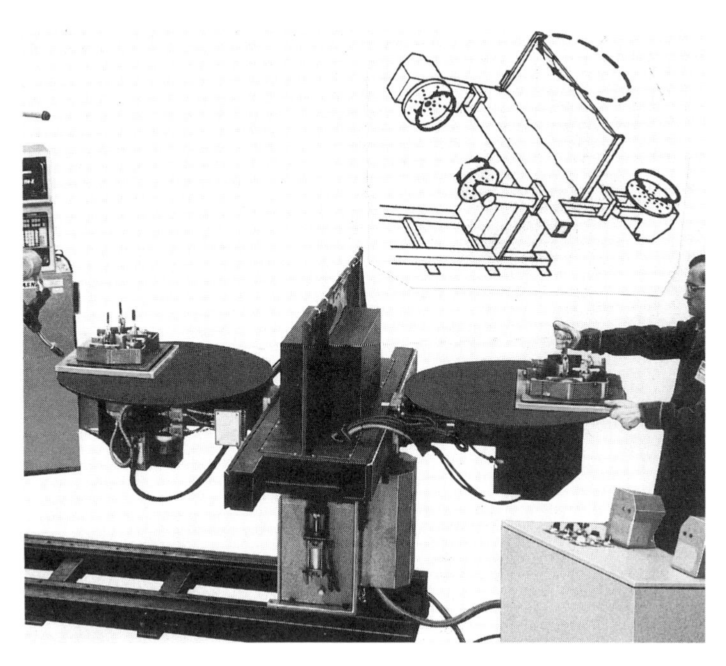

Figure 12.27 Turnaround positioner with servo table at each end.

axes of motion to the robotic system. An axis of rotary motion can be added to each end. Again, the two sides do not have to carry the same workpiece. In this illustration a complete welding cell is shown, and the control station for the turnaround fixture is shown in the right foreground.

A shuttle or transporter has many uses in robotic welding. Normally it does not have coordinated motion and moves in a straight line from one end of the track to the other. The track is usually 6 or 12 ft long. The transporter can be used to carry the robot, in which case when it arrives at a position it stops and is locked at that precise location. The robot provides all necessary welding motions. An optional version of the shuttle has servo-coordinated motion so it can make welds while the shuttle is traveling. This greatly extends the longitudinal axis of the robot's work envelope (Fig. 12.28).

Another use for the shuttle is to carry the work positioner and fixture into and out of the robot's work envelope barrier. This is primarily for safety but also adds to the productivity of the system. This application is shown in Fig. 12.29. The fixtures are loaded and unloaded outside the barrier while the welding is done inside.

A shuttle can also enable two robots to be used on one work fixture (Fig. 12.30). In this case, the motion of the shuttle is synchronized and coordinated.

Figure 12.28 Robot on shuttle.

Another common type of welding positioner is known as the turnstock, a headstock/tailstock positioner. This is a single indexed axis rotational unit as shown by Fig. 12.31 and is usually used for fairly long slender parts. It improves safety and productivity because one side is inside the robot envelope and the other is outside for loading and unloading. A turnstock positioner can include servo-coordinated motion on both sides. It can also accommodate multiple fixtures

Figure 12.29 Two work-holding shuttles.

on each side. A typical installation is shown in Fig. 12.32. In this case, each side of the turnstock has a separate coordinated axis that moves during the welding operation.

Figure 12.33 shows a turnover fixture, a type of positioner used for larger and heavier weldments. Turnover fixtures come in various styles, some have the main axis horizontal, while others have the main axis in the vertical position.

Another heavy-duty positioner, known as a headstock/tailstock positioner, can accommodate fairly large and heavy weldments. This positioner, shown in Fig. 12.34, has only one servo-coordinated axis.

Electricity for welding must be carried to the workpiece being welded. This means that the work lead of the welding circuit must be connected to the work and/or to the fixture holding the weldment. Electric welding power must not be transmitted through bearings of any type but transmitted through rotating joints. The most common device is a rotary copper connector that rubs against carbon brushes on the stationary side. These are normally copper slip rings with carbon brushes. Normally 1 square inch of carbon brush is required to transmit 100 A of power. For better conduction, most connectors require the application of an

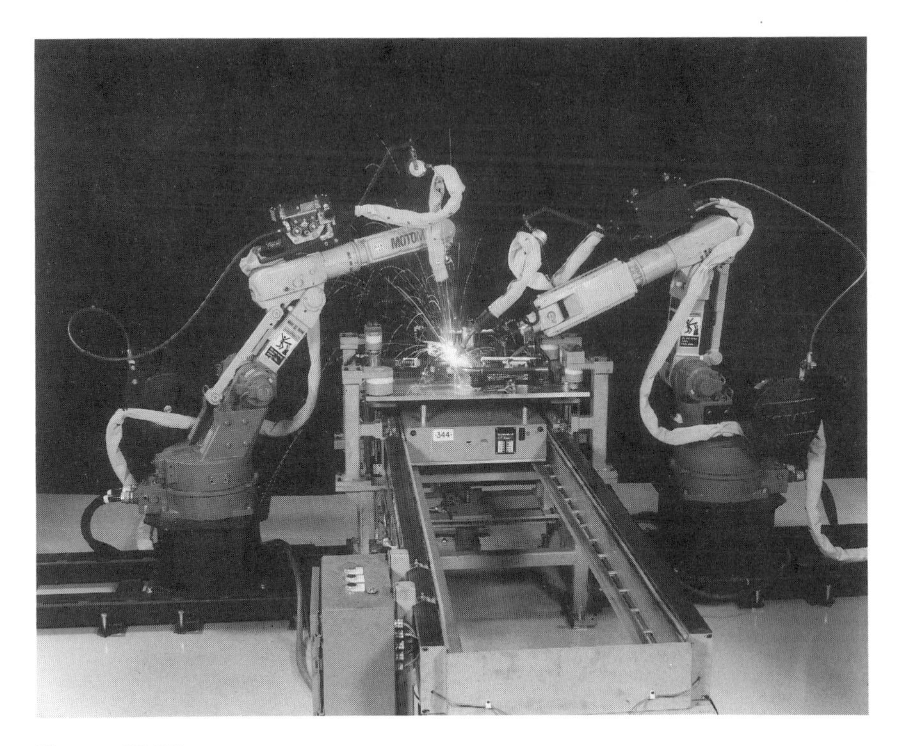

Figure 12.30 Two robots welding on a shuttle.

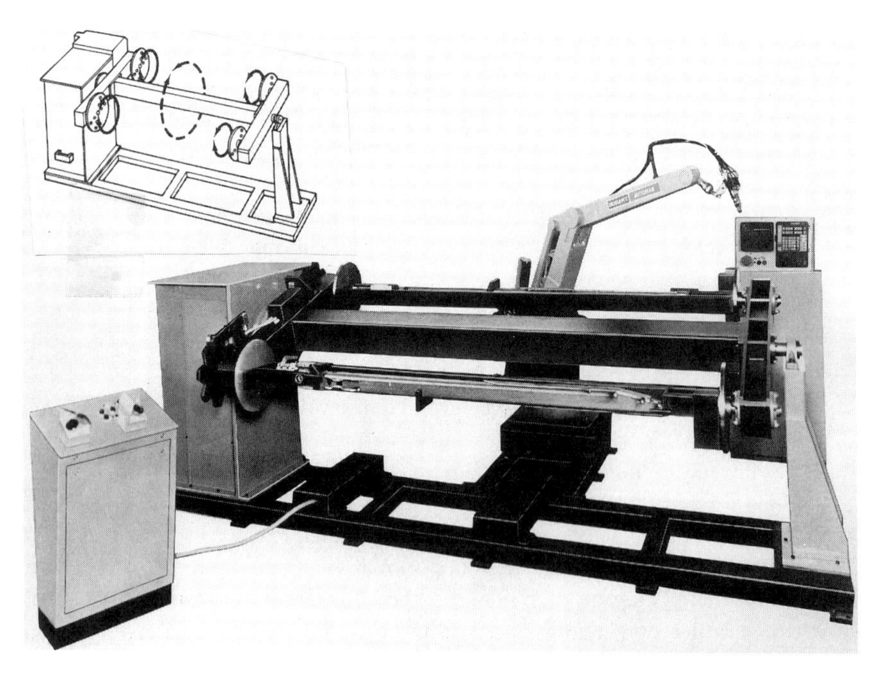

Figure 12.31 Turnstock positioner.

Figure 12.32 Turnstock with multiple jigs.

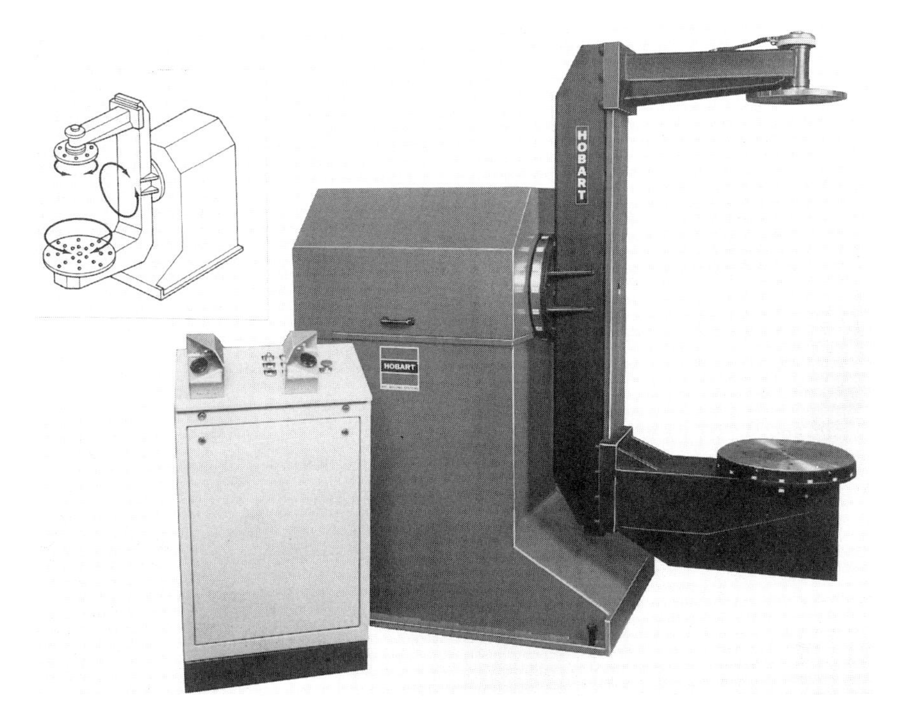

Figure 12.33 Turnover positioner.

Figure 12.34 Head stock/tail stock positioner—one axis coordinated.

electrically conductive grease so that they do not dry out. Trailing cables or electrically flexible conductors are usually used for sliding or linear motion, and trailing cables are sometimes used for headstock/tailstock positioners.

Undoubtedly other combinations of work motion fixtures are available. The selection must be based on the welding requirements.

All of the positioners and work motion devices mentioned in this chapter are available in a choice of sizes with various specifications. Vendors provide specification sheets for each type that indicate weight and size capacity as well as the motion motors and the type of controller that is required. It is imperative that a single controller control all coordinated motions in the entire robotic system.

12.5 Teaching the Robot

For more efficient and better-quality welds, the robot should be programmed by a human welder. This means that programming must be straightforward and logical and follow the normal mental processes of a welder welding a new product. The involvement of a welder is important. It has been found that it is quicker and easier to teach a welder to program a welding robot than it is to teach a programmer about welding. Most robot suppliers offer training programs to teach programming of robots.

In some advanced systems, off-line programming is used. This is accomplished by the general computer system. For off-line programming the entire program is prepared on a computer, using an appropriate programming language. The program is entered into the robotic memory very quickly. This increases the utilization of the robot, since lead-through teaching ties up the robot during programming. Off-line programming is difficult and requires experienced personnel. It has not yet become popular.

There are at least four methods used for teaching or programming a robot controller: manual, walk-through, lead-through, and off-line programming. The manual method is used for pick-and-place robots and is not used for arc welding.

The walk-through method requires the operator to move the torch manually through the desired sequence of movements. Each move is recorded into memory for playback during welding. The welding parameters are entered at appropriate points during the weld cycle. This method was used with early welding robots.

The lead-through method is the most popular way of programming a robot. The robotic welding programmer accomplishes this by using a teach pendant. Three typical teach pendants are shown in Fig. 12.35. By means of the keyboard on the teach pendant, the torch is power-driven through the required sequence of motions. In addition, the operator inputs electrode wire feed speed, arc voltage, arc-on time, counters, output signals, job jump functions, and much more. All of these functions are related to a particular position along the taught path, not to time. If the robot's speed is changed, it is not necessary to change the program for certain actions to take place because actions are sequential and position-related rather than time-related. The travel speed of the torch is independently programmed between specific points by the keyboard. This allows higher motion speed during "air cut" time, that is, when not welding.

The path of the arc or tool center point (TCP) is taught by moving the TCP to a particular location using the teach pendant keyboard. The machine axes locate the torch, and the wrist axes control its angle. There is a control for each drive motor (i.e., one for each axis). When the desired position is reached, it is recorded in memory by pushing the record button. This operation is repeated for the next location point and the next until the complete path is taught. For straight-line motion, location points are required only at the beginning and end. This is true also for arcs. The robot controller is coordinated to control all the axes simultaneously.

In the playback mode the robot will follow the path between points according to its interpolation function. Normally, linear interpolation is used,

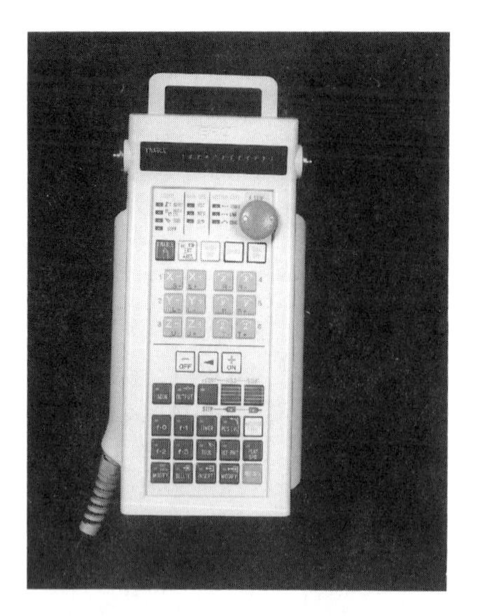

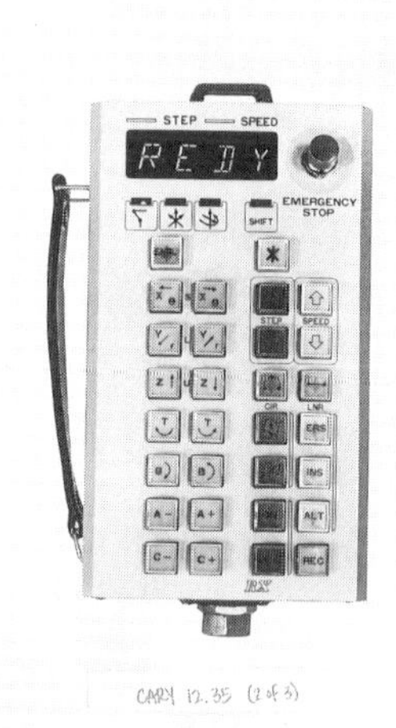

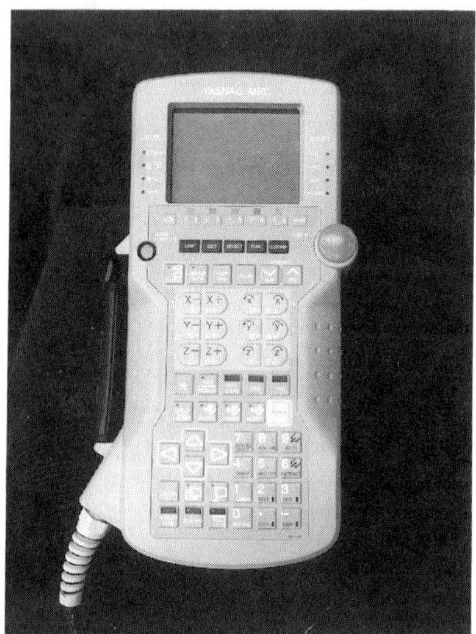

Figure 12.35 Typical teach pendants.

which means that the arc or TCP will move in a straight line between two taught points. Circular interpolation means that the TCP will move in a circle designated and located by three points. It is used for developing a curved path and reduces the number of location points required. The playback mode gives a continuous path.

The controller can be set to an edit mode that allows taught points to be revised, deleted, or added without reteaching the entire path and also allows changes in travel speed or welding parameters. The programmer can check the taught path and welding parameters without actually welding. The speed of the arc may be set in absolute values or by transverse run time (TRT) or time between points.

The amount of controller memory is usually indicated by the number of steps and instructions that can be programmed with the number of axes involved, say a memory capacity of 2200 steps and 1200 instructions. Memory should have 32K bytes with battery backup. There should be a programming terminal with keyboard and monitor (screen) display in addition to the teach pendant.

Welds can be made only when the power is on for all components, the electrode wire is installed, and the controller is in the playback or operate mode. The material to be welded must be in the fixture and in position and all sensors interlocks closed. Pushing the start button will initiate the welding operation. The robot will move the torch to the start point. The welding equipment will begin its cycle of operation (gas preflow, start the arc, etc.). The robot controller will determine that the arc has started and then start motion. Points along the taught path will initiate other programmed activities. At the end of the taught path, the welding equipment will terminate the weld program, and the robot controller will determine that the electrode wire has separated from the work. After this the robot will return to its home position, ready for another cycle. After the initial run the weld should be checked for quality. If improvement is needed, the program should be checked and edited. It is important to minimize the air cut time and increase air cut speed. When the weld quality is acceptable and the cycle is as fast as possible, it is time to freeze the program and start production.

12.6 Specifying the Welding Robot

There are many choices when it comes to buying a robot or robot welding system. Before considering the purchase you must have accurate information concerning the goals of the project—safety; productivity; volume; quality; expected production rates; the size, weight, and material of weldments to be processed; the welding process and amount of welding on each part; the size and length of each weld; the welding position; tolerance of parts and of the weldment; weld distortion and spatter allowed; whether a code is involved; expected payback; and the budget for this capital expense. When all of these questions are answered and the specifications have been agreed upon, the cost and risk will help decide how to buy the system.

The least risky way is to buy a custom-designed total integrated system from a single vendor. The vendor has total responsibility, but the cost will be the highest, especially if tooling is included. However, the components will work together to produce quality weldments soon after installation.

The least expensive system with minimum risk is an off-the-shelf prepackaged integrated welding system. The advantage is that the system is a package of matched components guaranteed to work together. Tooling is usually purchased separately but specified to match the robot cell; thus there can be split responsibility.

Perhaps an ideal solution, which might be most cost-effective, is to buy a semicustomized system, which is a modified standard off-the-shelf preengineered cell. This is composed of modular units, which greatly reduces the risk and is less expensive than the custom turnkey system. Tooling could be purchased from the same vendor. Such a system might be more difficult to install, and responsibility is split.

Those who are buying their second or later system may choose to buy the robot arm and its controller directly from the robot vendor; buy the work-holding positioner, safety equipment welding package, and tooling from different vendors; and integrate the parts to form the total system. This might be the least expensive, but the buyer has different responsibilities and the highest risk.

To obtain the exact robot system that you want, you must supply or reference the specifications for the robot. It is advantageous to combine the robot manipulator, the controller, and the welding system. The specifications describe the basic robot system. For example, we want an industrial machine, either the vertical articulated jointed arm type or the Cartesian type. We want the electrical drives to be a servoamplifier for each axis with position feedback sensors. The motion drive system and position feedback sensors are extremely important. The motion system must be smooth at all times in all positions. The hardware, software, and program language for the robot controller must be specified.

In describing the manipulator, the total number of axes, including the wrist axes, must be specified. In general, a five-axis jointed arm robot is the minimum requirement. The six-axis robot is better suited for welding applications. We then want to specify the maximum reach, the minimum reach, and the size of the work envelope, which relates to the arm length. It is important to compare the envelope size in both the horizontal and vertical planes when comparing robots. It is then necessary to determine the travel velocity while welding and while not welding. The welding speed must be compatible with the welding process and procedure to be used.

The next important feature is the payload or weight-carrying capacity of the robot. This influences the selection of the welding torch, the torch breakaway device, the mounting of water and gas hoses, and current-carrying cable. In some cases the electrode wire feeder is mounted on the robot, and its weight must be considered. The total weight of these items must be accommodated by the robot. The normal operating envelope must not be disrupted at normal travel velocities at the end of the wrist.

A very important factor is the repeatability of the robot. This is the closeness of agreement of repeated position movements under the same conditions, moving the welding torch to the same point every time it goes through its program. Most electric robots provide a maximum variation of ±0.015 in. in robot movement for repeated returns to a programmed point. This variation is affected by operating speed, and the value quoted is acceptable for gas metal arc welding. For gas tungsten arc or plasma arc welding, a tighter tolerance is required, and a repeatability of ±0.008 in. is desired. Repeatability information should be provided by the manufacturer's specifications.

Accuracy and resolution are both very important. Accuracy is the degree to which the actual position reached corresponds to the desired command position. This is measured by comparing the command position to the actual position. Resolution is a measure of the smallest possible increment of change in variable output of the robot. It is determined by the ability of the position feedback encoders or resolvers to determine the location of a particular joint and the position of the end point, called the *tool center point.*

Consult the specifications to determine the mounting position possible, base height adjustments, manipulator weight, environmental limits, and approvals. Next, review the robot controller specifications. The following should at least be considered and covered by the specifications; the robot controller teach pendant, type of keyboard, type of display, memory (RAM), read-only memory (ROM), type of processor (CPU), number of digital input/outputs, number of analog input/outputs, communication ports, power requirements, cabinet size and weight, number of axes that are fully coordinated, power failure recovery method, teach program, operating system, and application software and its language. There are specific features desirable for the arc welding robot. Many controllers have these built in, but they may be optional. The following is a brief review of some of these more important desirable features.

Linear interpolation is necessary. This means that the arc, or TPC, will move in a straight line between taught points. Circular interpolation indicates that the arc will move in a circle if three points are designated and located. It is useful in developing a curved path and reduces the number of points required. Software weave is extremely important for welding robots. Figure 12.36 shows how this works. It is useful for making welds when the joint can vary slightly, and it can be used instead of complex seam trackers.

Other normally standard features include, among others,

Automatic acceleration and deceleration
Three-dimensional shift
Simultaneous control of extra axes
Scale-up and scale-down
Mirror image

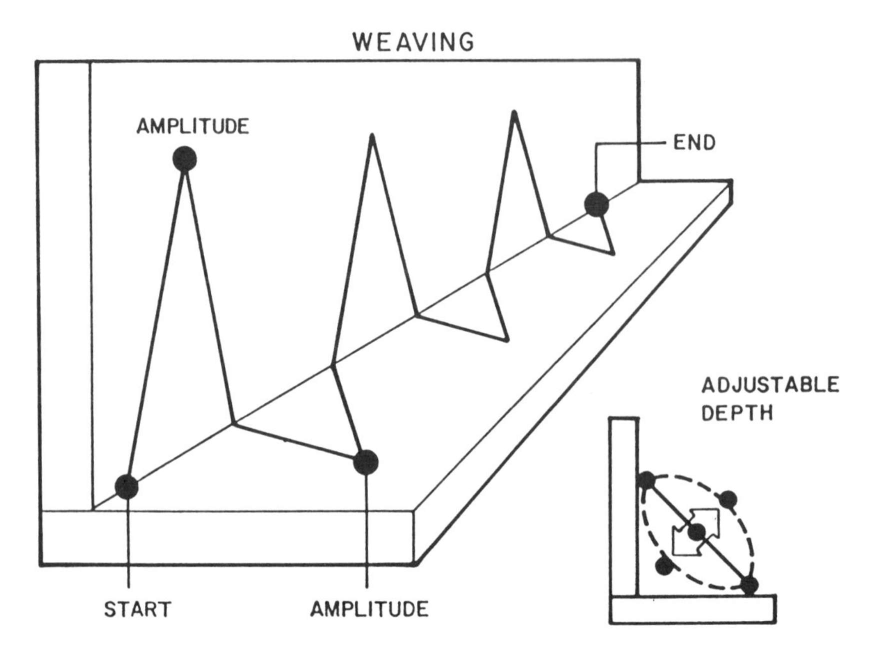

Figure 12.36 Software weave.

The information discussed above will help in comparing robots and is also necessary to obtaining the most suitable robot for your purposes.

The automotive and special machinery industries have formed a nonprofit organization known as the Automotive Industry Action Group (AIAG). The AIAG is dedicated to strengthening the relationship between the automobile companies (the customers) and the special machinery companies (the suppliers). One of their activities has been to publish the *Tooling and Equipment Supplier Quality Assurance Manual* [1]. This manual is intended to help foster continuous quality improvement among the suppliers and provide guidelines for customer product acceptance. It includes the prior to delivery 50/20 dry run program, which is a 50 hr quality test that applies to robots, a 20 hr continuous reliability run, the 10 piece preliminary evaluation, and process stability. The manual also describes the 20 hr continuous reliability run to be carried out and evaluation of short-term and long-term process performance, with acceptance criteria and a description of the test of the supplier's and customer's responsibilities. It is suggested that you obtain this publication, review it with your robot supplier, and come to an agreement to use these recommendations in your relationship.

Finally, it is important to determine whether you received the robot you ordered. You may want to make tests to determine that the robot meets your specifications. The following will assist you in this.

The Robotic Industries Association (P.O. Box 3724, Ann Arbor, MI 48106) cites three American National Standards Institute standards that help facilitate understanding between robot users and manufacturers by defining the performance criteria and a method of evaluating these criteria and testing requirements. For specific information, see the following American National Standards for Industrial Robots and Robot Systems:

ANSI/RIA R15.15-1 Point-to-Point and Static Performance Characteristics, Evaluation
ANSI/RIA R15.05-2 Path-Related and Dynamic Performance Characteristics—Evaluation
ANSI/RIA R15.05-3 Reliability Acceptance Testing—Guidelines

Reference

1. AIAG, *Tooling and Equipment Supplier Quality Assurance Manual,* AIAG-CQ1-2, Automotive Industry Action Group, 26200 Lahser Road, Suite 200, Southfield, MI 48034.

13

Case Studies of Robotic Welding Applications

13.1 Straightforward Robot Applications

Industry today is continuing its efforts to improve manufacturing quality and productivity through automation and systems integration. Probably no other piece of industrial machinery, during the past 25 years, has received as much scrutiny as or had a greater impact than the industrial robot. Robotic applications are progressing at a rapid pace in all types of industry and in all sizes of companies, small, medium, and large. Robots have proven to be accurate and reliable, to produce consistent quality, and to reduce manufacturing cost. They are justifying the investment in advanced production technology. Today companies are moving from the single robotic work cell to the fully automated flexible manufacturing cell, which includes multiple robots, positioners, and fixtures providing a variety of operations to complete a total part or subassembly and operate with simultaneous programmable motion control.

Robots are being assigned more and more welding jobs. Increasingly, they are being used to relieve monotony, remove the welder from the arc area, improve quality, and greatly improve productivity. They are justifying the systems investment and are being paid off in approximately one year. Robots are now a standard manufacturing tool in many industries. They are helping companies to become competitive worldwide.

One of the best ways to describe these different applications is through the use of case studies. This chapter contains 19 case studies of robotic welding applications. Some of the earlier case studies are of simple, straightforward applications in companies attempting to become familiar with the capabilities of robots. The studies become increasingly more complicated. Some are very new, but some are not. Each illustrates a point and introduces a new situation or new conditions.

A review of these case histories will help you decide whether the robot is applicable to your particular welding applications. All of these applications use a five- or six-axis articulated robot, and many use multiaxis positioning equipment.

It is realized that one or two pictures cannot explain the complex motions involved in robotic welding. Therefore the final section of this chapter provides a list of videotapes of automated and robotic welding applications available from manufacturers and integrators.

Note: In many of the pictures that follow, the protective equipment has been removed to better illustrate the welding operation.

Case Study 1

One progressive agricultural equipment manufacturer adopted robotic welding for short production runs on subassemblies of their equipment. They were actively searching for improvements that would help get the job done better, faster, and more cost-effectively. The company installed two jointed arm robotic work cells, each with a 180° rotary indexing table. Each end included positioners that added five axes of movement including the indexing action. The short production runs required a change of tooling once or twice a day for welding different parts. The robots fit the company's objective of just-in-time manufacturing, reducing parts inventory and minimizing storage space and materials handling. Parts to be welded were selected based on a comparison of load/unload time with arc welding time. It was found that if the arc-on time was less than the loading and unloading time, the operator had time to inspect the parts, which was desirable.

The robots were equipped with an adaptive through-the-arc seam-tracking system. The seam tracker enables the robot to adjust the programmed path automatically to changes in weld path location, maintaining the correct relationship between the torch and the joint centerline. The robotic welding operation is shown by Fig. 13.1. The robot is welding a knife support for a hay baler. The

Figure 13.1 Robotic welding operation.

rotating indexing positioner allows the part to always be placed in the best position for welding and provides easy access for the torch. The workpiece consists of 13 separate parts, with only two parts that need to be tack welded. The knife support is approximately 18 × 18 × 6 in. and is made of 1/4 in. steel. The robot produces 50 separate fillet welds with a total weld length of 70 in. The weld cycle time is 5 min 50 sec with an arc-on time of 74%. The part is shown in Fig. 13.2. Figure 13.3 shows the welding system layout and the location of the various components of the welding cell. This application met the company's objective.

Case Study 2

A small subcontractor to the automotive industry received a large order from a major automobile producer that required more production than the subcontractor had available. The company met this need by purchasing four robotic arc welding cells with 180° rotary index tables. Two are used to produce a steering hanger beam assembly that structurally supports the automobile steering column that contains the air bag system for the driver's side. The other two weld the beam assembly that supports the air bag system on the passenger's side of the car. The robots are six-axis articulated manipulators. The robot operation is integrated with the other plant manufacturing operations. In the case of the hanger beam, a CNC tube bender is used to bend the beam. It is then moved to a punch press, which

Figure 13.2 Knife support assembly.

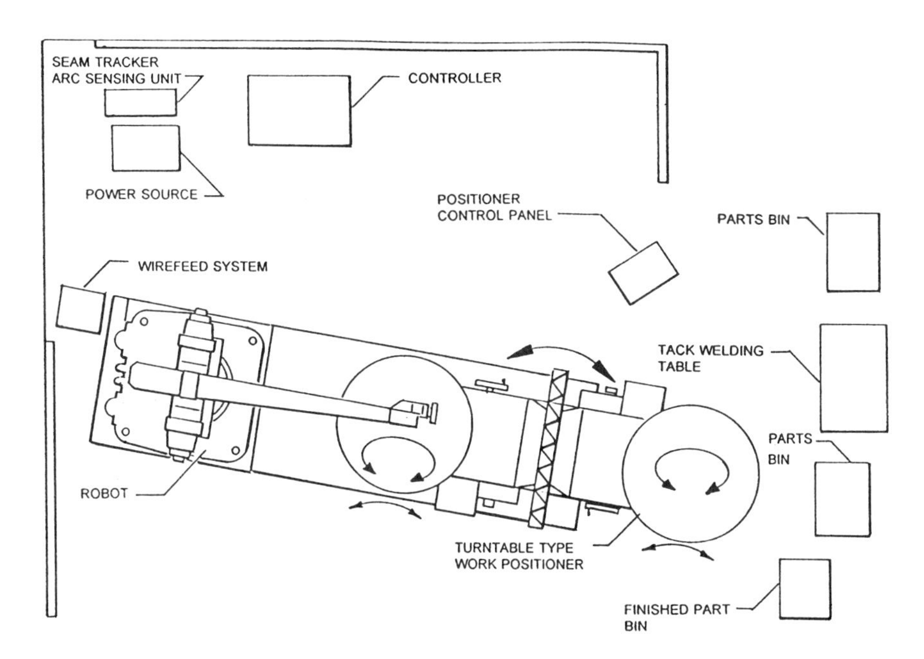

Figure 13.3 Welding system layout.

performs coining, flattening, and piercing operations. The operator loads the welding fixture with the beam and three brackets. The positioner is rotated into the robot's work envelope, and welding begins while the operator unloads the previously welded beam and reloads the necessary components for the next weldment. The two beam assemblies are similar, and the welding operations are about equal. Mild steel electrode wire from a 1000 lb coil is supplied via a wire dereeling system. A shielding gas of argon and CO_2 produces almost spatter-free welds. This new system, shown in Fig. 13.4, enabled the company to produce the required number of components just in time. Since the weld cell is self-contained with protective wall curtains, the operator does not need to wear a helmet and is not exposed to the sparks, gases, and splatter that come with manual welding.

Case Study 3

Management of another subcontractor to the automobile industry reviewed their operation and decided that automation would yield the greatest improvement in quality and production costs and would also provide greater flexibility and versatility for bidding on a wide range of jobs without the expense of continual retooling. They selected a five-axis jointed arm robot to weld an automotive brake pedal that they produce in large quantities. The robotic cell is equipped with two stationary work tables to hold the fixtures. A variety of fixtures were built that

Figure 13.4 Welding air bag support assemblies.

were designed to handle different types, sizes, and models of brake pedals. A typical brake pedal is shown in Fig. 13.5. Fixtures were designed to hold three or four brake pedal assemblies. Each fixture can accommodate three, four, or more of the same model. The operator places the foot pad and the brake arm plus the stub and the tube in the fixture. No tack welding is required. As soon as the fixture is loaded, the robot moves to the first pedal of the first fixture and makes the welds using gas metal arc welding. It then moves to the next pedal and makes the welds. This is repeated for each pedal. Figure 13.6 shows the welding robot in operation. The operator moves to the second fixture and loads it with the same parts. The robot finishes the welding on the first fixture, and the operator returns to the first station, unloads and inspects the brake pedal assembly, and reloads parts for the next welding cycle in the first fixture. Welding time is 9 sec for each assembly. The layout of the welding cell is shown by Fig. 13.7. The operator controls each weld cycle through a dual palm control panel, which opens and closes the clamps holding the parts in place. Use of the robot increased productivity and improved and maintained high quality welds on every pedal.

Case Study 4

A major automobile manufacturer reduced welding costs by replacing manual electrode welding with robotic arc welding. This program was welcome to the employees because the robots replaced "stick" welding, which exposes welders to hot, dirty, unpleasant, monotonous, tedious, boring, and fatiguing operations.

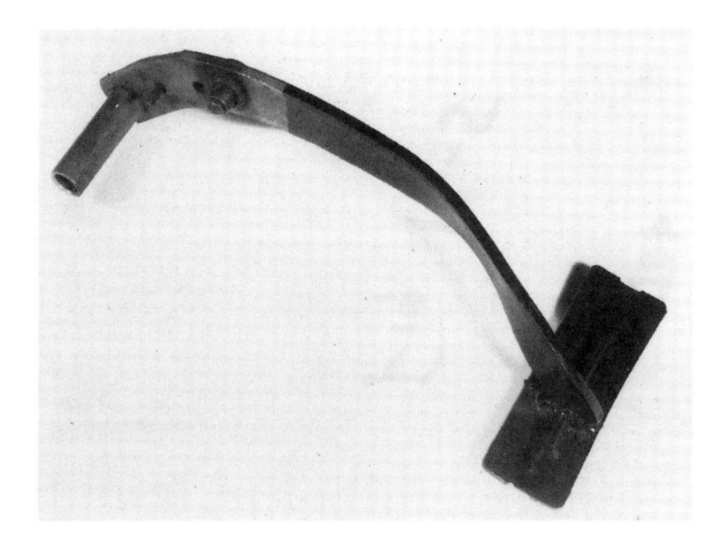

Figure 13.5 Brake pedal.

Figure 13.6 Welding operation—welding four assemblies.

At the same time, they give better weld quality, eliminate reworking, and greatly increase production. The project was initiated to reduce the cost of the engine mount bracket shown in Fig. 13.8. This assembly consists of two stampings of ¼ in. thick steel. The stampings are semi-self-jigging and are tack welded together with two resistance spot welds. The spot welded assembly is brought to the robot workstation and eight-position indexing table. Each workstation has a fixture to hold the assembly in the precise location needed for robotic welding.

Figure 13.9 shows the weld system layout. The operator loading the rotary table is outside the work envelope of the robot. The operator may load two or more fixtures at a time. The robot has a nozzle-cleaning station and is programmed to take the nozzle to the cleaning station periodically. The rotary table

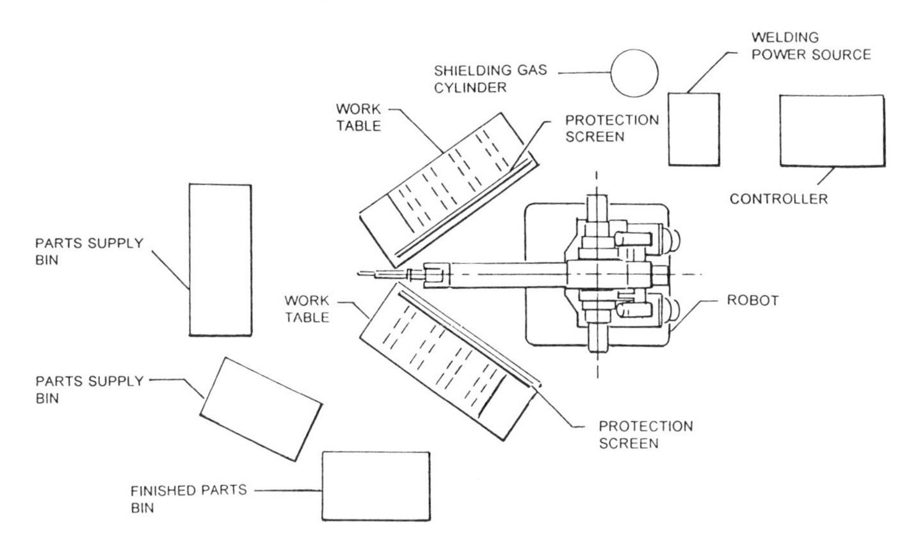

Figure 13.7 Layout of welding cell.

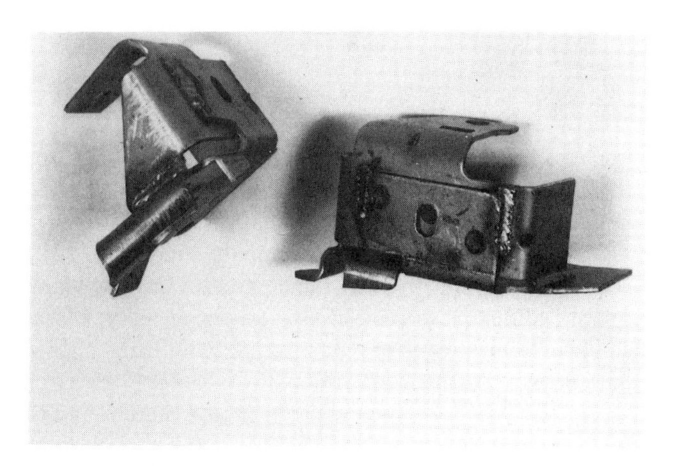

Figure 13.8 Engine mount bracket.

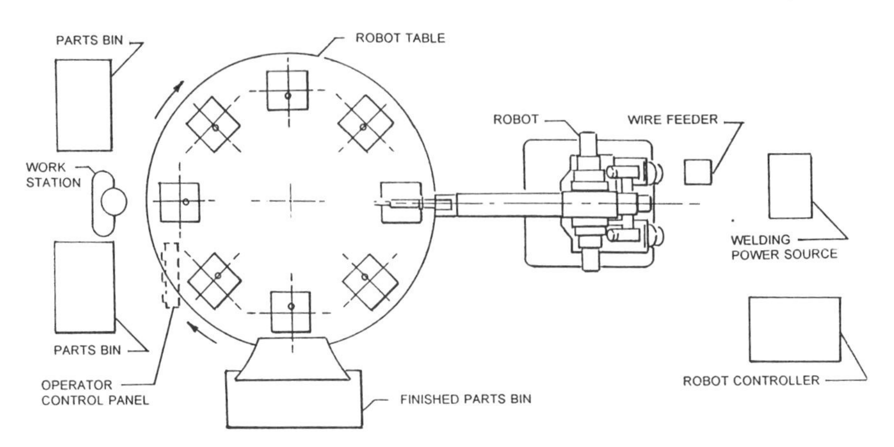

Figure 13.9 Layout of workstation.

is mechanically unloaded, and the welded bracket is placed into a finished parts bin for removal to the next department. The robot makes four 1/4 in. fillet welds 1 in. long on each assembly, and the welding time is 18 sec. The resulting benefits met the company's expectations.

Case Study 5

A prominent elevator manufacturer sought to accommodate shorter production runs for a greater variety of parts along with just-in-time manufacturing. Their investigations led to the purchase of a robotic cell containing a six-axis jointed arm robot with a five-axis rotary indexing turntable positioner. Figure 13.10 is a picture of this operation. Since putting the robot into production, they have selected and programmed 70 different workpieces for robotic welding. These parts all require short production runs of 4–6 hr. The operator will usually change tooling once during a shift. However, some parts are small, and the operator often sets up tooling to weld a different part on each end of the indexing positioner. The robot supplier provided the tooling for the original installation, but the manufacturer has made the additional tooling. The operator monitors the operation, inspects the welded parts, and unloads and reloads the parts behind the protective screen. The use of the robot has reduced manufacturing costs, increased production rates, and improved weld quality. The system payback was approximately 1 year. A variety of robotically welded parts are shown in Fig. 13.11.

Case Study 6

A manufacturer of large liquid fertilizer spreaders for agriculture has a product line that includes a large number of piping assemblies for applying liquid

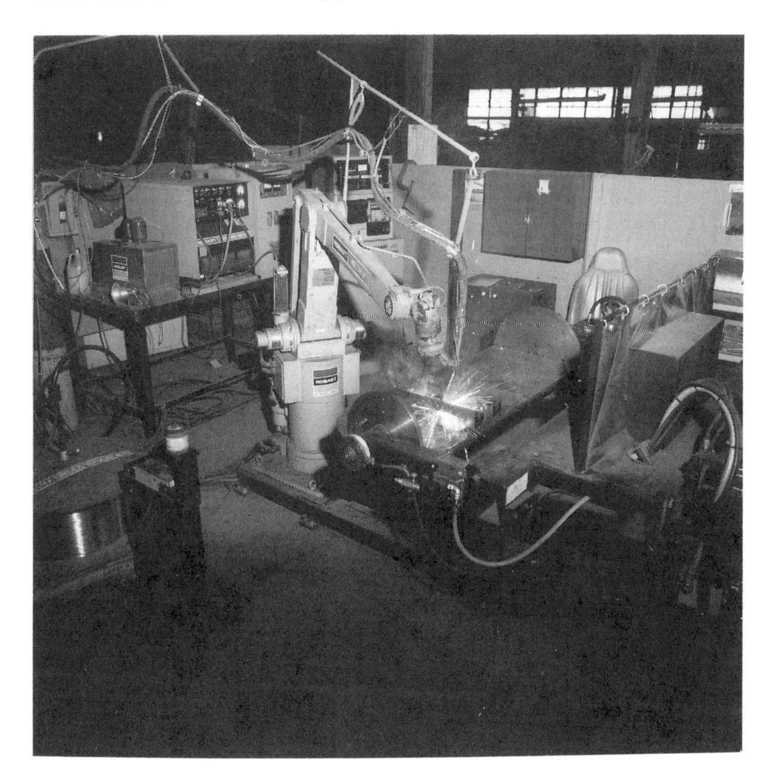

Figure 13.10 Welding an elevator part.

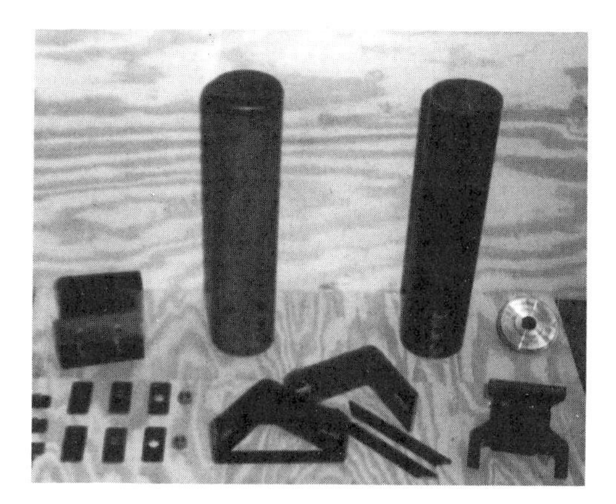

Figure 13.11 Variety of robotic welded parts.

fertilizers. Production runs of 25–200 parts were normal. Their investigation indicated that robotic welding could provide improvements in both production and quality, but they needed to overcome the problems of short production runs and variations in joint fit-up. The installation of a five-axis jointed arm robot in combination with a five-axis indexing positioner with tilt and rotate capabilities on each end helped them achieve their requirements. Figure 13.12 shows the robot and the positioner welding a typical pipe assembly. Figure 13.13 shows a typical pipe assembly that must contain high-quality welds. To ensure quality, the COM-ARC seam-tracking system is employed. Full-penetration V-groove welds are required on pipe assemblies. Production runs are all short, and tooling is normally changed once or twice a day with different setups at each workstation. Pipe assemblies are tack welded together and manually placed on the positioner. They are also manually unloaded. Arc time was reduced, which reduced manufacturing costs and increased production rates while improving weld penetration and quality.

Figure 13.12 Robot and positioner welding assembly.

Figure 13.13 Typical pipe assembly.

Case Study 7

An enclosed welding cell (Fig. 13.14) is designed to weld rod hanger brackets to an automobile exhaust muffler. Three rod hanger brackets are welded to each muffler tail pipe assembly. Three intermittent lengths of welds are applied by a jointed arm robot to weld each rod hanger to the muffler. Two welding stations are employed, with the robot welding first at one workstation, then the other. Material is loaded manually at the nonwelding station. The muffler assembly is shown in the foreground in Fig. 13.14. The cell employs an automatic nozzle-cleaning station and a semiautomatic unloading device. It also incorporates automatic arc shielding at each workstation and safe light curtains. The power source is mounted overhead to conserve floor space. The system's production rate ranges from 90 to 95 assemblies per hour.

Case Study 8

This case study concerns a much heavier weldment than any of the previous ones—a bolster assembly for a railroad tank car. The bolster assembly is made of heavy flame-cut mild steel plates that are, in some cases, formed on a press brake. This assembly was previously welded with semiautomatic self-shielded flux-cored wire. Approximately eight welders were required on two shifts, with setup men providing the production required. To reduce costs, the railroad car manufacturer installed two robot systems with positioners. The new system is operated by one man with several setup people. The pieceparts are accurately tack welded together in a setup fixture, then taken to the holding fixture, which is

Figure 13.14 Enclosed robotic welding cell.

mounted on the positioner. Another difference from previous case studies is that 3⁄32 in. (2.3 mm) diameter flux-cored wire with CO_2 shielding is employed. The flux-cored arc welding procedure operates at 450 A, and production has been greatly increased over that achieved by semiautomatic welding. The bolster assembly mounted on a positioner is shown in Fig. 13.15.

Case Study 9

In this case the weldment is made of aluminum, and GMAW with inert gas is used. A subcontractor producing ordnance material for the government was studying ways to reduce welding costs. They manufactured products of various geometries and configurations made of aluminum. A five-axis jointed arm robot was installed along with a torch-cleaning station and a double-ended 180° index positioner with positioners on each end. The product is an aluminum louvered grill assembly, shown in Fig. 13.16. This assembly measures 23 × 16 × 4 in. and weights 43 lb. It consists of 22 louvers attached to two side frames and top and bottom base plates. The side frames have rectangular slots stamped in them, and a louver is mounted in each slot. Each louver has a machined tab with 45° bevels,

Figure 13.15 Railroad car bolster.

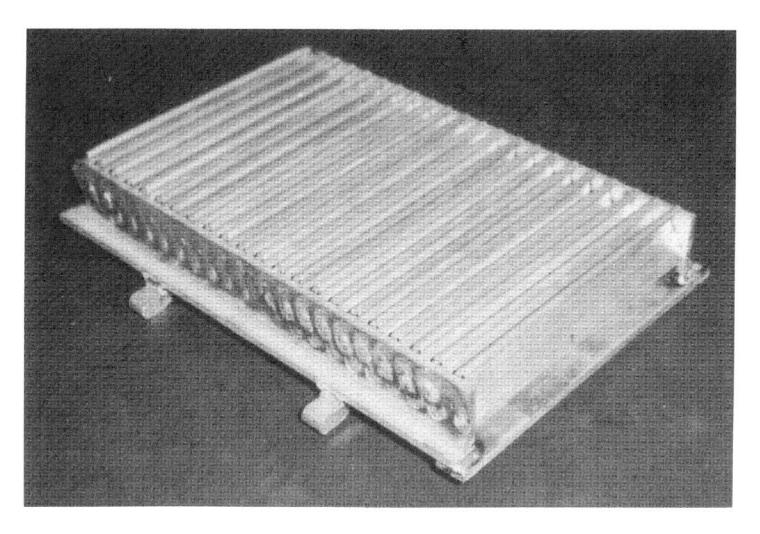

Figure 13.16 Aluminum grill assembly.

and the resulting joint consists of two single-bevel joints for joining each louver to the side frames. There are 21 louvers, and the welding amounts to 165 linear inches. The robot welds the assembly in approximately 10 min. The previous method using semiautomatic welding required 1 hr 15 min. Since the welds are located very close to each other, the robot is programmed to avoid welding adjacent louvers and thus virtually eliminates distortion. The welding operation is shown in Fig. 13.17. The introduction of robotic welding eliminated reworking and led to an appreciable reduction in cost.

Case Study 10

A food equipment manufacturer is using a robot to weld a stainless steel meat saw frame, shown in Fig. 13.18. This saw frame is a complicated part, difficult to fixture and difficult to provide torch access to, and must have accurate mounting locations. A five-axis jointed arm robot with a 180° rotary index positioner is employed. A fixturing concept was developed in which four fixture plates can be used to weld the frame and subassemblies. A fixture is used to weld the right and left sides of the frame separately. After a quantity of side subassemblies are made, a different fixture is installed on the positioner. This allows the right and left subassemblies to be welded together (Fig. 13.19). The accessibility of the welding torch to the final assembly is illustrated. In this operation, specific locating points are held constant. The joint fit-up varies according to tolerances of the pieceparts. This allows weld distortion of the side subassemblies, which allows excessively large root openings of up to 0.60 in. (15 mm). The welding procedure was designed to accept this imperfect fit-up. The welding electrode is a 0.045 in. (1.1 mm) diameter 308LSI. The shielding gas is a mixture of 90% argon, 8% CO_2, and 2% oxygen, which provides short-circuiting transfer to successfully weld

Figure 13.17 Welding the aluminum grill.

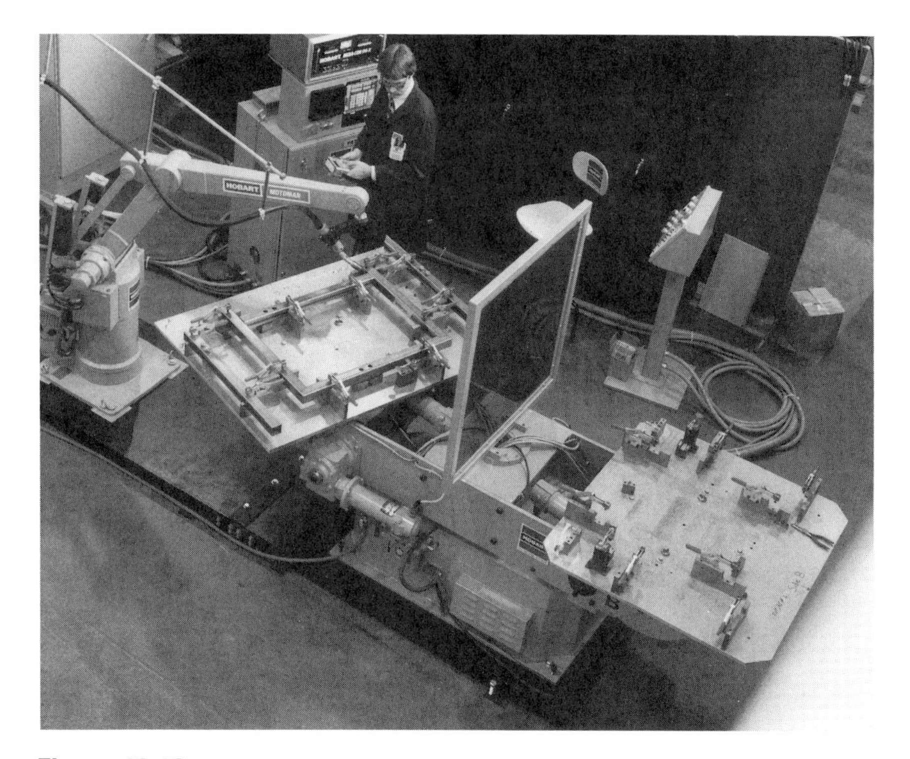

Figure 13.18 Meat saw frame.

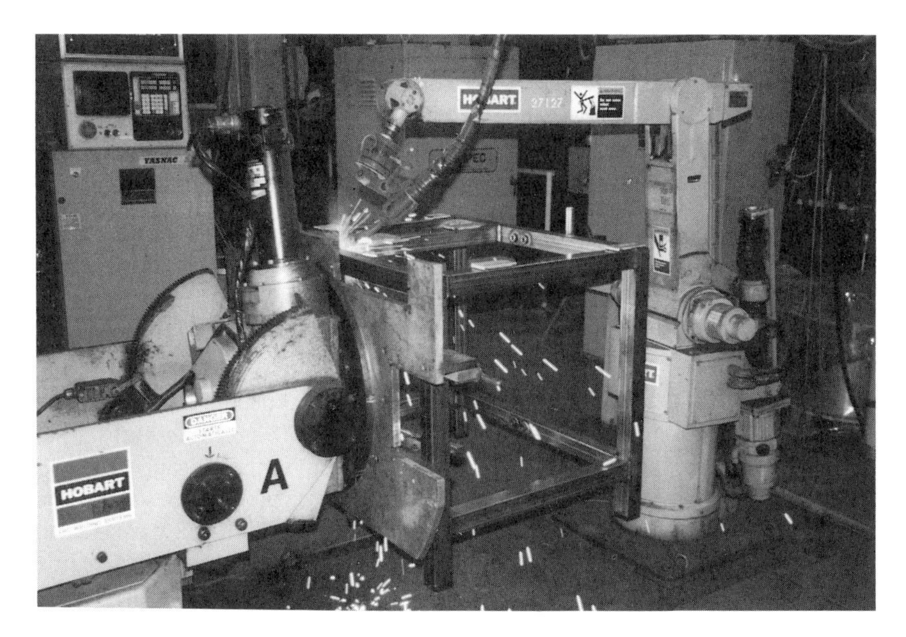

Figure 13.19 Welding the frame together.

joints that have gaps. This procedure produces welds with a smooth surface and very low spatter level. The robot operation greatly reduced the amount of scrap and reworking previously encountered with semiautomatic welding. The welding sequence as programmed by the robot controller is always the same, so distortion is minimized. This eliminated the need to straighten the frames after welding because the holes in the frame line up with the holes in the saw base. Another benefit is the weld surface quality consistency, which has greatly reduced the amount of metal finishing needed.

13.2 Complex Applications

The next five robot applications are complex compared to the previous case studies. In general, they involve more than one robot in a manufacturing cell, or more than one welding or cutting process. As more companies acquire experience with robots, robots will be used for more complex jobs. These case studies provide insight into complex robot jobs and ideas for robotizing your more complex welding or manufacturing operation.

Case Study 11

This case study is unique in that it involves both gas metal arc welding (MIG) and gas tungsten arc welding (TIG) on a stainless steel component. In addition, the welding cell contains other innovative features. The product in question is an undercounter dishwasher for restaurants. It is produced by a major company in the restaurant equipment supply business. The manufacturer initiated a redesign project for this product to improve its durability, reliability, and to facilitate servicing. At the same time, they wanted to bring in house several major components furnished by outside vendors. It was decided to design a manufacturing cell for the main components of two different models. The main components are the stainless steel 16 gauge dishwasher chamber and the 14 gauge stainless steel heater booster tank. To be most cost-effective, coengineering was employed, with the design engineers, manufacturing engineers, tool makers, and robot vendor all working closely together.

Numerically controlled punching equipment and bending equipment was also included in the new manufacturing cell. Restaurant dishwashers must have a cosmetically clean and smooth surface on both the inside and outside to meet food industry requirements. Based on their experience with a previous dishwasher model, the company selected gas tungsten arc welding for the 16 gauge stainless steel dishwashing chamber. This item measures approximately 30 in. square. The design includes many square and sharp corners. Discoloration and warpage had previously been experienced at these points. To overcome warpage, a very sturdy weld fixture was designed that included both inside and outside braces. It also had copper backing bars that provided backing gas. The inner portion of the tooling expanded, and the outer portion clamped, so that the stainless chamber was held rigidly in the exact location with perfect joint fit-up. Robotic welding was faster and more uniform in speed than manual welding, and warpage was reduced to a bare minimum. No filler wire was used, and the shielding and backing gas was argon. Direct current electrode negative was used with 3⁄32 in. tungsten electrodes. The tooling for the box chamber was affixed to one side of the indexing turnover positioner.

The booster tank was made of 12 and 14 gauge stainless steel on the front, back, and sides, and a 16 gauge baffle. Fixturing was very rigid, and the front and back pieces were clamped to the baffle that runs through the middle of the booster tank. The clamps holding the parts in the fixture are automated by an air-driven sequential cylinder, on both the inside and outside and for both weldments.

The booster tank uses gas metal arc welding with direct current electrode positive and shielding gas of 98% argon and 2% oxygen. This provides a short-circuiting metal transfer mode using 0.045 in. diameter electrode wire. The high silicon content of the ER 308LSI electrode improved wetting and reduced spatter. The tooling for the booster tank was attached to the opposite side of the indexing turnover positioner.

The robot welding power source is an inverter type with a programmable output characteristic curve. When the gas tungsten arc process is used, the welding power source has a constant current or fairly vertical output characteristic. When gas metal arc welding is used, a constant voltage with a fairly flat characteristic curve is used. The controller commands a reversing switch to change welding current polarity when the positioner indexes. With GMAW welding on the booster tank, a wire feeder and gun are employed. This is shown by Fig. 13.20. In this mode a nozzle-cleaning station, which is a vibrator rather than a reamer type, is used. This is sufficient for the amount of spatter collected. When the gas tungsten arc process is used, a capacitor start system is employed to prevent high-frequency stabilizing current from potentially harming the robot computer. When GTAW is employed for welding the thinner dishwasher chamber, a GTAW torch is used. This is shown in Fig. 13.21. The robot changes to the MIG electrode gun or to the TIG torch as required. The cables for both torches are shown in Fig. 13.21. The controller changes the output characteristics of the power source and the polarity as required and will also change to a completely different welding procedure. The cycle time for the washer chamber is 12 min, and the cycle time for the booster tank is 4 min. Figure 13.22 shows the overall picture with the robot working on the washing chamber side of the positioner and the booster tank in the other side of the fixture ready for welding. This manufacturer is extremely pleased with the results, which produce more accurate weldments, more uniform welds that require less finishing, and weldments that are produced much quicker.

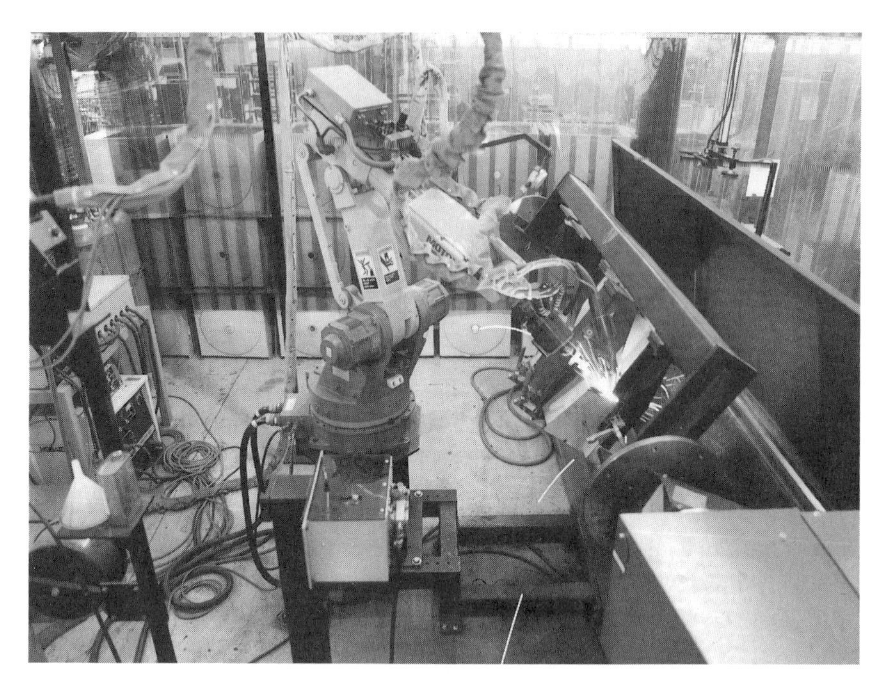

Figure 13.20 Welding a booster tank with GMAW.

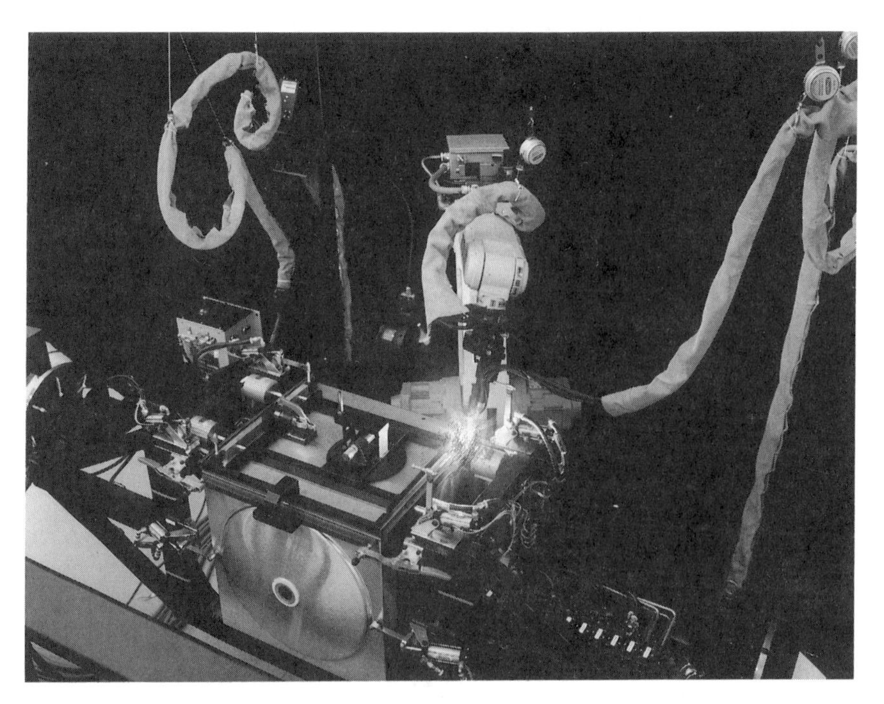

Figure 13.21 Welding a washer chamber with GTAW.

Figure 13.22 Overall view of welding operation.

Case Study 12

An overall view of one of the largest automatic robot cells for production of a single part is shown in Fig. 13.23. The part in question is an agricultural irrigation pipe assembly. This unusual operation employs two welding processes and a cutting process to produce these giant parts. The cell has the capability of welding pipes of different diameters to produce assemblies for different applications. It also possesses massive material-moving capabilities, moving material to the cell, moving it during production, and removing the finished parts. The unit is fed by a supply of pipe of various diameters from the stockpile. The pipe is rolled to the robot workstation, where the first operation is to push the pipe to the fixed end of the cell and qualify it. At this workstation the length of the two pieces of pipe must be as specified. If they are too long or too short the operation will not proceed. The next operation is to weld the two lengths of pipe together with a roll weld using gas metal arc welding as shown in Fig. 13.24. Then the flange plates are picked up and placed on each end of the welded pipe assembly. Figure 13.25 shows the robot picking up the flange plate from a special precision carrier. Welding is done using the submerged arc welding process with twin electrode wires in the roll welding position. Subarc welding is used because the owner

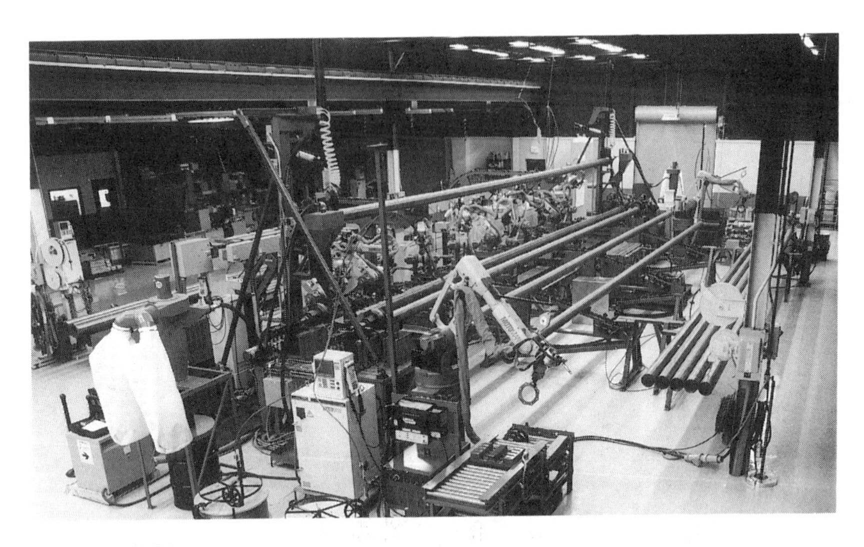

Figure 13.23 Overall view of manufacturing cell.

Figure 13.24 Welding pipe together.

Figure 13.25 Subarc weld—flange to pipe.

wanted to use it to weld the flanges to the pipe and wishes to maintain the high quality and uniformity of this operation. After the weld is completed, the solidified slag is automatically removed and a flux recovery unit picks up the unfused submerged arc flux.

The next operation involves eight robots. Each robot has a special plasma torch with a gripper. The plasma torch cuts a hole in the pipe where water nozzles are to be installed. The cut-out part is discarded into a scrap bin. The gripper then picks up a nozzle from a feed conveyor line and inserts it in the hole in the pipe. A GMAW torch tack welds the nozzle in the pipe, as shown in Fig. 13.26, then makes a weld around each nozzle to the pipe, as shown in Fig. 13.27. Brackets are picked up by the gripper from another feed conveyor line, and a GMAW torch tacks it to the pipe. There are two brackets per pipe assembly. Other robots with GMAW torches completely weld the bracket to the pipe. The finished pipe assembly is then moved overhead for unloading.

Case Study 13

A subcontractor to a foreign auto manufacturer makes extensive use of robotic welding. One weldment is a front axle support arm, shown in Fig. 13.28. Both right-hand and left-hand arms are made on the same equipment. An overhead conveyor line takes the tack welded assembly through the operation from start to

Figure 13.26 Inserting nozzle into hole in pipe.

finish. The system is programmed to produce right-hand arms for 4 hr, then left-hand arms for 4 hr. The overhead conveyor carries a scissors arm mechanism that lowers and raises the fixture holding the axle support arm. This is shown in Fig. 13.29. The scissors arm mechanism holds the weldment and lowers it at each robotic welding station for welding. It then picks up the arm and carries it to the next station, where it is lowered and released to the holding device at that welding station. The conveyor line indexes, and all weldments move to the next robot workstation. Figure 13.30 shows the robotic welding operation and also shows

Figure 13.27 Welding the nozzle to the pipe.

the carrier in the upper or withdrawn position without the arm assembly. The carrier then lifts the assembly to the upper position and moves to the next workstation, as shown in Fig. 13.31.

At each workstation two robots, one from either side, make a specific weld or several welds while the weldment is held in the work-holding device at that station. In some stations both the work and the robots move. The conveyor system continues through all of the welding stations, and the welds are completed on the front axle arm. The carrier on the conveyor line then goes to the discharge or

Figure 13.28 Front axle support arm assembly.

Figure 13.29 Lowering the assembly into the welding station.

Figure 13.30 Robot welding assembly.

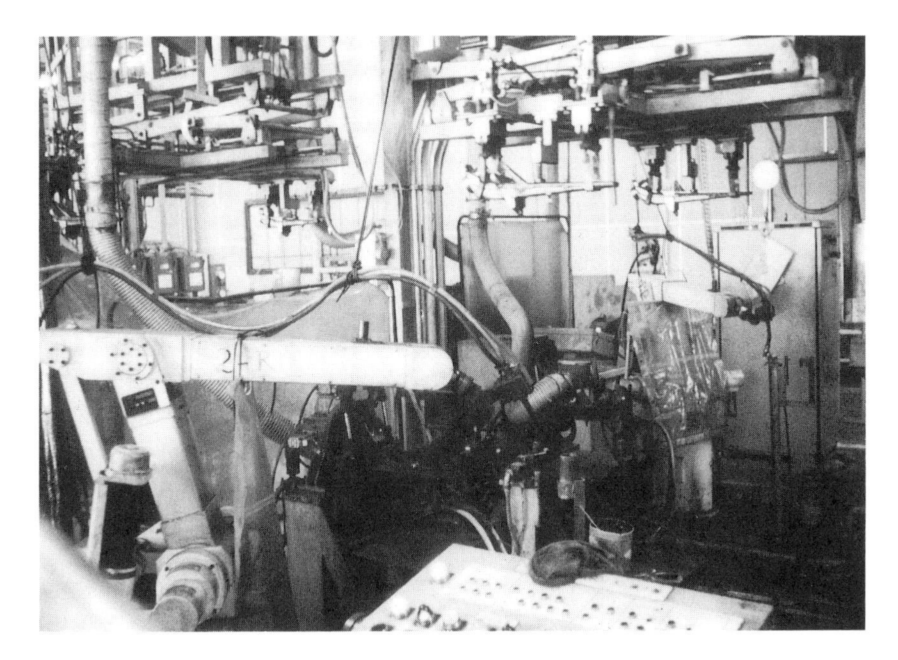

Figure 13.31 Conveyor line moving assembly to next welding station.

unload point, where the welded assemblies are placed in a bin for shipment to the auto manufacturer. Production is completely unmanned, except for a roving technician who quickly corrects any problem that surfaces. This type of production can be used only where continuous production is justified.

Case Study 14

An automated manufacturing system that makes subassemblies for automobiles from stamped and formed sheet metal uses a special moving assembly line containing workstations that employ GMAW to spot weld, punch, and extrude parts of the assembly. At some workstations loading is manual, but at others components are fed automatically. The total manufacturing cell (Fig. 13.32) is 148 ft (45 m) long and 10 ft (3 m) wide and made up of modular components. It produces 120 units/hr. This high-speed production manufacturing system produces automobile rear suspension assemblies for new front wheel drive luxury cars made by a major automobile manufacturer.

The overhead transfer system allows the entire manufacturing process to take place in 20 workstations. The main overhead line transports the suspension components from workstation to workstation by means of a special carrier working in unison with a pneumatic lifter device. The overhead transfer system will cycle as soon as all work is completed at each workstation. At one station the lifters pick up loose component parts and lower them into the next station.

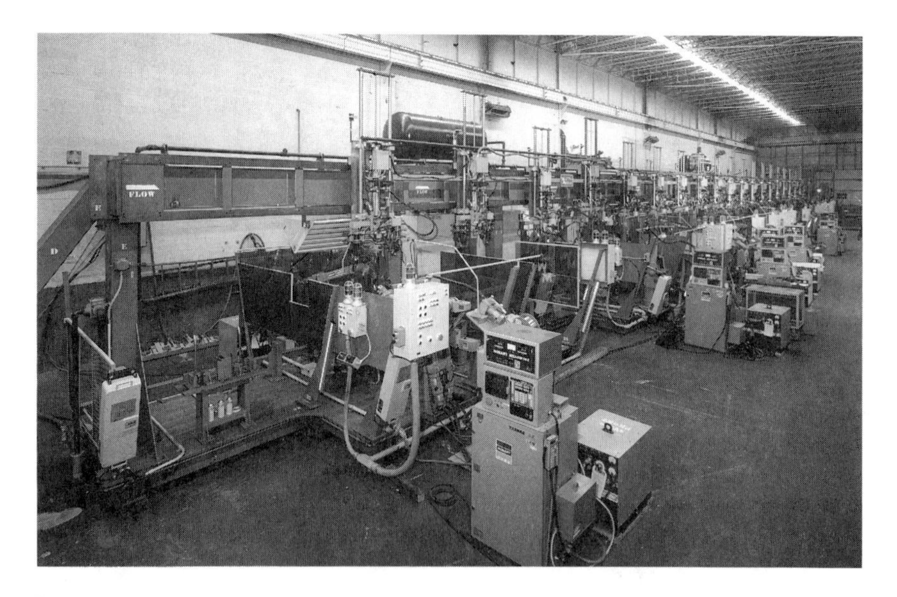

Figure 13.32 Overview view of total cell.

The cross member, which is the largest stamping, contains two locating holes that match reference points in the transporter and are used at each workstation to locate the components in the same relative position every time. The main line has 15 jointed arm robots for GMAW welding.

The assembly contains 30 formed sheet metal components, shown in the foreground of Fig. 13.33, which are welded into the rear suspension assembly, shown in the rear. This total assembly weighs 50 lb (22.6 kg) and has 236 in. (6 m) of gas metal arc welds made at an average speed of 60 in./min (1524 mm/min). The manufacturing system has five subassembly operations including two GMAW welding stations. One with a rotary index table is shown in Fig. 13.34. The subassembly operation requires the operator to load and clamp component parts in the fixture. The parts in the fixture move to a robotic GMAW station or to a projection welding station. After the welding is completed, the operator removes and inspects the finished subassembly and transfers it to the main welding line. At some workstations, as shown by Fig. 13.35, the main line robots are located across from each other in the same work envelope, working

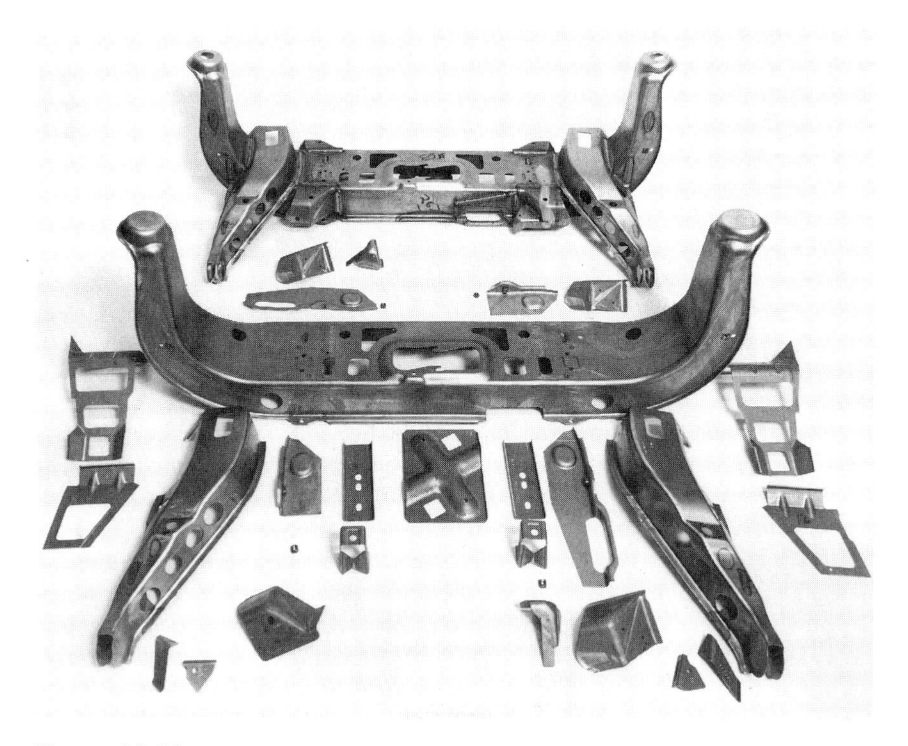

Figure 13.33 Rear suspension assembly and component parts.

Figure 13.34 Subassembly workstation.

together to minimize the required length of the line and maximize the utilization of equipment and space. The shielding gas used is an argon–CO_2 mixture.

To remove accumulated weld spatter, each robot is programmed to periodically go to a torch-cleaning station.

The rear suspension assembly is completely welded prior to the piercing and extrusion of 20 holes, slots, and extrusions, which requires precise locations. To provide flexibility for design changes, the welding robots can be reprogrammed. If the redesigned assembly will fit within the existing work envelope, the manufacturing system can be quickly retooled for a different product line. Each assembly must be correctly welded or the line will shut down until the proper adjustments are made. This manufacturing system produces perfect, accurate assemblies.

Figure 13.35 Two robots welding on the same subassembly.

Case Study 15

This case study describes a completely automatic manufacturing cell used to produce a welded engine manifold for a major automotive company. The new automotive engine exhaust manifold is made in a very large, completely enclosed and completely automatic work center. The manifold, shown in Fig. 13.36, is designed for a four-cylinder engine and is to replace a cast iron manifold. It is made of 403 stainless steel and consists of seven parts—the tank or hood, the base plate, three flanges, the tube, and the collar. It is completely welded and tested in this master robot work cell, which contains 12 articulated six-axis robots, two testing stations, eight welding workstations, a conveyor system that extends the length of the work cell, and four offloading conveyors for removing the finished product. One operator controls the master cell and loads all parts onto the precision carrier supplying the parts to the robots. At this control center, a data panel informs the operator of any problems within the master work cell such as parts missing, parts misalignment, or robots out of position. The operator also has complete control including start, stop, hold controls, and a method for stepping the system to realign all units and functions

Figure 13.36 Welded exhaust manifold top and bottom.

prior to start-up. The master work cell is fully enclosed with sliding panels, with interlocks at every access opening.

All parts for a complete manifold system are placed on a single carrier by the operator. An inspection station monitors each carrier to be sure all parts have been loaded and properly placed in position before the carrier is moved into the cell. One side of the cell handles the tank welding, and the other side handles the welding of flanges.

The base plate and the tank, or hood, are placed in a fixture that is in a water tank. Automatic clamps close and hold the tank in proper alignment and in tight contact with the base plate. The robot then comes in and makes the weld completely around the joint, joining the tank to the base plate (see Fig. 13.37). After the tank is welded to the base plate, the water in the weld fixture tank rises and quenches the assembly, and after a short delay the clamps retract. The welded assembly is then moved by another robot, turned over, and placed in another fixture. This fixture locates the tank base plate assembly, and three flanges are

Figure 13.37 Joining the tank to the base plate.

then placed and located on top of this assembly. Note that the two center ports are one flange piece. A hold-down fixture tightly clamps the assembly together prior to welding. The top is water-cooled copper and has access holes that allow the welds to be made. These welds join the inside diameter of the holes in the flange to the base plate. This is shown by Fig. 13.38.

This assembly is then removed by a robot and placed in a new fixture. The robot then adds the collar and the tube assembly. These are clamped in the fixture. Another jointed arm robot welds the collar to the tank and the tube assembly to the collar. This finished assembly (shown in Fig. 13.39) is removed and taken by a robot to the test station, where the manifold is tested for leaktightness. If it passes, it is placed on the conveyor by a robot and taken to the outside of the cell. If it fails the test, it is placed on a different conveyor and taken out of the cell to

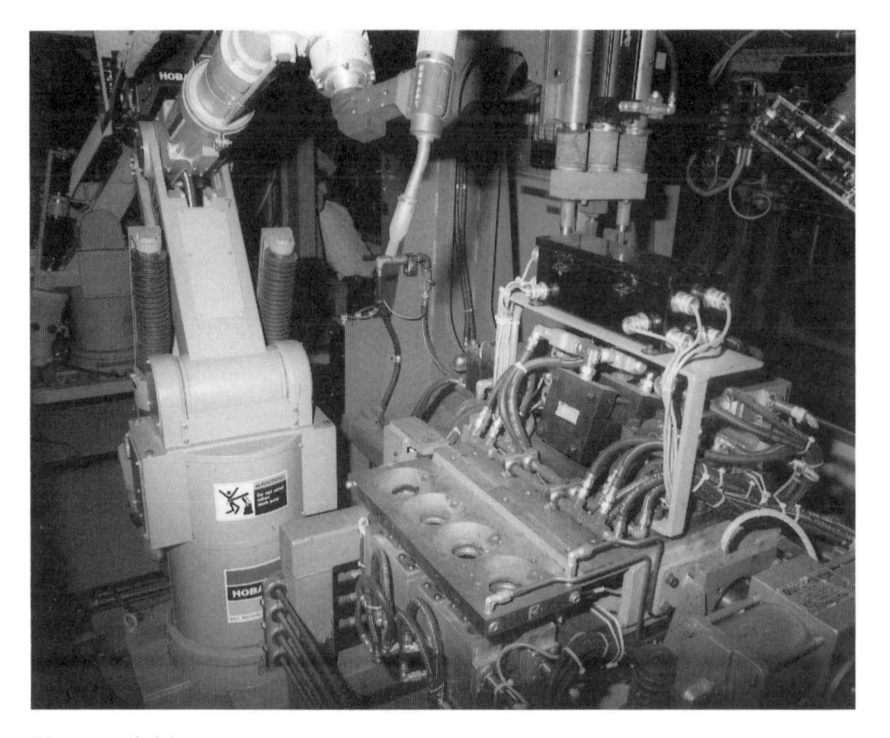

Figure 13.38 Joining the flanges to the assembly.

Figure 13.39 Joining the tube and collar to complete the assembly.

a repair station. All robots are back at their starting positions ready to start the next complete cycle.

Sensors are used extensively to determine whether the fixture is loaded and the parts are in their proper location. They provide information to the master control panel, and all sensors must be satisfied before the next indexed operation can occur.

This manufacturing system with the robot's versatility provides many benefits, including increased productivity, minimum scrap and rework, and reduced manufacturing cost.

13.3 Related Processes

Robots can be used for many metalworking activities, including machine loading, die casting unloading, press loading and unloading, and robotic grinding and deburring. These types of robotic applications are outside the scope of this book.

Except for one, the preceding case studies concerned gas metal arc welding or flux-cored arc welding. Gas tungsten arc welding, plasma arc welding, and laser beam welding are becoming robotized, and the following two case studies deal with these processes.

Case Study 16

Gas tungsten arc welding can be applied in two different modes. One uses filler wire to provide weld metal for making the weld. In the other mode, known as autogenous welding, filler metal is not employed, and the heat of the arc is used to fuse edges together. When filler metal is used, a delivery cable is attached to the welding torch as shown in Fig. 13.40.

In the welding operation seen in an overall view in Fig. 13.41, gas tungsten arc welding is used to weld jet engine components such as heat exchanger cores, louvers, and exhaust housings. Materials range in thickness from 0.032 to 0.215 in. and include two different high-nickel alloys. The robot is a six-axis jointed arm articulated type plus an indexing positioner with coordinated rotation and tilting on both ends. The welding machine is an inverter-type constant-current power source. It has an automatic voltage control (AVC) system that works through the robot's software. This measures the arc voltage as the torch and workpiece move and adjusts the path to maintain the correct torch-to-work distance.

The use of the indexing positioner with a work load station allows the operator to load and unload one part while another part is being welded. Figure 13.42 is a close-up view of the weld being made. Note the filler wire conduit feeding the wire into the arc. The company modified the GTAW torch to allow

Figure 13.40 GTAW torch with cold wire feed.

the tungsten electrode to be changed without moving or relocating the torch from the holder. This provides weld repeatability. The use of the robot and its programmed motions has greatly reduced distortion and has improved the appearance of the finished product. The company is so satisfied with the results that more welded assemblies have been given to the GTAW robot station to improve productivity and reduce heat buildup and distortion.

Case Study 17

A company that manufacturers turbines ranging from small steam turbines to large gas turbines made a study of the fabrication of turbine compressor diaphragms. They decided to eliminate manual GTA welding to reduce welding time and improve weld quality, accuracy, and reliability and invested in a six-axis jointed arm robot and two indexing positioners. The unique feature of this installation is the location of the robot and the positioners. They are located in a rectangular area 8 ft below the plant floor level. This allows the workpiece to be at floor level when the positioner table is horizontal and facilitates parts loading and unloading. The robot manipulator is attached to a vertical transport system that moves the robot up and down so the welding can always be accomplished in the flat

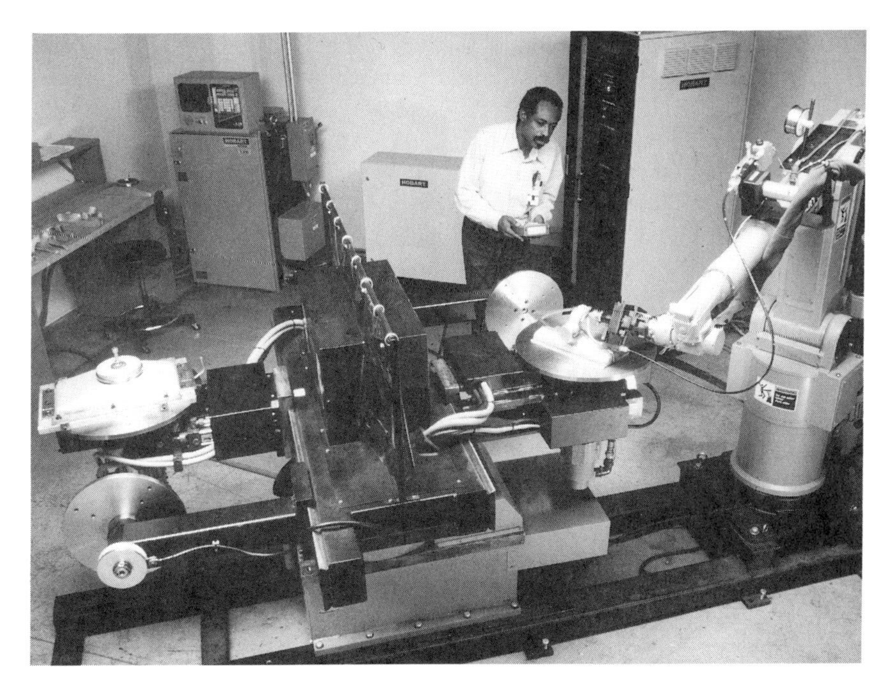

Figure 13.41 Overall view of robot and positioner.

Figure 13.42 Close-up view of weld being made.

position. The parts for the turbine diaphragm are fabricated in two identical semicircular sections. They are assembled and tack welded. The holding fixtures consist of two semicircular plates that also act as heat sinks. They hold the inner and outer shrouds but do not interfere with the torch positioning during welding. The airfoil-shaped vanes are tack welded to both the inner and outer rings. The robot welds each vane to the outer ring and then to the inner ring. To reduce the possibility of distortion, the robot is programmed to sequentially weld every other vane.

With manual welding, welding speeds averaged 6 in./min with an average of 30% arc-on time. The robot welding speed varies between 10 and 20 in./min with 90% arc-on time. Figure 13.43 shows an overall view of the robotic welding station in the pit. The metal thickness on the diaphragm averages 5⁄16 in. (8 mm), and the fillet size varies from 1⁄16 to 1⁄32 in. (1.6–0.8 mm) with 50% penetration required.

The two-axis positioners are automatically placed in a horizontal position at floor level for easy parts loading and unloading. They swing down below floor level into a vertical position during the welding operation, as shown by Fig. 13.44. The positioners are integrated into the welding robot controller

Figure 13.43 Overall view of weld cell in pit.

Figure 13.44 Position for welding.

so that each vane-to-shroud frame is manipulated for welding in the flat position. Figure 13.45 shows the vanes being welded to the inner frame of the diaphragm. The airfoil-shaped welds can be seen in the center of Fig. 13.46. The workpieces average between 48 and 72 welds, with a weld length of 2 in. Stainless steel 0.035-in. filler wire is used. The robot's performance has met company expectations.

Case Study 18

A manufacturer of textile machinery was modernizing production facilities and replacing semiautomatic gas metal arc welding with robots. The same piece of machinery, however, required stud welds for the side walls. It was decided to integrate stud welding with the gas metal arc welding on this assembly. It was further decided to use the same robots for both operations. Figure 13.47 shows the robot at the tool-changing area, where the stud welding gun is attached. The robot picks up the gun, takes it to the workpiece, and welds the studs to the sheet metal part, as shown in Fig. 13.48, employing the drawn arc stud welding process

Figure 13.45 Shape of weld path.

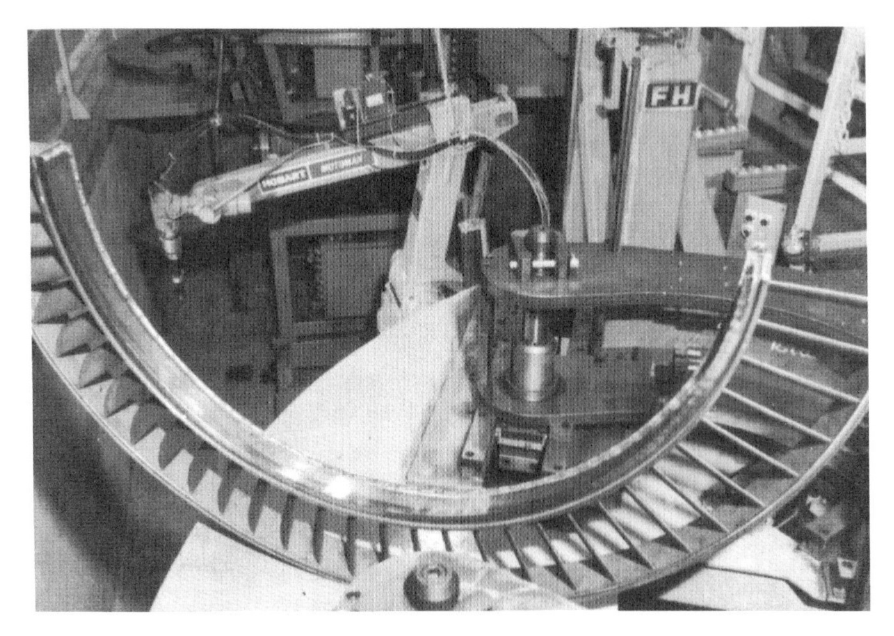

Figure 13.46 View of half of shroud.

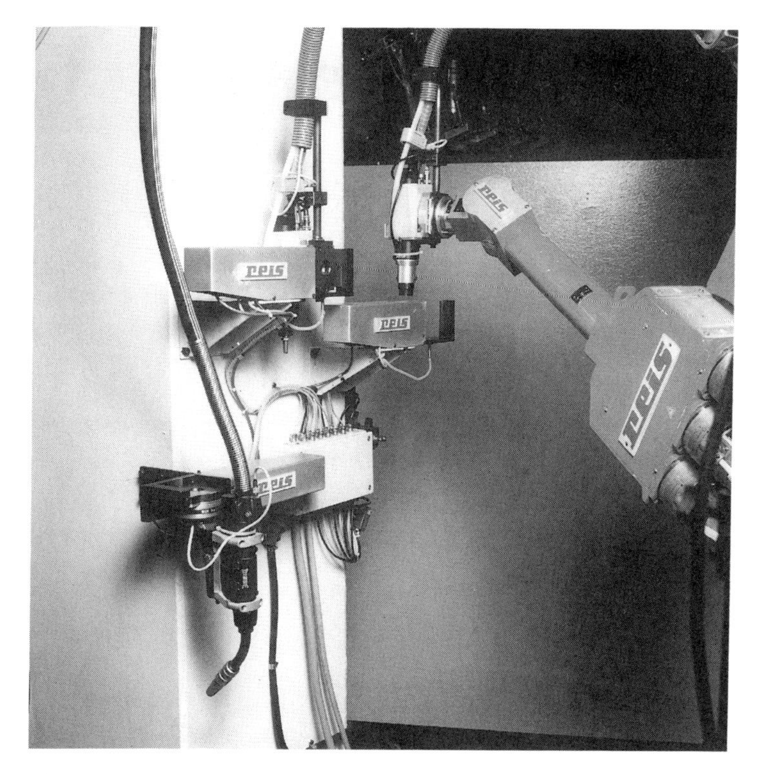

Figure 13.47 Gun-changing station.

with inert gas shielding. The studs are fed to the stud gun through a conduit by air from the magazine. The conversion from semiautomatic welding to robots greatly improved the quality of the product, reduced warpage, and reduced the weld time.

Nonwelding Applications

So far our case studies have related to welding; however, another important activity for the robot is thermal cutting. This was briefly discussed in Chapter 5. Until recently, oxyfuel gas cutting was the most popular thermal cutting process. Plasma is replacing oxyfuel gas cutting for many applications, especially for robotic applications. Figure 13.49 shows a plasma arc torch cutting a thin deep drawn part to a special contour to fit the adjacent component. This is a very common application for plasma cutting.

Laser beam cutting, which was briefly described in Section 5.2 of Chapter 5, has become extremely popular, particularly in the automotive industry. The Nd:YAG laser has become widely accepted for sheet metal cutting, particularly for auto body panels. The Nd:YAG laser uses a fiber optic cable to

Figure 13.48 Robot making stud weld.

transmit the beam to the cutting nozzle. This is very advantageous because reflective mirrors, which have adjustment problems, are not used. Figure 13.50 shows a continuous-wave (CW) solid-state laser cutting an access hole in the floor panel of an automobile. Different variations of cutouts are required for different types of transmissions. It is uneconomical to provide a special floor pan piercing die to handle each variation. This can be best done by use of the laser. Laser cutting has become very popular for making variations in specific stampings.

Another fairly new cutting method not considered an allied process to welding is *water jet cutting*. This employs a very high pressure fine stream of water directed toward the part to be cut. It is widely used for cutting nonmetal material, especially by the auto industry. One of the most common applications in the auto industry is the cutting of floor carpeting. Figure 13.51 shows the water jet cutting head attached to a jointed arm robot. Pressurized water is fed through the hose to the cutting head. When metals are cut, fine abrasive particles are added

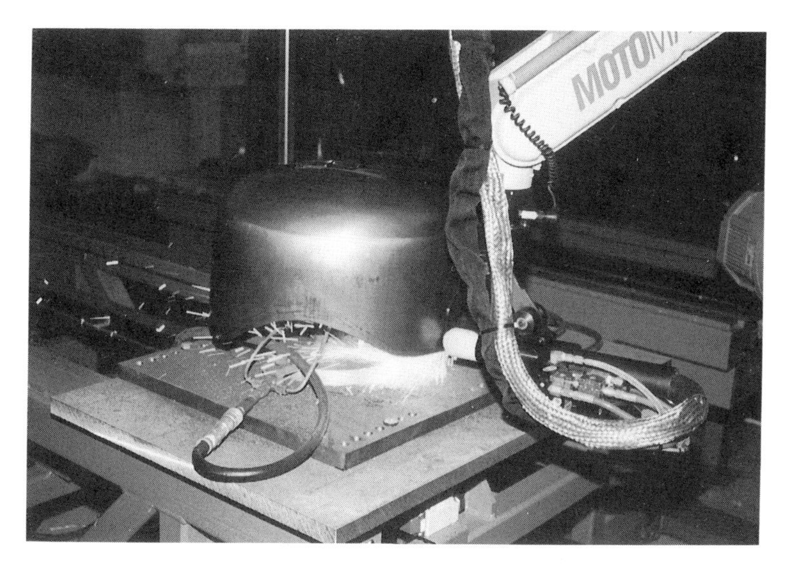

Figure 13.49 Plasma cutting.

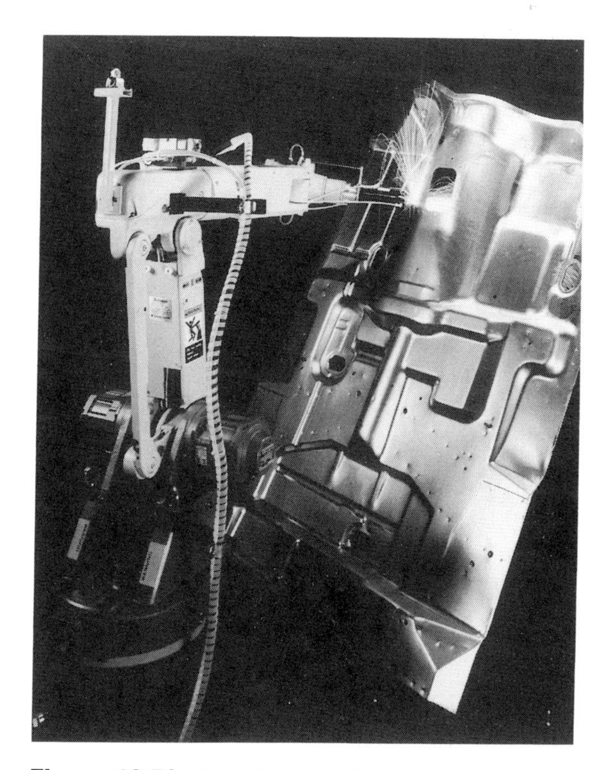

Figure 13.50 Laser beam cutting.

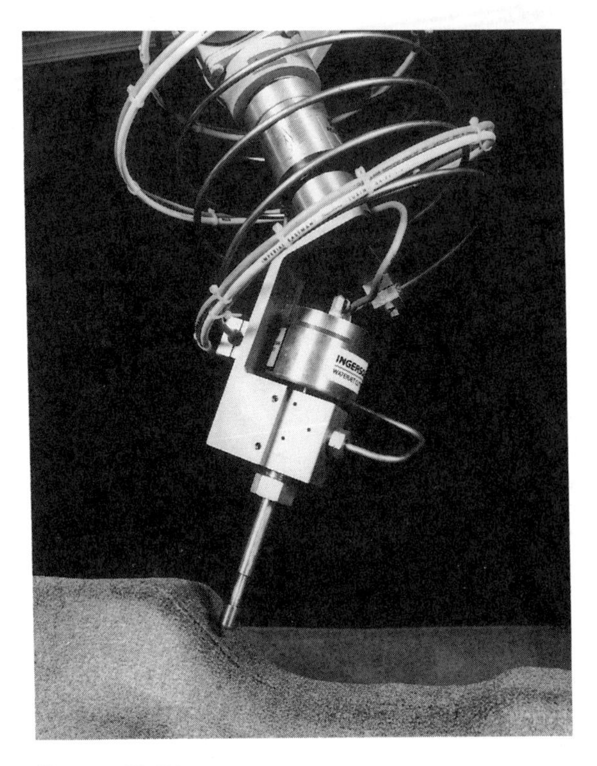

Figure 13.51 Water jet cutting.

to the water to provide additional cutting power. Figure 13.52 shows some parts cut by the water jet and used in automobiles.

Another popular application is the cutting of plastics. Figure 13.53 shows a water jet being used to cut holes in a helmet. Note that in this picture the water jet device is held differently in the arm of the robot.

Case Study 19

The final case study concerns a nonwelding robot application that is extremely important in the automotive and woodworking industries. Productivity improved and labor costs dropped because less cleanup was required after a window manufacturing company installed a robot to apply a butyl mastic sealant in the production of window units. Window frames are made of wood clad with an aluminum skin to withstand weather extremes and to minimize maintenance. In the assembly process, sealant is applied in a ⅛ in. (3.2 mm) bead directly to the wooden sash at the point where the aluminum cladding and the wood meet the glass (Fig. 13.54). This bond prevents moisture from reaching

Figure 13.52 Auto parts cut by water jet.

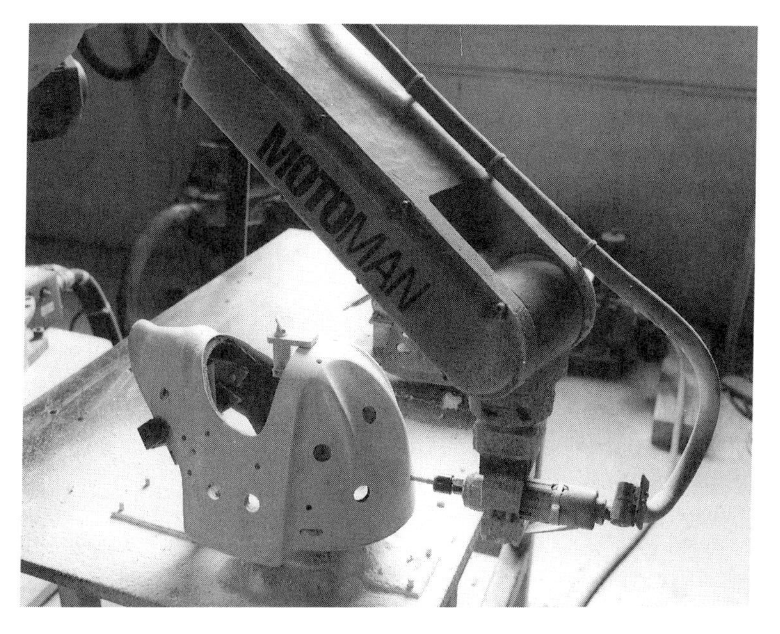

Figure 13.53 Water jet cutting of plastics.

Figure 13.54 Applying adhesive to house window assembly.

the wood. Prior to robot dispensing, the operator manually applied the sealant. When the robot was installed, the assembly process was modified and a cleaner delivery was attained. This is partially because when the bead shut-off is called for, a pneumatic "snuffer dispensing" valve quickly sucks back a small amount of material, eliminating the little "whiskers" that can develop. The dispensing speed is 20 in./sec (508 mm/sec), and the operator can specify the bead size, tip speeds, and other process parameters. This has become a very successful operation.

The automobile industry also uses sealants for attaching rubber molding strips to metal parts. Figure 13.55 shows the adhesive dispenser dispensing a small bead on a stamped panel. A similar operation is shown in progress

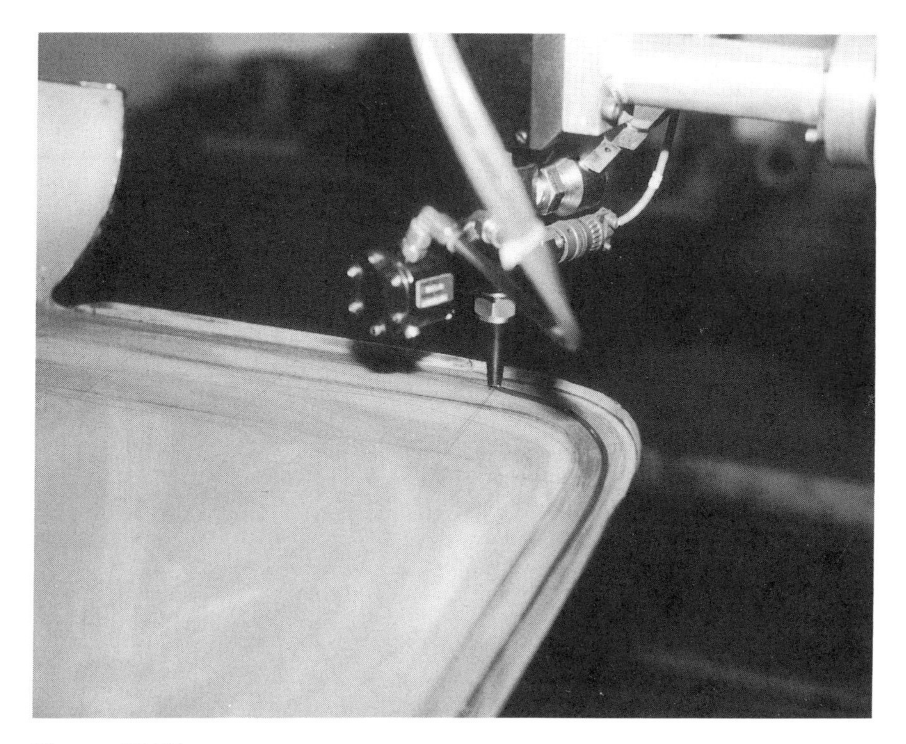

Figure 13.55 Applying adhesive to automobile component.

in Fig. 13.56. In these cases the dispensing head is attached to the robot and carries the adhesive, which can be a single-component or multicomponent type for the specific application. Travel speeds are much higher than those used in metalworking.

13.4 Videos of Applications

It is said that one picture is worth more than 1000 words, and this chapter contains many pictures. It is true also that one video is perhaps worth more than 10,000 words, because individual pictures in the book cannot show motion, and motion is so important in automation. To help overcome this problem, various automation companies were contacted and asked if they would provide videos of their products for review by readers. The following is a list of companies that provide videos and the subjects of each. Contact them directly, indicate your interest, and request the video.

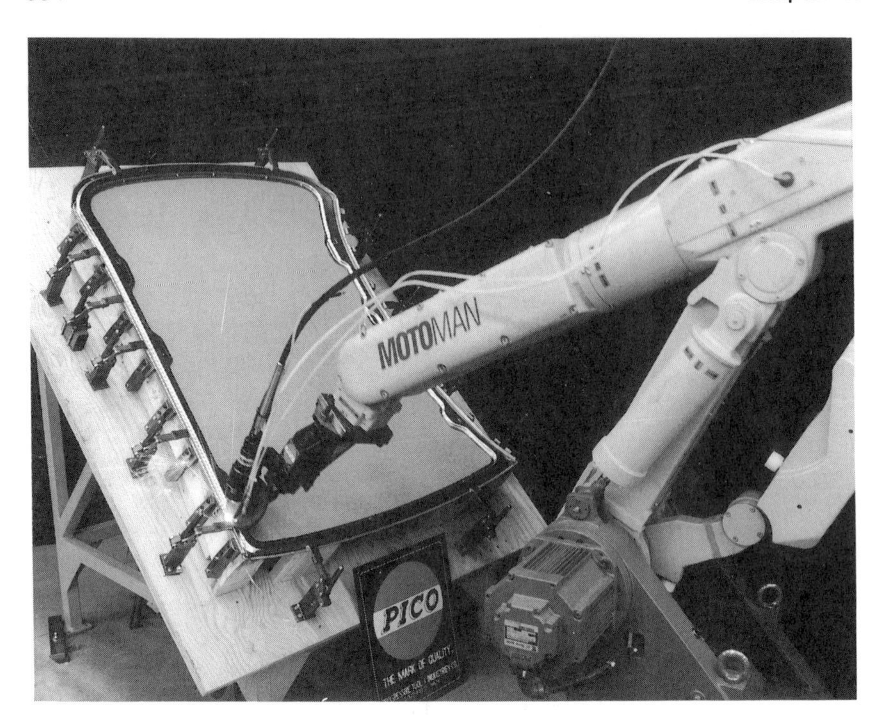

Figure 13.56 Applying adhesive to auto body panel.

Company	Title of video
Alexander Binzel, Corp. 650 Research Drive, Suite 110 Frederick, MD 21701-8619	Robot Uptime
Bortech Corporation BoreWelder P.O. Box 440 Chesire Turnpike Alstead, NH 03602	Demonstration of 306-P
Cybo Robots 2701 Fortune Circle, E. Indianapolis, IN 46241-5519	Cybo Robots
Fanuc Robotics 2000 South Adams Road Auburn Hills, MI 48326-2800	Robots, Systems: Solutions for Welding and Cutting
Genesis Systems Group 481 Tremont Avenue Davenport, IA 52807	Picture Book Family of Robotic Welding Workcells

Company	Title of video
Hobart Brothers Company 600 West Main Street Troy, OH 45373	By specific application
igm Robotic Systems Inc. W133 N5138 Campbell Drive Menomee Falls, WI 53051	Large Weldments—Two-Robot GMAW Oxyfuel Gas Cutting—Mechanized Fixtures
Jenzano Incorporated 820 Oak Street Port Orange, FL 32127	Company Introduction
KOHOL Systems, Inc. 980 Senate Drive Dayton, OH 45459	Safety Weld Systems
The Lincoln Electric Company Attn: Automation Division 22801 St. Clair Avenue Cleveland, OH 44117-1199	Application-specific video tapes
Melton Machine & Control Co. 1600 West Main Street Washington, MO 63090	Automatic Welding Systems
Miller Electric Mfg. Co. Attn: Audio Visual Department 1635 W. Spencer Street Appleton, WI 54912	Explore the Advantage #1
Motoman, Inc. 805 Liberty Avenue West Carrollton, OH 45449	By type of application
Pandjiris Inc. 5151 Northrup Avenue St. Louis, MO 63110	Various titles available
Redman Controls & Electronic Brick Kiln Industrial Estate Malders Lane, Pinkneys Green Maidenhead, Berkshire SL6 6NQ England	Various titles available (P.A.L. format)
Weldex, Inc. 23780 Hoover Warren, MI 48089	Various applications available

In North America, video cassettes are produced in the S.T.S.C. format for use on 60 Hz power. In Europe and most parts of the world, a different format is used. All of the above video cassettes are for the S.T.S.C. format unless noted otherwise.

14

Controls and Sensors for Automated Arc Welding

14.1 Controlling the Welding Arc

The objective of any welding system is to always produce a high-quality weld. A quality weld must have the proper joint fill, size, reinforcement, penetration, heat-affected zone, and surface smoothness and contour and an absence of surface or subsurface defects. A human welder doing manual or semiautomatic welding normally produces a high-quality weld. The transition from manual welding to a fully automated welding system appears to be a simple matter because few people realize how total the involvement of the human is in the production of a quality weld. The human welder is a closed-loop welding control system. The welder's senses—visual, tactile, and audio—develop information and transmit it to the brain, which causes the motor reflex motions necessary to make a quality arc weld.

A human welder learns through training and experience how to start the arc, control the molten weld pool, and obtain a quality weld of the proper size and penetration, for example. This knowledge and acquired skills allow the human welder to produce high-quality welds in spite of difficulties such as poor parts preparation and joint fit-up. Obtaining weld penetration means directing the arc to wet and melt the base metal. Controlling the weld pool means observing and understanding molten metal, moving the arc, and making compensating changes.

The welder maintains control over the molten metal and obtains the desired weld deposit with proper surface shape while avoiding defects. This is done by increasing or decreasing the arc length, increasing or decreasing travel speed, changing the attitude of the electrode, oscillating or weaving. To do this, the human welder must have a good understanding of the arc as well as experience.

The welding arc is a sustained electrical discharge through a highly conductive plasma that produces sufficient thermal energy to join metals by fusion. It is a localized source of heat that is moved to make welds. The welding arc has a point-to-plane geometric configuration, the point being the arcing end of the electrode and the plane the arcing area of the workpiece.

There are two basic types of arcs. The consumable electrode arc, in which the electrode melts in the heat of the arc and the melted metal is carried across the arc to become the deposited weld metal, is used for GMAW, FCAW, and SAW. In the nonconsumable electrode arc, the arc is a source of heat and the weld metal is added by melting a filler metal wire in the arc. This is used for GTAW and PAW. These types of arcs were explained previously.

To weld automatically, a control system must substitute for the human brain and command the equipment to perform each necessary function. The human welder is a closed-loop system and has adaptive feedback. Most mechanized systems do not have sensors and adaptive controls, yet produce quality weldments time after time. However, if serious variations occur in joint location or fit-up, defective welds may occur.

There are four methods of application of arc welding ranging from total human control welding to total machine control welding. These methods are compared in Fig. 14.1. The required functions, listed in the left-hand column, are performed by either a person or machine as indicated for each method of application. Notice that on the left-hand side of the chart the individual performs all functions, whereas on the right-hand side of the chart the machine performs all of the functions to make a weld.

The most important and most difficult function is starting the arc. This becomes habit for the human welder, who performs this function without thinking. Normally the electrode is touched to the workpiece to complete the welding circuit and start the arc. The arc gap is first lengthened until the arc is established and the surface of the base metal is melted. It is then shortened to the correct length, and travel motion begins.

Starting and maintaining the arc automatically is different for the consumable and nonconsumable arc processes. It is also different for consumable electrode processes that use a constant-voltage (CV, flat) or a constant-current

Method of Application / Arc Welding Elements/Function	MA Manual (closed loop)	SA Semiautomatic (closed loop)	ME Mechanized (closed loop)	AU Automatic (open loop)	RO Robotic (open or closed loop)	AD Adaptive Control (closed loop)
Starts and maintains the arc	Person	Machine	Machine	Machine	Machine (with sensor)	Machine (robot)
Feeds the electrode into the arc	Person	Machine	Machine	Machine	Machine	Machine
Controls the heat for proper penetration	Person	Person	Machine	Machine	Machine (with sensor)	Machine (robot) (only with sensor)
Moves the arc along the joint (travels)	Person	Person	Machine	Machine	Machine (with sensor)	Machine (robot)
Guides the arc along the joint	Person	Person	Person	Machine via prearranged path	Machine (with sensor)	Machine (robot) (only with sensor)
Manipulates the torch to direct the arc	Person	Person	Person	Machine	Machine (with sensor)	Machine (robot)
Corrects the arc to overcome deviations	Person	Person	Person	Does not correct, hence potential weld imperfections	Machine (with sensor)	Machine (robot) (only with sensor)

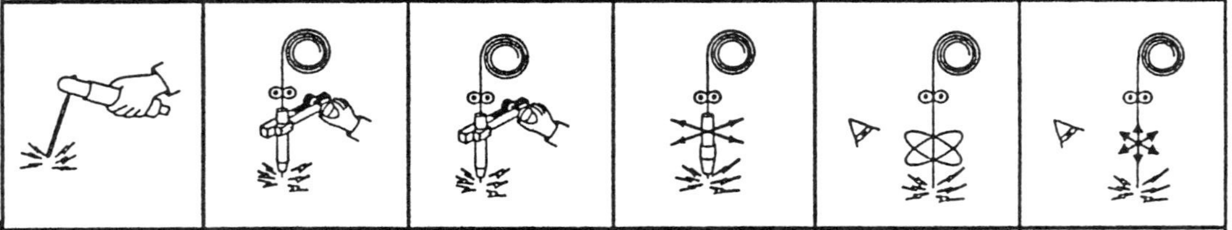

Figure 14.1 The manual versus mechanical methods of arc welding.

(CC, drooping) characteristic power source. With a CV (flat characteristic) welding machine, the welding electrode is fed until it strikes the workpiece. At the instant of contact the circuit is completed, and a large initial surge of current causes the arc to start. This type of start, called a *fuse start,* causes spatter. A better way is to feed the welding electrode into the work at a slower than normal rate; then the arc is formed without much spatter. The best start occurs when the base metal is clean and the electrode is freshly cut or has a point.

The submerged arc welding process normally uses CC (drooping) power and larger diameter electrodes. In the early days, to start the arc a small ball of steel wool was placed between the electrode and the work. The steel wool would immediately disintegrate when power was applied, and the arc would be quickly established. A more refined starting method used an electrode retrack system in which the electrode would feed until it touched the work. At the instant of contact, the wire feeder would reverse, pulling the electrode back to establish the arc, and then reverse again to feed the electrode into the arc.

Arc starting with gas metal arc welding using small-diameter electrode wire with a drooping (constant-current) power source uses a complex starting circuit and fast-reacting wire feed motors. Foolproof arc starts are required for automated arc welding systems.

The nonconsumable arc welding process, especially for gas tungsten arc welding, uses a constant-current or constant-power power source. There are three ways to start the gas tungsten arc. The most common method is to momentarily touch the electrode to the work and quickly withdraw it to start the arc. This is used for noncritical work but should be used only when the electrode and base metal are cold. If the tungsten electrode is hot, small bits of tungsten may transfer to the weld deposit.

A second way of starting the gas tungsten arc is by superimposing high-frequency current on the welding current. This hf current will cause a spark to jump across the arc gap, ionize the area, and allow the welding current to follow and establish the arc. The use of high frequency for arc starting has two serious drawbacks. The high-frequency current will create disturbances in the welding control system, particularly computer controls. Efforts to shield the current from computer controls are not always foolproof. Second, the high-frequency current may jump across insulation in mechanized welding machines such as small orbital tube heads and damage them.

Starting an arc with high-frequency current is not foolproof. If the torch is located a great distance from the power source, the arc may not start each time. Atmospheric humidity, shielding gas type, length and location of cable, and type of tungsten electrode all affect arc starting. A remote high-frequency unit placed near the arc will overcome this problem.

A third way to start the arc is to use a superimposed high-voltage impulse of short duration that creates a spark that will cross the arc gap. This is followed by the welding current, and the arc is started. This method can be used for an automated system provided the computer controller is disconnected momentarily when the arc is started.

The plasma arc welding process is easy to start because it employs a pilot arc within the welding torch. The arc is transferred from within the torch to the work by initiating the welding circuit. This is the plasma transferred arc variation that is the most common.

Once the arc is started, it is maintained by the machine. In consumable electrode welding, the wire feeder feeds the electrode wire into the arc at the same rate it is melted in the arc. This ensures a uniform arc gap, which is measured by the arc voltage across the gap. For nonconsumable electrode welding, the arc gap is fixed. If variations occur, automatic arc length control devices are used. These are known as automatic voltage control units. The power source also is more tolerant and able to maintain a stable arc.

The electrode wire is fed into the arc by a wire feeder in both the consumable and nonconsumable electrode welding systems. The wire feed speed is controlled to provide the desired weld deposit.

The heat of the arc must be controlled for proper penetration. The heat required for melting is related to the current in the arc circuit, which in turn depends on the type and adjustment of the power source and the current control adjustment.

Most weld joints, with the exception of spot welds, require the arc to move along the seam to make a deposit along the entire length of the joint. This requires arc travel, which is provided by a motion device. Travel speed is adjustable and is part of the welding program. In manual welding, the travel speed depends on the welding current and is controlled by the human to produce the size and shape of weld required. The direction of the arc is changed as needed to follow the work. This is done by the human operator or, for mechanized welding, by the machine.

The torch must be manipulated to direct the arc in such a way as to provide the desired penetration, size, and shape of weld. In manual welding, the human welder will change the travel angle and work angle of the electrode with respect to the axis of the weld. An experienced welder performs this function on a routine basis. In the case of robotic arc welding, the change of direction can be programmed into the machine to provide the desired contour.

The final function is to correct the arc to overcome deviations. Deviations are variations in the joint configuration such as changes in the included angle of a groove weld, the root opening, or the position of the joint. Again, the experienced human welder makes these corrections on a routine basis. In the case of mechanized welding, special controllers and sensors are required to accomplish the same action.

14.2 Functions for Mechanized Welding

To manufacture a weldment, certain things must be done. With mechanized welding, a program is required to substitute for the human operator's brain. It includes a specific sequence of events to weld an assembly of parts. Table 14.1 shows the functions involved in a mechanized welding program. This list is not all-inclusive and will vary according to the welding process and the weldment. The total program can be on a time basis or based on position or a specific action. Time is most often used for relatively simple welding programs. The control system can be very simple or very complex depending on the number of functions that must be controlled and the interdependency of these changes. An analysis of these functions follows.

1. *Make preweld decisions.* This must be done by a human welding expert for the weld or weldment to be produced. The decision is based on such factors as the welding process, the material, production quantity, the complexity of the weldment, and the tooling used. It involves establishing the welding procedure. Theoretically it could be done using a welding databank or expert system.
2. *Set up and adjust.* This assumes that the weldment is completely assembled and is in the work-holding fixture. The design of the weldment and the fixture indicates whether or not tack welding is required. This involves certain movements and adjustments. It also involves teaching if this capability is available.
3. *Preweld.* This is the point where welding parameters are selected. This may come from the welding procedure mentioned above. If adaptive controls are available, the machine may analyze the weld joint using sensing devices. If this is not possible, preweld parameters must be adjusted for the weld conditions.
4. *Start the weld.* This is the actual starting and making of the weld. This function is directed by the programmer-controller and is related to the welding process being used. For example, if plasma arc or hot-wire filler metal is being used, additional functions would have to be included, such as manipulating the torch and modifying the weld path trajectory. The weld portion of the total sequence may have different options; for example, more than one arc or multiple passes may be required. In addition, for quality control the weld parameters can be measured and periodically printed out for a permanent record. Parameter limits may be established so that welding will stop if any parameter exceeds its limits.

Table 14.1 Functions Involved in a Total Welding Program

1. Make preweld decisions
   a. Select welding process.
   b. Select welding procedure.
   c. Select welding electrode filler metal and size.
   d. Select shielding gas or flux and flow rate.
2. Set up and adjust
   a. Load workpieces and adjust to start point.
   b. Move torch to weld start point (teach or find weld joint).
   c. Adjust torch for proper angles.
   d. Teach weld path (trajectory).
3. Preweld
   a. Select weld parameters.
   b. Analyze weld and modify weld parameters.
   c. Turn on power to equipment.
   d. Turn on shielding gas and cooling water supply.
4. Start weld (depends on process)
   a. Start shielding gas flow.
   b. Start cooling water flow.
   c. Start arc weld power.
   d. Feed electrode or filler wire.
   e. Control current (heat).
   f. Start travel (relative motion); follow taught path.
   g. Manipulate torch oscillation; modify arc motion.
5. Terminate weld
   a. Crater fill at end of weld.
   b. Postflow shielding gas time required.
   c. Stop weld.
      Stop travel.
      Stop arc weld power.
      Stop shielding gas flow.
      Stop cooling water flow.
   d. Terminate travel.
      Move torch to home position.
      Move work (weldment) to home position.
6. Postweld shutdown
   a. Move equipment to home position.
   b. Remove weldment.
   c. Turn off power to equipment.

5. *Terminate the weld.* The welding operation is completed by means of a subroutine that terminates all factors in proper sequence and shuts down the machine.
6. *Postweld shutdown.* The equipment returns to the home location, the weldment is removed, etc., so that the welding machine is ready for the next piece to be welded.

The control functions fall into various classes. An analysis of these functions, excluding those necessary to assemble the weldment, indicates that some are preweld adjustments but most are motion-related functions. Many of the functional changes are variable and must be changed continually during the welding operation. Some functions relate to electrical welding power and its characteristics and would be assigned to the welding power source and the electrode wire feeder. The automatic welding system must include a control system capable of handling all control functions.

An adaptive control automated welding system will handle all control functions. It will include sensing devices and adaptive circuits in the controller (closed loop) to compensate for changes and produce a high-quality weld.

A welding program can be displayed as a line chart with time as the horizontal axis. A separate line is used to indicate each function. Each has its own baseline, and in some cases if amplitude changes it is shown. Each line indicates when the function starts and ends and when there is any change in its amplitude. Figure 14.2 shows a typical program for automated welding with a nonconsumable electrode welding process. Figure 14.3 shows a typical program for automated welding with a consumable electrode process. These charts do not include parts loading, indexing fixtures, fixture unloading, or similar operations.

To better understand a welding program it is necessary to know the terminology used. Common terms are listed alphabetically below, in general agreement with the American Welding Society definitions.

Arc current See *Welding current.*

Burnback time See *Meltback time.*

Crater fill current Current value during the crater fill time.

Crater fill time Time interval following weld time but prior to meltback time during which arc voltage or current reaches a preset value greater or less than welding current values. Weld travel may or may not stop at this point.

Crater fill voltage Arc voltage value during crater fill time.

Cooling water On/off control of circulation of cooling water.

Cycle start Start of welding cycle.

Cycle stop End of welding cycle.

Downslope time Time during which the welding current is continuously decreased.

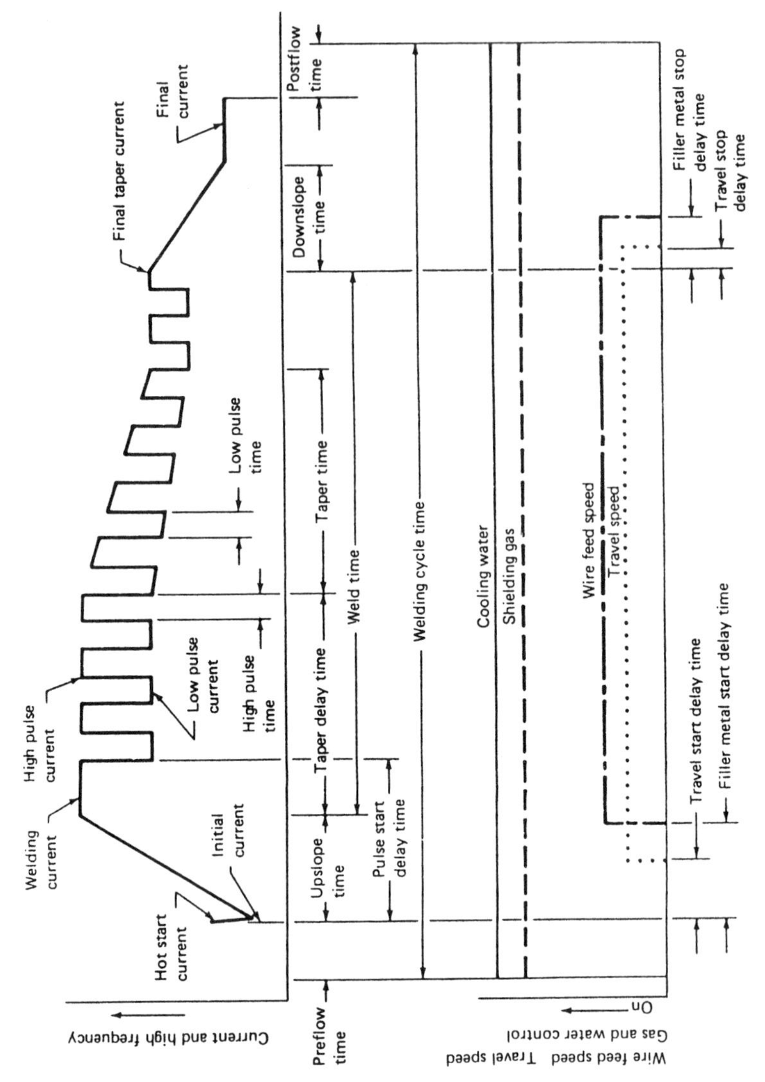

Figure 14.2 Typical program for automated welding with a nonconsumable process.

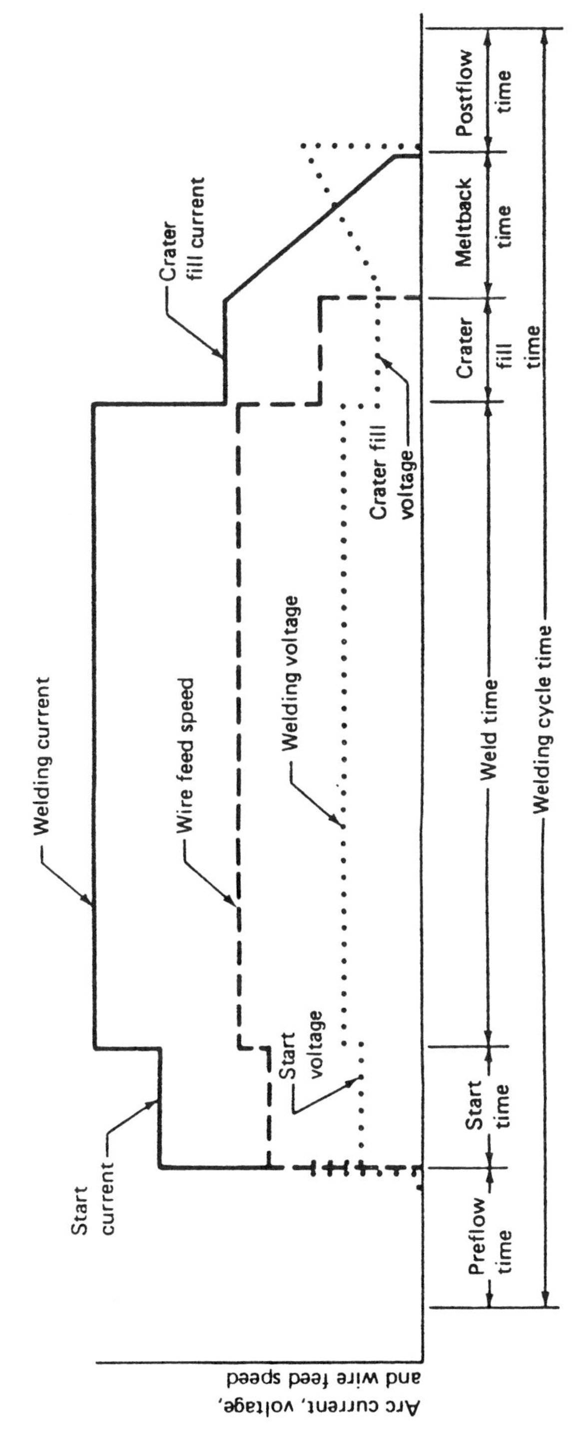

Figure 14.3 Typical program for automated welding with a consumable electrode process.

Filler metal start delay time Time interval from arc initiation to the start of filler metal feeding.

Filler metal stop delay time Time delay interval from beginning of downslope time to when filler metal feeding stops.

Final current Current after downslope but prior to current shutoff.

Final taper current Current at the end of the taper interval prior to downslope.

High pulse current Current level during the high pulse time that produces the high heat level.

High pulse time Duration of high current pulse time.

Hot start current A very brief current pulse at arc initiation to stabilize the arc quickly.

Inch Filler metal—forward/reverse, up/down.

Initial current Current after starting but before establishment of welding current.

Low pulse current Current level during the low current pulse.

Low pulse time Duration of the low current pulse.

Meltback time Time interval from end of crater fill time to arc outage during which electrode feed is stopped. Also called *burnback time*.

Preflow time Time interval between start of shielding gas flow and arc starting.

Postflow time Time interval from current shutoff to either shielding gas or cooling water shutoff.

Pulse start delay time Time interval from current initiation to the beginning of current pulsation.

Purge Preflow of shielding gas.

Ramp down Decrease the value of the factor.

Ramp up Increase the value of the factor.

Retract Electrode wire reverse, i.e., opposite to electrode wire feed.

Shielding gas Protective gas used to prevent atmospheric contamination.

Start current Current value during start time.

Start time Time interval prior to weld time during which arc voltage and current reach a preset value greater or less than welding values.

Start voltage Arc voltage during start time.

Taper delay time Time interval after upslope during which the current is constant.

Taper time Time interval when current changes continuously from the welding current to the final taper current.

Travel speed Velocity of the arc relative to the work.

Travel start delay time Time interval from arc initiation to the start of the travel.

Travel stop delay time Time interval from beginning of downslope time or crater fill time to shutoff travel.

Upslope time Time during which the current changes continuously from initial current value to the welding value.

Welding current Welding current during welding period. Also called *arc current.*

Welding cycle The complete series of events involved in the making of a weld.

Weld program Sequence of events based on time or position (or both) to accomplish the weld.

Weld time Time interval from the end of start time or end of upslope to beginning of crater fill time or beginning of downslope.

Welding voltage Voltage across the welding arc.

Wire feed speed Velocity at which electrode wire is fed into the arc.

Wire stop delay time See *Filler metal stop delay time.*

14.3 The Welding Control System

The controller for an automated arc welding system must control all weld functions. The overall welding program must provide the sequence of events and control each function so that it has the correct value and occurs in the proper sequence. This is necessary to provide the selected welding procedure and produce a quality weld.

The simplest welding program for mechanized welding would start the arc, provide one direction (axis) of motion, and stop the arc. Since it is under the observation of the welding machine operator, this program is considered a "closed-loop" system. In such a simple system, a skilled welding operator is required because the machine and control system have no intelligence to provide a quality weld. If the weld is long, the operator can make changes by adjusting the welding power source, the travel speed, or torch position and angle. A high level of welding skill is necessary to understand the relationship between the welding parameters and the weld and adjust parameters to produce the desired weld.

Early welding programmers used relay ladder diagrams as a basis for programming. The circuits consisted primarily of command switches, relays, timers, motor speed controllers, and limit switches. The need to improve productivity and quality quickly led to more complex controls. As electronic circuits became more capable, they provided a more complete welding control system but still used relay ladder programming. These systems were actuated on a time basis. Timers opened and closed relays that caused a travel motor to start, stop, or change speed. The programmer commanded the power source to start or stop. The welding power source often had its own programmer to control welding parameters. This led to automatic welding systems that were open-loop systems, as they operated without operator observation. They can be categorized according to the level of complexity of the welding procedure. The type of circuitry employed depends on the process, type of power source, and complexity of motion. Simple mechanized systems have controllers that do not include a memory system or a display screen. A controller for a basic single-axis automatic welding system is shown in Fig. 14.4A. The controls for making adjustments (Fig. 14.4B) can be locked.

Complex welding programs require computers with memory, keyboard input, and monitors and accommodate subroutines. For example, a subroutine to provide simple weaving of the torch may be desired to control the speed and width of oscillation and the dwell time at either end of the stroke. Another subroutine may provide automatic arc length control (ALC) or arc voltage control (AVC). This is usually provided by a separate controller that uses the arc voltage as a signal to provide a constant arc length for precise gas tungsten arc welding. These and other subroutines can be included in the total complex controller. Some controllers contain menu-driven computers with a keyboard, a color monitor, and auxiliary drives that allow programs to be uploaded and downloaded. They can be used for different welding processes and provide up to six channels of analog output that are capable of pulsing and sloping. Some units provide for process recording and monitoring and allow limits to be set for warning signals. A controller for a complex automatic welding system is shown in Fig. 14.5.

Complex controllers are also required for robotic or automated arc welding systems. Controllers of this type include a high-speed microprocessor to enable them to provide coordinated, simultaneous, continuous motion of up to eight axes and all welding parameters. As the number of axes increases, the computer capacity must increase.

The machine tool industry introduced numerical controls (NC) years ago. Automatic flame-cutting and plasma machines use numerical controls for controlling the path of the cutting torches. These are known as point-to-point (PTP) control systems. The points are all in one plane. For arc welding robots, the arc is moved from one point to the next in space. A typical robot controller for arc welding is shown in Fig. 14.6. The location of the arc is known as the "tool center point" (TCP). It is the path of the TCP that is programmed and stored in memory. For flame cutting, pick-and-place, and machine loading, PTP playback is used. For an arc welding robot, playback of the arc motion is a continuous path in three dimensions. The robot controller must coordinate arc motion with respect to the axes so that all movement begins and ends at the same time. It is the function of

(A)

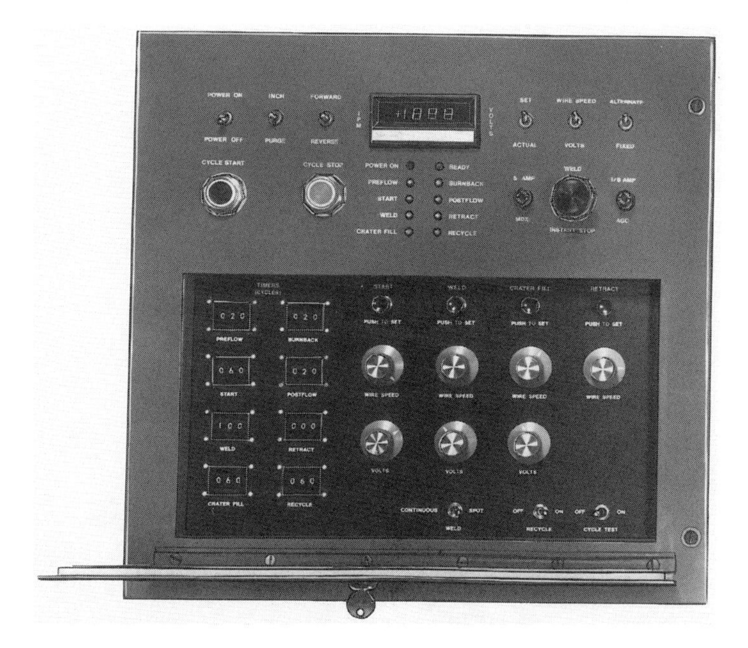

(B)

Figure 14.4 Typical controller for simple automatic welding system.

Figure 14.5 More complex controller.

the programmer to accept the input of many point locations, relate welding parameters to the path taught, store this information in memory, then play it back to execute a welding program.

Welding parameters may change at various points along the path of the torch. The travel angle and work angle of the torch with respect to the axis of the weld may change continuously. The wrist motions of the robot may also change continuously. The motion of the work-holding positioners and the travel speed must be coordinated with the motions of the torch. This requires more axes of motion and a more powerful computer.

Welding system and robot controllers are available for any arc welding process. They control the welding parameters and all axes of motion, employ a high-speed microprocessor, and provide accurate repeatable welding procedures. A welding program can be inputted from a keypad in response to a menu

Figure 14.6 Robot controller.

on the monitor, allowing the operator to preset all welding parameters for each weldment. During welding the monitor displays a readout of actual values.

An interface controller (Fig. 14.7) is required for some welding systems. The interface controller takes the robot controller analog signals and provides the welding power source and the wire feeder with information to supply the correct wire feed speed and arc voltage. It also allows the operator to adjust the current and voltage with respect to the command reference supplied by the robot controller. In addition, the interface controller may contain subroutines such as burnback control, electrode wire forward/reverse capabilities, purge switch for shielding gas, and an arc fault timer. The interface controller allows the robot to make simulated welds and display the welding parameters without striking an arc.

A remote video system can be used for some welding operations. It consists of a color video camera, a color monitor, and a camera control station. The camera is attached to the torch assembly so that it focuses on the arc. The welding operation control panel is placed at the operator station with the monitor so the operator can observe the welding operation and make adjustments as required. This system is often used for welding in radioactive areas.

Figure 14.7 Interface control.

14.4 Sensors and Adaptive Control

The ultimate automated welding system will simulate the human manual welder and provide a closed-loop system that compensates for all variations to produce a high-quality weld. This is true adaptive control welding. The components of a welding system of this type are diagrammed in Fig. 14.8. Adaptive control welding can be applied to robotic welding systems or to complex automatic welding systems, which are open-loop systems until the adaptive control is added.

True adaptive control for automatic or robotic welding closes the loop because sensing devices can replace the human operator for almost every function. (See Fig. 14.1.) In practice, however, sensors are provided only for the functions that require surveillance and control. The number of sensors indicates the level of completeness of a "closed-loop" system.

Two components must be added to a system to provide adaptive control:

1. A controller with a high-powered processor that will accept signals from sensors and make the necessary changes in the welding parameters in real time
2. Sensing devices that provide real-time information to cause parameter changes

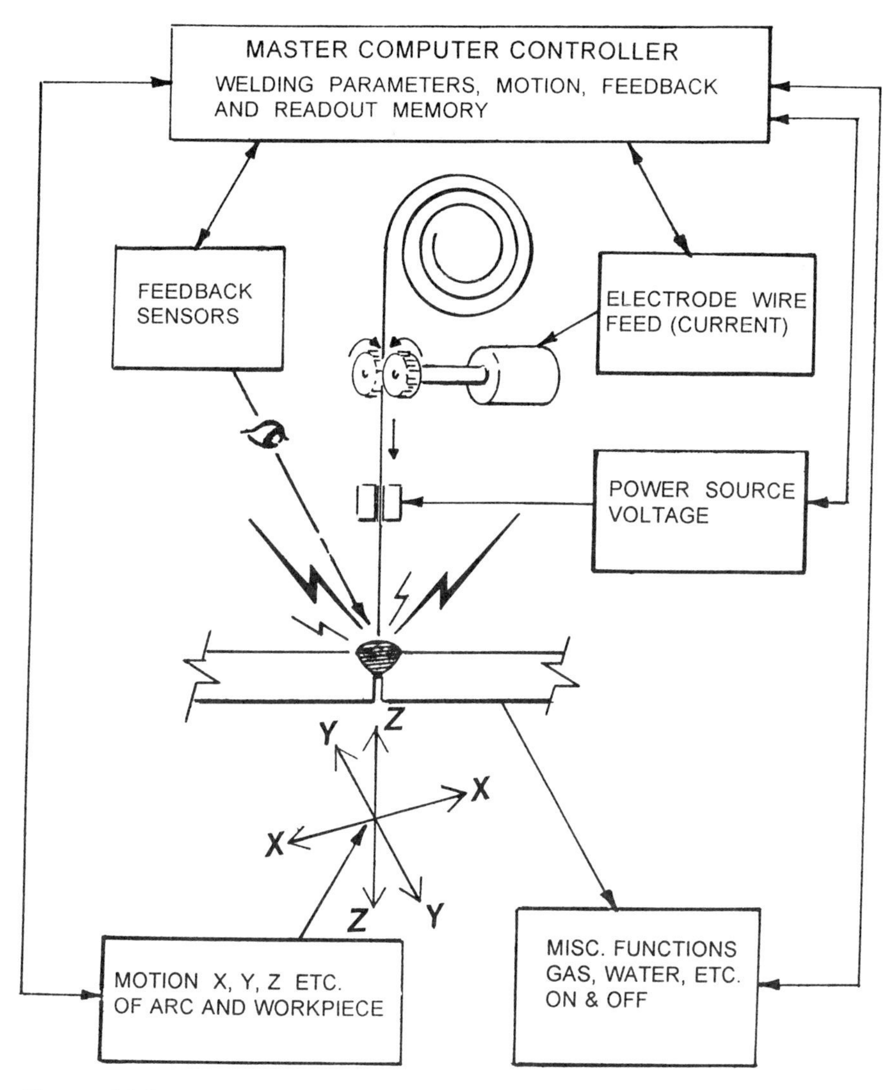

Figure 14.8 Adaptive control for an arc welding system.

The controller for a welding robot or for a multiaxis automated welding machine was described in the previous section. An advanced controller coordinates additional movements and has the ability to accept feedback sensor signals.

An adaptive controller for a robotic arc welding system must control all functions and accommodate feedback signals. The controller's central processing unit (CPU) controls and monitors arc motion and workpiece motion, torch location and angular position, welding parameters, parts location, shielding gas supply, etc. The advanced controller eliminated the need for an interface controller and provides any necessary analog-to-digital conversions for inputs and outputs. The processor must have very high speed with the ability to process vast amounts of data. The unit should have at least 4 MB of RAM and a 20 MB hard disk for program storage and weld history. A floppy drive or a tape drive should be included for making copies of welding procedures or taking history data. The robot controller should communicate with other factory processors and fit the CAD equipment used. It should provide a warning and alarm system adjustable for variations in parameters and shut down the equipment if values exceed the set tolerances.

Configuration editors should be included for different procedures and applications. It should have a color display monitor with a touch-sensitive screen so the operator can correct operations easily and accurately. The controller should have a logical menu-driven program with color symbols and sounds and written in plain English to lead the operator through the weld procedure. Only selections that make sense at a given time should be presented to the operator. If an error occurs, the controller should suggest solutions. The controller should contain hundreds of welding schedules in its library that can be recalled and applied as required. The controller should allow procedures to be checked by the system for logical consistency before the actual weld is made.

Password routines should be used to allow a procedure to be viewed by many but changed only by personnel with the authority to do so. A lightweight portable pendant with a single connector to the main processor console should be provided. The pendant should contain an emergency stop button, individual trim knobs for each weld parameter, jog buttons for each motion axis, wire feed, and subroutines. It should have a display screen that displays messages and actual weld parameters. The programmer controller should be able to gather data on all activities and provide printouts of procedures, parameter values, etc. Printouts should include the date, time, operator identification, weld procedure, and part identification. Programmers should be sufficiently flexible that the operator working at the main menu can touch the screen and review the welding procedure or select a new one from the library. It should be possible to make a dry run to determine the procedure without the arc on. The operator's pendant, which displays parameter values, can be used to modify and control the weld. The master robot controller should be able to communicate with other computers in the factory. Some typical modern controllers are shown in Fig. 14.9.

The other components necessary for adaptive control are sensors. A sensor is a device that determines or measures a function in real time during the welding operation. Sensors are used to determine actual conditions so the welding procedure can be modified if necessary. They provide signals that are used to modify the motions of the arc as well as for changing welding

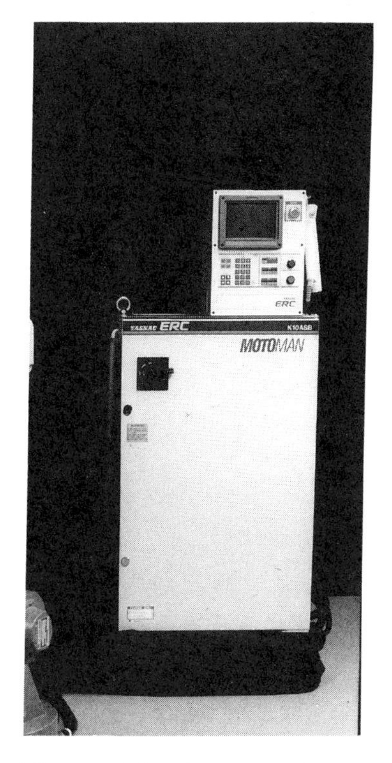

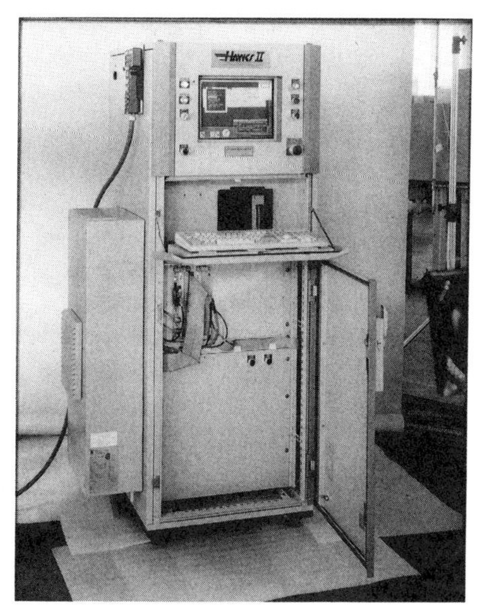

Figure 14.9 Advanced robot controllers.

parameters. Sensor feedback on parameter variations causes the adaptive controls to change the parameters and travel path to produce a quality weld in spite of problems that may be encountered. Sensors close the loop and make truly automatic welding possible.

Sensing devices are available from a variety of sources. Special software or a special computer may be required to match a sensor to the robot controller. New and improved sensors are continually being developed, and their use becoming more widespread.

Sensors are difficult to categorize because they can be classified according to the physical principles employed or by the purposes for which they are used. The principles employed may involve actual contact or optical or laser viewing of the arc or weld pool. They may depend on magnetism, capacitance, or impedance or may relate to arc characteristics. The purpose of the sensor may be seam finding, seam tracking, welding parameter control, arc length control, real-time arc monitoring, etc. One classification system is outlined in Table 14.2 and is briefly described below.

Table 14.2 Classification of Sensor Systems

I. Contact or tactile (for seam tracking)
   A. Mechanical—roller spring loaded with floating torch
   B. Electromechanical—probe with torch on motorized cross sides
   C. Intermittent contact probe—electromechanical
   D. Electrical—electrode extension probe with complex control
II. Noncontact (for various activities)
   A. Physical characteristics relationship
      1. Acoustical—for arc length control
      2. Capacitance—for proximity control
      3. Eddy current—for seam tracking
      4. Induction—for seam tracking
      5. Infrared radiation—for penetration control
      6. Magnetic—electromagnetic
      7. Ultrasonic—for penetration control and quality control
   B. Through-the-arc (electrical contact)
      1. Arc length control (arc voltage control)
      2. Oscillation with electrical measurements—GMAW
      3. Oscillation with electrical measurements—GTAW
   C. Optical-visual (image pickup and processes)
      1. Reflected light with photodiode detection
      2. Viewing the welding arc
      3. Viewing the molten weld pool
      4. Viewing the joint ahead of the arc
      5. Laser shadow technique
      6. Laser range-finding technique (rastering)
      7. Optoelectronic
      8. Laser or standard light
      9. Other systems

The two major categories here are contact (tactile) and noncontact. Tactile sensors have been used for joint or seam tracking for many years. They range from simple mechanical systems to complex electrical-mechanical contacting sensors. The simplest seam tracker is a spring-loaded roller with a floating welding torch. The roller fits against a reference surface and causes the head to maintain a specific dimensional relationship with the joint. The head will follow the motions generated by the roller.

The electromechanical system is more versatile. In this system, a wheel or a stylus probe will contact the surface, which can be the plate surface, the edge of a groove, the edge of a tee joint, or similar surface, and provide a signal that operates a motorized cross slide to adjust the torch for making the weld. A second axis can be provided to maintain accurate torch-to-work dimensions. The probe and torch are mounted on the carriage (Fig. 14.10), and the system is used for long straight seams. Probes wear and must be replaced. They are connected to switches that provide the correction signal. The probe must be sufficiently distanced from the arc to prevent spatter buildup. Tack welds and the start and end of welds pose a problem. This type of equipment is not suited for robotic arc welding.

An intermittent contact seam follower, shown in Fig. 14.11, can be used for robotic arc welding. In this case the sensor is attached to the torch. A pneumatic vibrator causes the probe to intermittently touch the work four to five times each second. This eliminates the drag problem associated with the normal probe. This system can be used for fillets in tee and lap joints with a minimum material thickness of $\frac{1}{16}$ in. (1.6 mm). The probe feeds signals to the computer to control the motion of the torch. It is also used to search for the beginning of the joint. The search function and the changed path are stored in the robot controller memory. The seam follower will maintain a constant

Figure 14.10 Continuous-contact sensor.

Figure 14.11 Intermittent contact sensor.

torch-to-work distance as well as following the joint. This device will take care of gradual changes of direction but cannot accommodate a sharp 90° turn. It has not become popular.

The distance from the arc to the sensing location can pose a problem for a mechanical probe or wheel. If the distance is too great, deviations can occur; if it is too short, the arc will interfere with the probe and cause rapid wear and deterioration. Another problem is the inability of these systems to accommodate abrupt changes of direction at welding speeds.

A different type of touch system is employed in conjunction with through-the-arc tracking systems. This system uses the electrode wire, which protrudes beyond the current pickup tip, as the contact. The robot is programmed to move the electrode wire and touch the work surface at different points along the joint to determine the location of the start and end of the joint. It can also be programmed to measure the weld geometry and establish the size of the weld groove. It utilizes a complex motion system. This computer-driven system may employ an expert system with a memory. It is capable of sensing the joint path in three dimensions and storing it in memory. The calculation of the weld joint detail in connection with the expert databank will establish a new welding procedure and modify the welding parameters.

Noncontact sensor systems have become the most popular. There are three basic types: (1) sensor systems that rely on physical characteristics of materials or energy output relationships, (2) through-the-arc systems that use electrical signals generated in the arc, and (3) optical/visual systems that attempt to duplicate the human eye.

Acoustics can be used to control the length of a gas tungsten arc. The sound energy is linearly proportional to arc voltage. An acoustical waveguide close to the arc leads to a microphone. The signal is amplified, filtered, and rectified and is used to control the torch movement and thus to control the arc length. It is used for pulsed current gas tungsten arc welding but is not widely used.

Capacitance is the property utilized by some proximity switches. The capacitance limit switch has been used in automatic equipment for years. It can be adjusted for different distances. It is also used to detect the presence or absence of material.

Eddy currents are currents set up in the base metal by an adjacent alternating current field that is generated by a coil located close to the base metal. Another coil acts as the pickup and detects the eddy current. Electronic circuitry produces a voltage dependent on the distance from the base metal. The output changes when a joint interrupts the metal surface. The sensor is oscillated across the joint to produce control signals, which are processed to give the position of the joint centerline. Different types of sensors are required for ferrous and nonferrous metals. Thickness is not a major factor. This system is popular for noncontact seam-tracking systems.

Inductance or induced current in the base metal can be detected and measured and used for seam tracking. In this case the sensor contains two coils, which scan the seam and produce signals that provide information on the location of the joint. This is similar to the eddy current system. The sensor must be at a given distance above the base metal and placed ahead of the arc because of its sensitivity to heat, spatter, and so on.

Infrared radiation can be picked up by sensors that are used for penetration control. The infrared sensor is focused on the underside of the weld pool to detect the color of the metal under the weld. This system's accuracy is subject to surface conditions and exact target location. It is not considered extremely reliable as a penetration control system and has limited applications.

Through-the-arc seam tracking is a noncontact system with many advantages. It does not need accessory items attached to the torch. It is a real-time system that can be used for most types of welds. Monitoring occurs while the weld is being made. There are several types of through-the-arc systems, and they are used both when metal crosses the arc and when metal does not cross the arc.

The earliest through-the-arc sensing system was the arc length control system for gas tungsten arc welding. Such a system is called an *arc voltage control* (AVC) system; however, *arc length control* (ALC) is a more appropriate name. Some AVC systems have a starting mechanism such that when the cold tungsten electrode in the torch touches the work it initiates the arc and immediately withdraws to the preset voltage. Arc length control systems are very reliable and are widely used.

The major use of through-the-arc systems is for seam tracking, for which the welding torch is oscillated and the arc voltage and/or welding current are monitored before and after each oscillation. Mechanical oscillation is normally used; magnetic oscillation can be used for gas tungsten arc welding but not for gas metal arc welding. Through-the-arc systems can be used for fillet or groove welds. Figure 14.12 illustrates the principle of operation. Control circuits measure the voltage and/or current and reference the left-hand and right-hand values to equalize them. The control circuit moves the torch to the center point between the two equal points. This automatically adjusts the path. This system also has a corner recognition mode that allows tracking around a 90° change of direction and is capable of sensing the joint path in three dimensions. It can be used with all modes of metal transfer. Welding speeds of up to 40 in./min (1025 mm/min) can be attained. Oscillation can vary from 1/8 in. (3.2 mm) to 1 in. (25 mm), and the frequency from 1 to 4 1/2 Hz. The controls can be integrated into the controller. The final pass of a groove weld is attained by using the previous passes to establish the torch path in memory. This system can be coupled with the electrode contact system mentioned previously, where the electrode wire is used to find and measure the weld joint. If the root opening or gap in the groove joint is excessive, the machine can be programmed to select a different procedure from the memory bank and make alternate layers rather than a single pass.

Optical-visual sensor systems are based on an analysis of the manual

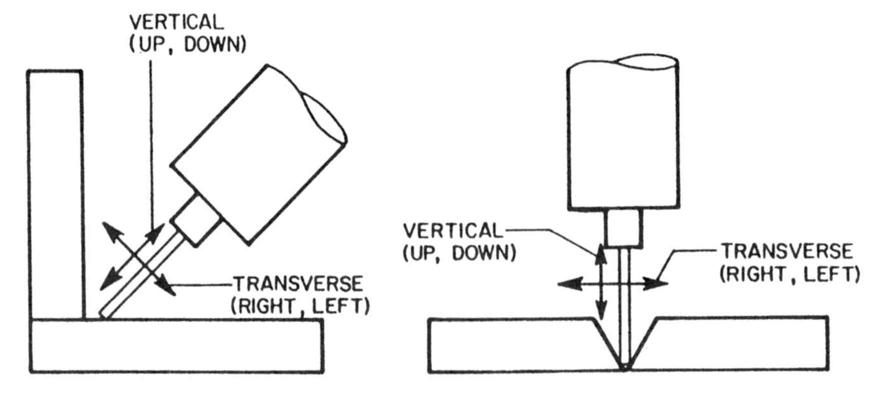

Figure 14.12 Through-the-arc sensing system.

welding operation, which reveals that the welder derives the bulk of the information required to make a high-quality weld through visual input. Optical-visual systems have the most potential for providing real-time signals for fully automated arc welding. Optical-visual systems find the seam, follow it, and identify and define the joint detail so that parameters can be adjusted to produce a high-quality weld. Optical-visual systems are extremely fast and do not become fatigued. However, they are extremely complex. A system flow chart is shown in Fig. 14.13.

Optical-visual systems have overcome the problem of viewing different colors and surfaces—bright, rusty, smooth, rough, and so on—that tend to confuse the sensor. They can pick up a very small joint in thin material, even when the joint separation is minimal and scratches on the surface are more distinctive than the joint.

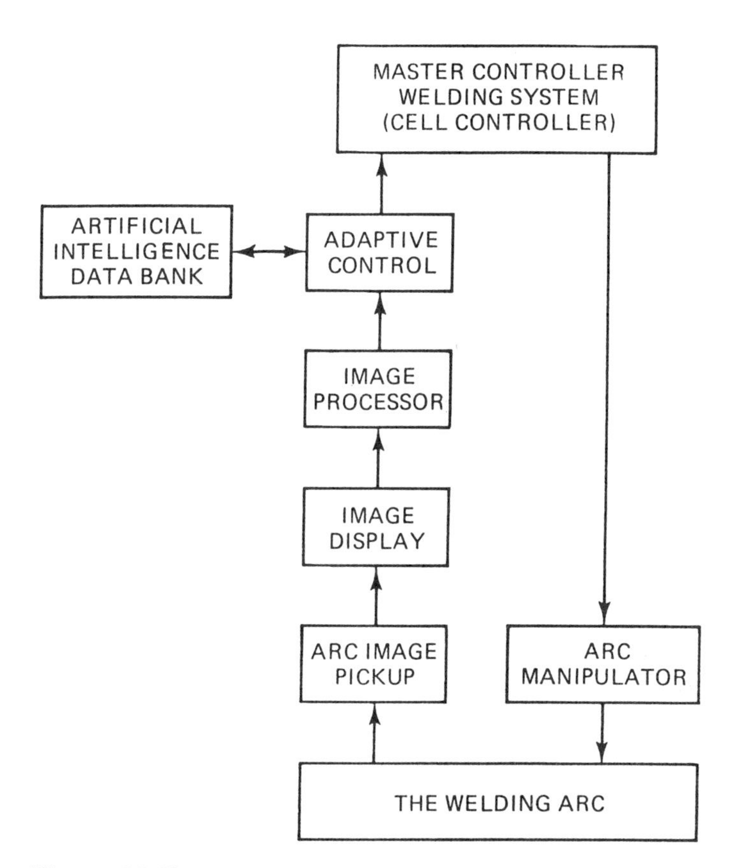

Figure 14.13 Components of an optical-visual system.

Many optical-visual systems are operating successfully, but there is no one system that can be applied universally to robotic welding applications. Different systems are designed for particular applications.

The image to be viewed can be the weld joint ahead of the arc, the arc itself, the weld pool under and behind the arc, or the light generated by the arc. The selected image depends on the viewing area and on how it is lighted. The image can be picked up by a TV camera as shown in Fig. 14.14 or by photodiodes arranged in a matrix array. The pickup method affects the image display and processing system. One or two images are required. In some cases, images are triangulated to determine exact location. Fiber optics can be used to transmit the image to a remote camera. The angle of viewing depends on the image processing method. Images can also be picked up by a system operating through the torch.

The picked-up image must be enhanced for better visibility. One method

Figure 14.14 TV camera to pick up image.

uses structured light, usually a pattern of bright light and no light that can be directed from an oblique angle as illustrated in Fig. 14.15. The light source can be an incandescent bulb or a laser. Laser light is more useful because it is monochromatic and can be highly focused. In addition, the incident arc light can be filtered out, which simplifies processing. If structured light from a point source is used, it is sometimes supplemented by a beam from another direction to facilitate triangulation to be used for precise positioning. The image from the pickup device must be processed to provide a display. Digitizing the image has certain advantages and disadvantages.

The most common image display device is the cathode ray tube, as shown in Fig. 14.16. Image analysis requires the use of high-speed microprocessors. It also requires an extremely complex program to analyze all the data received and put them into a useful form so they can be used to make real-time changes based on variations in the joint or weld.

Adaptive control systems require an interface between the sensor and the robot controller. It normally uses a database and an expert system to provide weld parameters when conditions change. The complete system will provide the necessary input and close the loop to produce the perfect weld.

Each optical-visual system has advantages and disadvantages. Each system is useful for certain applications. The following is a brief list of the uses of optical-visual systems.

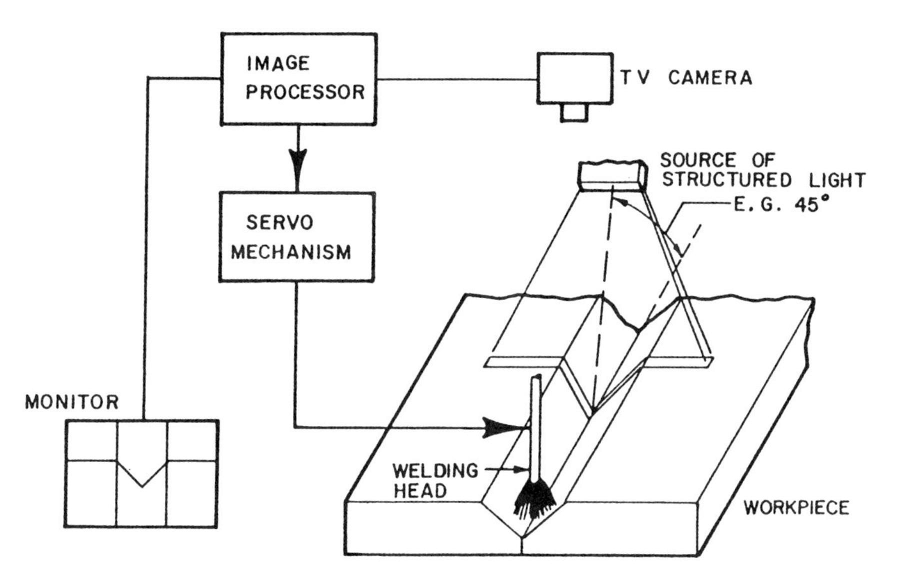

Figure 14.15 Structured light for seam tracking.

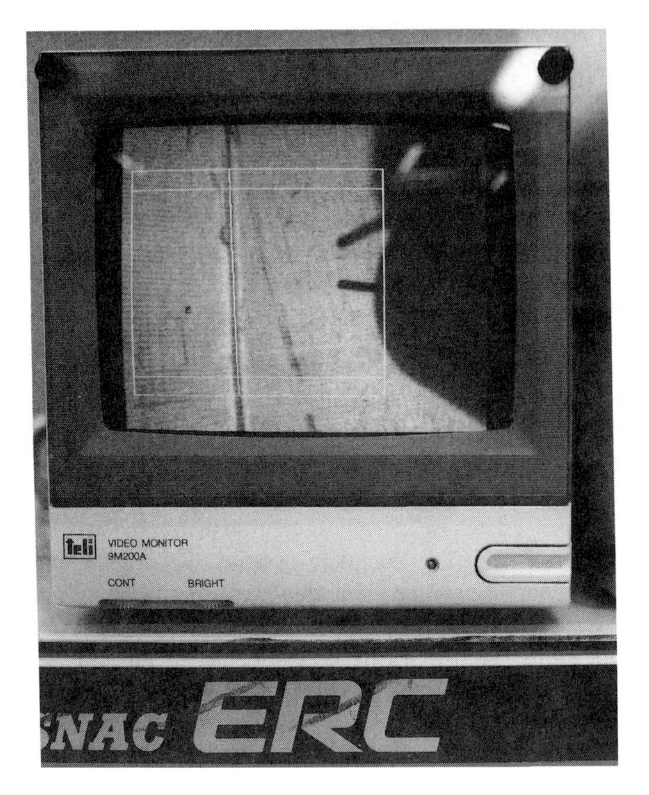

Figure 14.16 Image displayed on cathode ray tube.

Reflecting light with the photodiode pickup
Viewing the welding arc with a TV camera
Viewing the molten weld pool either through the torch or adjacent to the torch
Viewing the joint ahead of the arc
Laser range-finding (rastering)

Optical-visual sensing systems are being continually improved and are being more widely used for automated arc welding applications.

The economics of the use of sensors must be considered. Only sensors needed to detect a repeatable problem should be used. For example, if pieceparts are always made accurately and holding devices are accurate, there is no need for a seam follower. If the location of the equipment, tooling, and pieceparts are always accurate, there is no need for a seam-finding sensor. The more sensors involved, the more expensive and troublesome the system becomes. Sensors must be small, robust, and durable. They must be able to withstand the hostile

Figure 14.17 Torch-to-work distance for laser cutting.

environment near the arc. They must be easy to connect to the controller. They must immediately and routinely send a correction signal to the controller for immediate weld parameter correction.

The more commonly used systems are the electrode touch system for finding the joint combined with a through-the-arc system for seam tracking. An optical-visual system is used for seam tracking when welding with GMAW. A system for maintaining the torch-to-work (standoff) distance with feedback from a capacitance sensor is used for laser cutting (see Fig. 14.17).

Most robot systems will accommodate various types of sensors. Select the system most suitable for your work.

15

Design and Tooling for Automated Welding

15.1 Weldment Design

The design of a weldment, also known as a welded fabricated structure, is a very, very complicated process. It is basic that the weldment must perform the required functions. Many factors must be considered so that it meets its service performance requirements. It must use the least amount of material and the minimum amount of labor to meet a cost target yet have a pleasing appearance.

The success of a weldment lies with the design engineer. The designer must have a clear understanding of its expected service and be completely aware of the properties of the materials involved, how the parts are processed, and what welding processes and procedures will be used. He must also be aware of any codes or specifications that apply. The design is extremely difficult because weldments range from miniature parts in precision equipment to massive machines. It is the designer's responsibility to select and choose the correct weld joint configurations to carry the loads that will be encountered and to sustain the service life of the weldment. The following principles should be followed.

The total cost of the resulting weldment is of major importance. The initial cost is less important than the cost over the product's service life.

Minimum weight is desirable. This means that sections and welds may be stressed to their design limits based on appropriate safety factors.

Normal loading factors and possible overloads must all be considered. For some products, only static loading is involved. Often, however, dynamic loading must be considered. This relates to the fatigue life of the structure based on reversal of loadings and also on varying loads, frequency, impact loadings, and other factors.

Structures exposed to low temperatures must be designed to sustain dynamic loads. Cold temperatures magnify the problems of stress concentrations. Fracture toughness is extremely important and is related to impact loading and low-temperature service.

Exposure to high temperatures also poses design problems, especially if the temperature changes continuously. High temperature in conjunction with corrosion creates a difficult design problem.

The weldment environment affects design. A corrosive environment limits the choice of metals to be used, which must be selected on the basis of the environmental service.

Resistance to abrasion and corrosion combined requires attention. These and any other service-related factors must be considered in the overall design.

In order to meet service requirements, a composite weldment may be required, with different metals placed strategically to withstand different service requirements such as corrosion resistance and abrasion resistance.

The designer must have a clear understanding of the available fabrication equipment, procedures, and techniques. This includes the cutting and forming equipment used to produce pieceparts and the use of premachined parts versus large machining capabilities of the final weldment. The designer must understand the weldability of the base metal and the welding procedures. The designer must provide inspectors with guidance in selecting the type of inspection required and indicate special needs, the type of NDT specified, and tolerances for dimensions.

Very sophisticated techniques are available to the designer to help design for and predict the service life of the weldment.

Finally, the designer must transmit to the shop the exact specifications for the weldment employing recognized welding symbols. Designing for automated welding requires extra effort to be sure every weld joint is accessible to the welding torch.

It is beyond the scope of this book to cover how stresses are related to loading via finite stress analysis, sophisticated computer analysis or computer-aided design (CAD), and other techniques including fatigue factors, hot and cold temperature exposure, discontinuities, stress risers, etc.

The designer must remember that a weldment is a monolithic structure—

any load anywhere on the weldment creates stresses everywhere in the weldment, whether intended or not. Unintended or unexpected stresses that are not properly accommodated for by the design can cause unexpected problems. Consideration must be given to notches, discontinuities, and stress risers. Prototype designs are usually tested to destruction to determine problem areas so the design can be refined to provide the optimum structure capable of withstanding the normal anticipated loading.

In the following we will consider only those factors related to automatic or automated welding that influences the design of the fabricated structure. To help simplify the design process we will make some general assumptions that will apply in most cases. The material being used will be low-carbon mild steel or low-alloy high-strength steel. In these widely used materials the yield strength of the deposited weld metal exceeds the yield strength of the base metal. We will consider thin and medium thicknesses of steel, since these are the most usual thicknesses of steel and include flat hot-rolled sheet, plate and bars, hot-rolled structural members and pipe, low-carbon simple castings, and forgings. We will consider only gas metal arc welding, flux-cored arc welding, and gas tungsten arc welding, since these are the processes normally used for automated welding. The welding will be done using automated or robotic equipment without sensors or adaptive control.

The most important consideration in designing the weldment is to provide accessibility to every weld for the type of welding torch to be used. If the weld joint is not accessible to the torch, the weld cannot be made. Figure 15.1 illustrates welds that cannot be made because of torch inaccessibility. If a weld is inaccessible, redesign is required.

The weldment must be made with as few parts as possible. Large numbers of gussets and braces are normally not necessary. Slightly increasing the thickness of a part will eliminate the need for them. This also means that the weldment should be made with the fewest possible welds. Weld metal is by far the most expensive metal in any weldment. A simple bend, the use of a structural shape such as an angle, will reduce the amount of welding. Special attention must be given to the welding procedure if different types of metal are to be joined in a composite weldment.

The minimum number of thicknesses of metal should be utilized. At the same time the welds should all be the same size if at all possible or of the fewest different sizes possible. If bends are employed in the weldment, the minimum radius or the standard radius for the thickness involved should be employed. Bends that require more than one press operation should not be used. Parts made by rolling to form should be avoided unless accurate roll diameters or radii can be achieved in the parts preparation department.

Piecepart dimensional accuracy greatly affects the welding operation. Man-

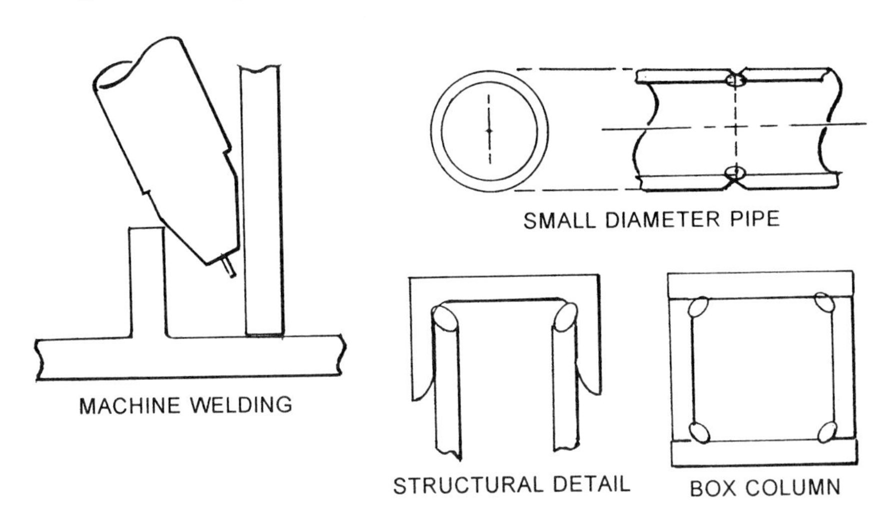

Figure 15.1 Welds that cannot be made.

ual flame cutting should not be used. When oxyfuel gas cutting is used, automatic tracing systems must be used for cutting. Dimensional tolerances should be established such as ±1/8 in. on thinner material. Plasma cutting is more accurate than oxyfuel gas cutting. Tolerance can be held to ±1/16 in. Laser cutting of thinner sections is even more precise and should be employed if possible. Sawing and machining can be used, but they are usually more expensive and their use must be justified.

The machining of pieceparts that go into the weldment versus machining the finished weldment should be thoroughly considered. It is much less expensive to premachine small parts such as bosses and journals than to machine the finished weldment. This is because machining of small parts is done much faster on less expensive machinery. If premachined parts are used, then special fixturing and welding procedures must be used to reduce or control warpage so that it does not interfere with the operation or with fitting the weldment to other parts of the structure. Many times, simple premachined low-carbon castings can be integrated into the weldment.

Where machining is required, allowances should be provided so that sufficient metal is available to clean up surfaces that are to be machined.

Designers must anticipate the inherent tolerances of pieceparts and possible increases in dimensions and potential problems. They should attempt to compensate with adjustments. The cost of preparing parts for the weldment should be given special attention. This is usually related to the quantity of weldments to be produced. Certain metal thicknesses can be sheared and formed in a press that produces more accuracy than thermal cutting. However, the cost of tooling to accomplish this must be considered. Forming tools and presses encounter the problem of material springback, and the piecepart tolerance may be exceeded.

The use of thin sheet metal may require the use of stiffeners. This can sometimes be overcome by a bend in the part. Stiffeners can be eliminated, thus reducing costs, by using a heavier gauge sheet metal. An analysis is needed to compare the cost of the thicker material versus the cost of adding braces and reinforcements.

The use of intermittent or skip welds on fillet welds should be carefully studied. Continuous welds can create warpage problems. On the other hand, a weld twice as large but half as long provides the same strength but requires twice as much filler metal and twice as much welding time. This is illustrated by Fig. 15.2.

Sharp changes in the direction of a seam should be avoided, particularly on thinner material where high-speed robot travel is used. If possible, increase the radius of corners or eliminate the need for continuous welds. The dexterity of the

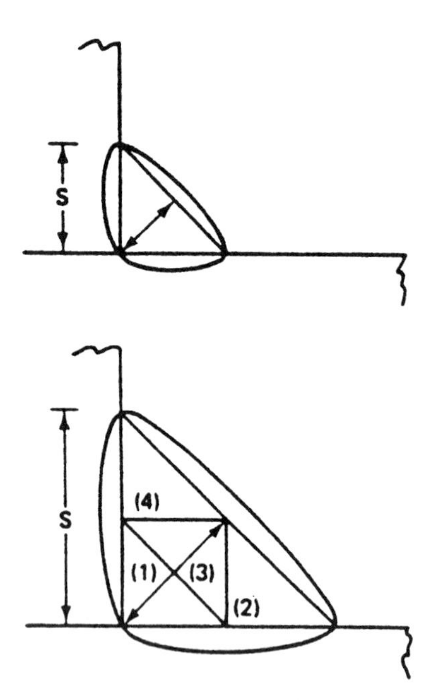

Figure 15.2 Intermittent fillet weld size versus continuous fillet.

robot in manipulating the welding torch has an influence on accessibility. It is wise for the designer to have a torch for reference but also to have exact knowledge of the wrist action of the welding robot.

A weldment is an assembly of parts. Automated assembly should be anticipated. Parts should be designed so they can be inserted with a straight-line up-and-down motion. Robots with appropriate grippers can be used to assemble weldment parts in the proper type of holding fixture. Tack welds should be employed only as a last resort. Additionally, if specific welds must be made prior to adding another piecepart, consideration should be given to building the weldment in subassemblies and using a final welding operation to combine the subassemblies.

Fixturing for weldments should be coengineered with the weldment design. The weldment design should allow building the weldment on a base with additional parts added to the top side. Self-jigging should be incorporated if at all possible. This depends on the gauge of the material. The addition of parts after a welding operation should be held to an absolute minimum.

The designer must take into consideration and anticipate the shrinkage and warpage inherent to weldments. When a weld cools, it shrinks. This will cause warpage of the weldment or lock up stresses in the weld if the fixturing is sufficiently heavy. One advantage of robotic welding is that distortion will normally be more uniform because the robot makes the weld in the same pattern every time. The designer should anticipate the sequence in which the welds will be made and attempt to balance welding to minimize warpage.

The position in which the weld is made has a bearing on the cost of making the weld. Flat-position welding is preferred because higher currents and faster travel speed can be used, which reduces weld time. Position, however, must be coupled with the use of fixturing and the use of positioners for mounting the fixturing. In sheet metal designs, position is not as important as with medium and heavy thicknesses. Welding position is better explained by referring to Fig. 15.3. See Ref. 1 for the official AWS definitions.

Precise fit-up of weld joints is necessary to employ robotic welding. Welding procedure, including electrode size, shielding gas type, metal transfer type, and welding current and voltage, is a factor in successful automated welding. It is wise for the designer to work closely with welding department personnel as well as with the pieceparts preparation departments. These personnel should review new weldment drawings during the design stage and be asked for suggestions. They should be involved in the production of the prototype. Producing the prototype will bring out many of the problems previously discussed and provide redesign suggestions for accommodating specific welds.

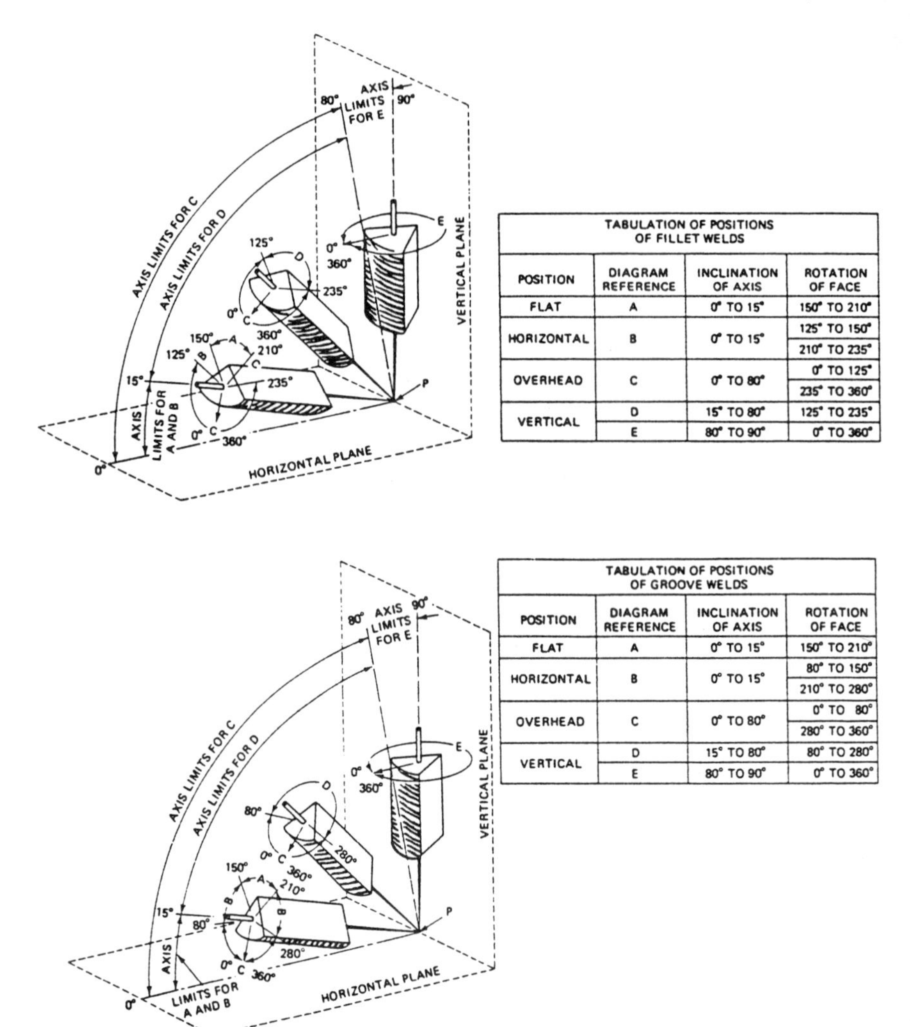

TABULATION OF POSITIONS OF FILLET WELDS			
POSITION	DIAGRAM REFERENCE	INCLINATION OF AXIS	ROTATION OF FACE
FLAT	A	0° TO 15°	150° TO 210°
HORIZONTAL	B	0° TO 15°	125° TO 150°
			210° TO 235°
OVERHEAD	C	0° TO 80°	0° TO 125°
			235° TO 360°
VERTICAL	D	15° TO 80°	125° TO 235°
	E	80° TO 90°	0° TO 360°

TABULATION OF POSITIONS OF GROOVE WELDS			
POSITION	DIAGRAM REFERENCE	INCLINATION OF AXIS	ROTATION OF FACE
FLAT	A	0° TO 15°	150° TO 210°
HORIZONTAL	B	0° TO 15°	80° TO 150°
			210° TO 280°
OVERHEAD	C	0° TO 80°	0° TO 80°
			280° TO 360°
VERTICAL	D	15° TO 80°	80° TO 280°
	E	80° TO 90°	0° TO 360°

Figure 15.3 Weld positions—fillets and grooves.

15.2 Weld and Joint Design

A weld joint is required to weld pieceparts together to produce the weldment. Welds, weld joints, and welding positions were described in Section 2.3. This information was provided early in the book to enable you to better understand the text. The use of the proper weld joint design affects the strength of the weldment and has a great influence on the cost of making it. The objective of any weld joint is to ensure the distribution of stress to each part of the monolithic structure. The five basic types of joint designs are illustrated in Fig. 15.4. These can be combined or varied in numerous ways. The five basic joint types are as follows.

Butt joint (B)
Corner joint (C)
Edge joint (E)
Lap joint (L)
Tee joint (T)

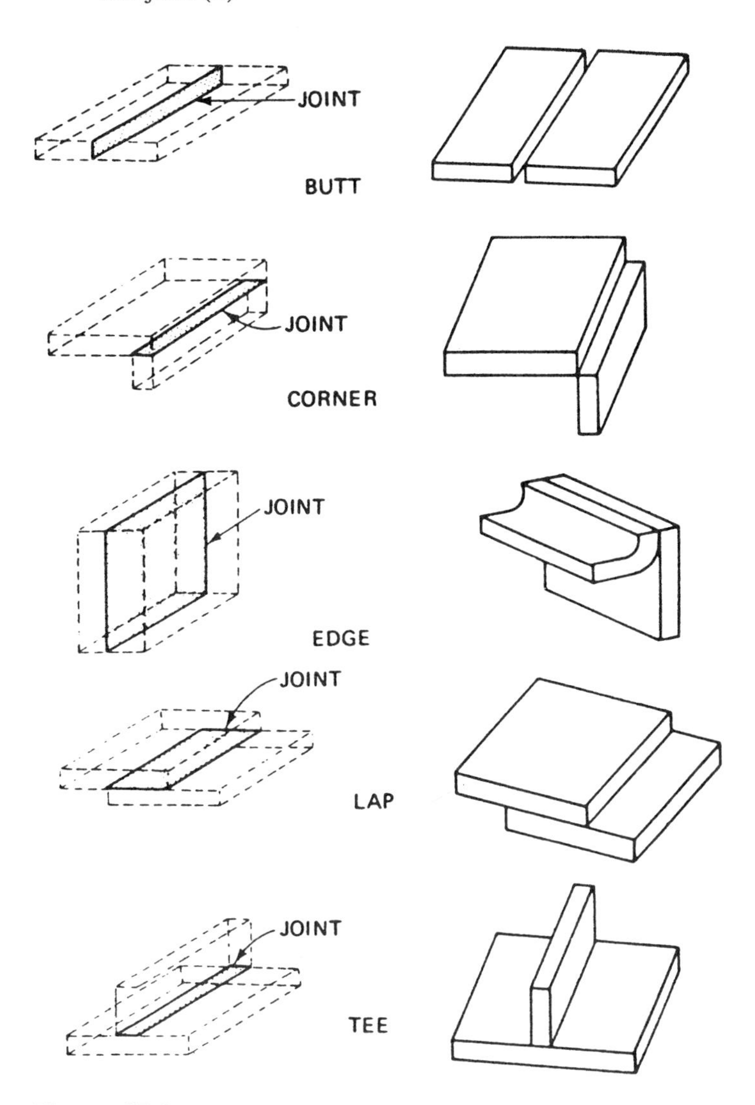

Figure 15.4 Five basic weld joints.

Two basic types of welds are used to make weld joints. By far the most popular is the fillet weld, illustrated in Fig. 15.5. The fillet is basically a triangular weld (in cross section). Fillets can also be made between parts that do not meet at 90°, as shown in Fig. 15.6. It is claimed that over 80% of all welds are fillet welds. It is further stated that most fillet welds are single-pass fillets. Fillet welds

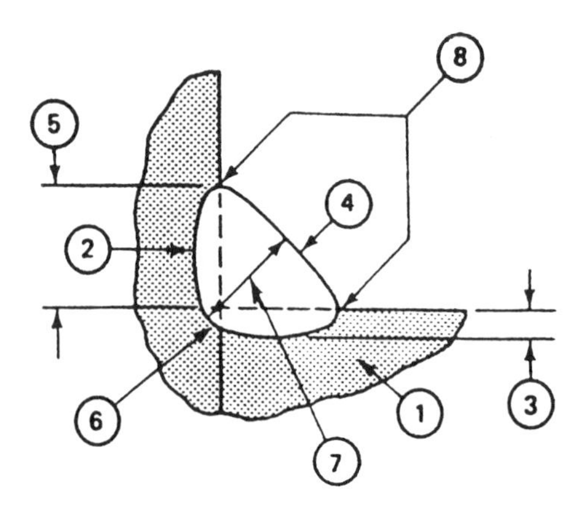

(1) BASE METAL	Metal to be welded.
(2) BOND LINE	The junction of the weld metal and the base metal.
(3) DEPTH OF FUSION	The distance that fusion extends into the base metal.
(4) FACE OF WELD	The exposed surface of a weld on the side from which the welding was done.
(5) LEG OF A FILLET WELD	The distance from the root of the joint to the toe of the fillet weld.
(6) ROOT OF WELD	The point or points, as shown in cross-section, at which the bottom of the weld intersects the base metal surface or surfaces.
(7) THROAT OF FILLET WELD	The shortest distance from the root of the fillet weld to its face.
(8) TOE OF A WELD	The junction between the face of a weld and the base metal.

Figure 15.5 Fillet weld, including size and definition.

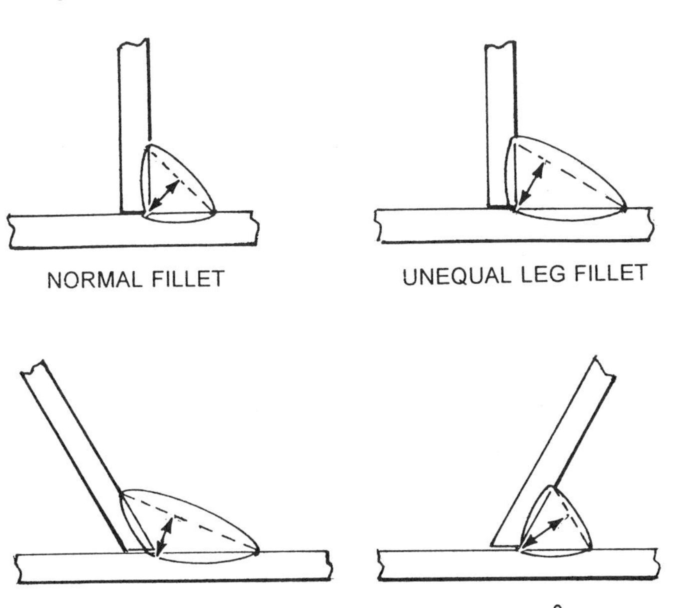

Figure 15.6 Variation of fillet welds.

can be made in any position; however, they can be made faster in the flat or horizontal position. When automated welding is used, the strength of the fillet is increased because deeper penetration is achieved. The fillet is measured by leg length without regard to penetration; however, for most mechanized welding greater penetration adds greatly to the strength of the fillet, and in many cases the fillet weld can be reduced in size to accommodate the greater strength due to the penetration. This is illustrated in Fig. 15.7. The strength of the fillet is its throat area, which is related to penetration depth. Normally the throat dimension is 0.707 times the size.

In spite of their many advantages, fillet welds have certain deficiencies. In the total joint there can be an unfused root. The fillet weld can be stressed, putting the root in tension, which has a negative effect on its strength when it is dynamically loaded. Figure 15.8 shows the high stress concentrations related to single fillet welds. Fillet welds can be used in pairs to avoid this deficiency. Fig. 15.9 shows the lower stress concentration of a double fillet weld joint. Note that each fillet of the double fillet weld protects the root of the other and improves the weldment's serviceability under fatigue loading. Fillets can be used to produce different types of joints, as shown in Fig. 15.10. Note that the joint is the geometric relationship of the parts being joined with the welds.

In some cases, the fillet weld size will be governed by a code or specifi-

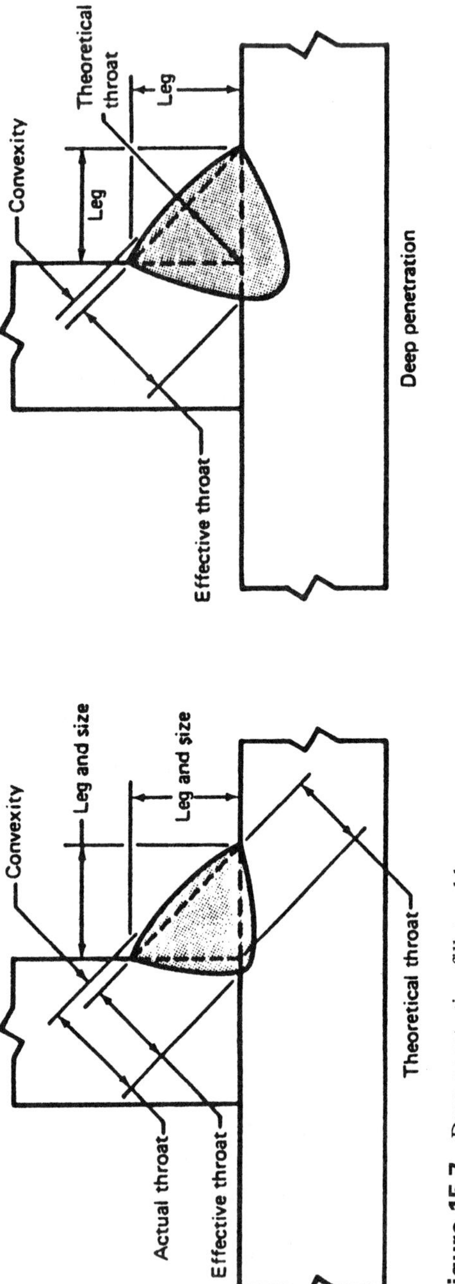

Figure 15.7 Deep-penetration fillet weld.

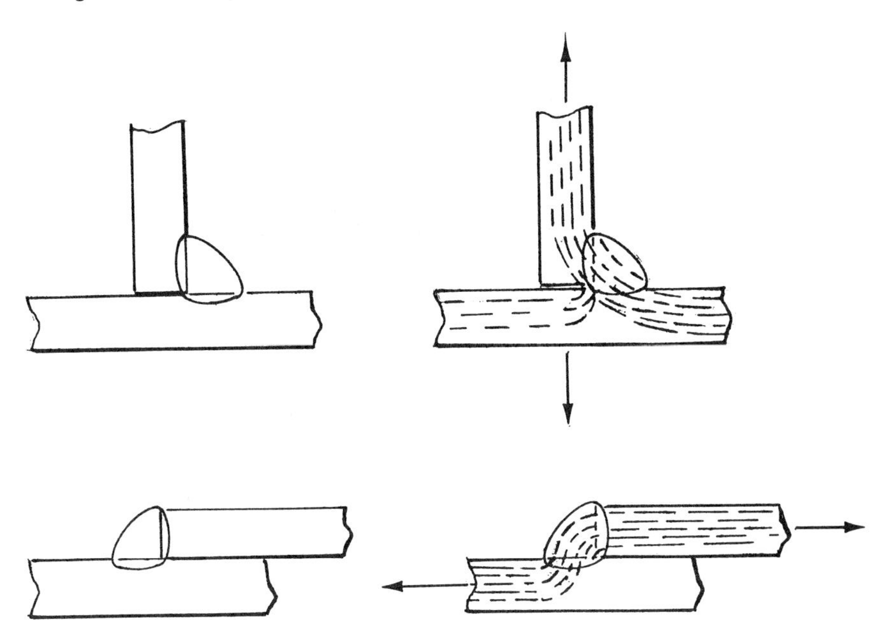

Figure 15.8 Stress concentration of a single fillet.

cation. In these cases, a minimum size is usually specified. Good engineering practice also dictates the minimum size of a fillet. The minimum size for various thicknesses of the metals being welded is listed in Table 15.1. This is based on the AWS document [2] concerning classification and application of welded joints.

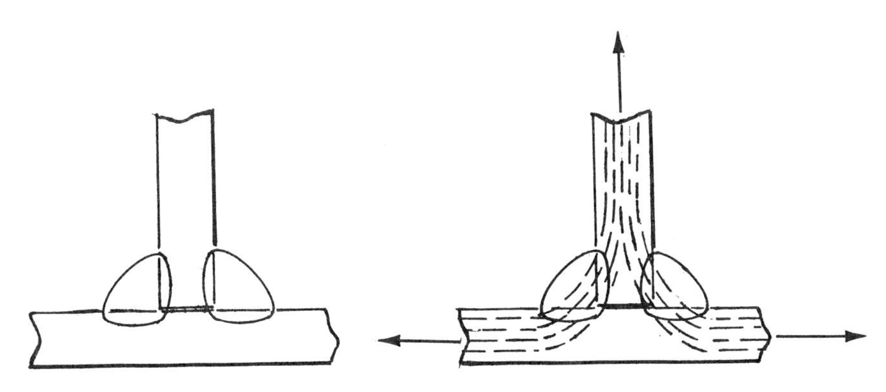

Figure 15.9 Double fillet weld stress pattern.

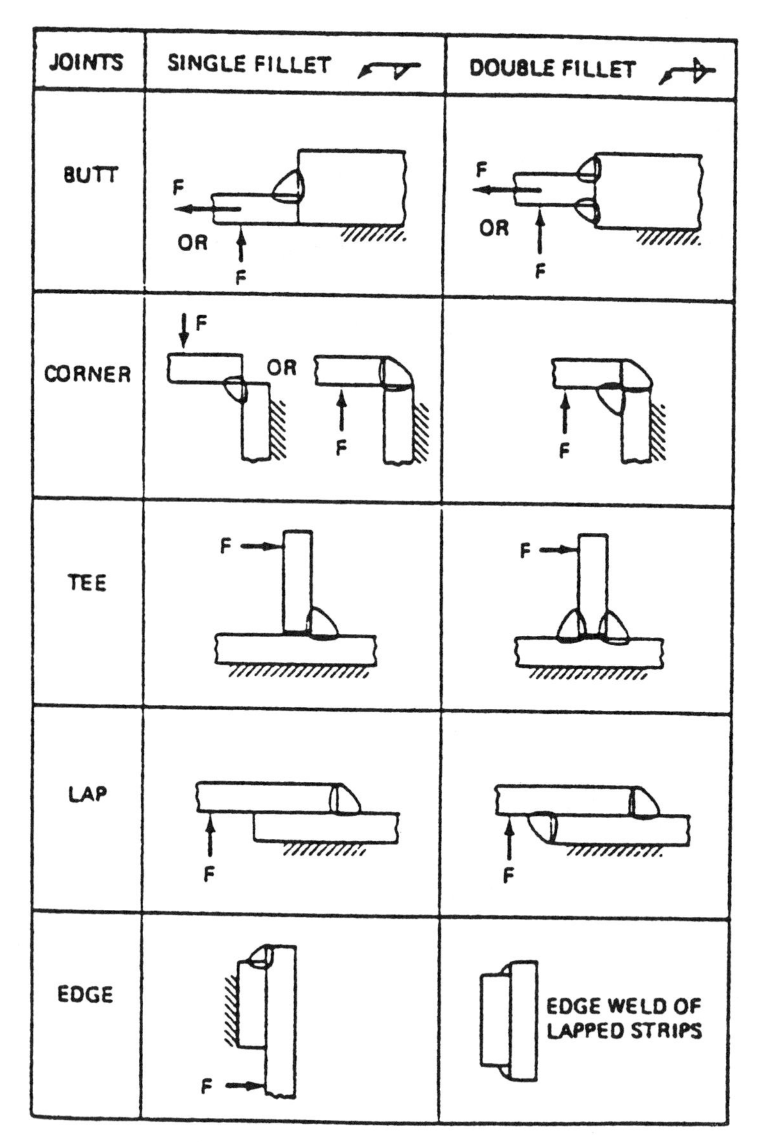

Figure 15.10 Double fillet welds used to make basic joints.

The groove weld is the second most popular type of weld employed. It is defined as a weld made in the groove between two members to be joined. The groove weld is regarded as being in the joint. The details of groove welds are shown in Fig. 15.11. There are seven basic groove weld designs, and they can be used as single or double welds. The difference is that single groove welds can be

Table 15.1 Minimum Fillet Weld Size

Material thickness of thicker part joined		Minimum size of fillet weld	
mm	in.	mm	in.
To 6 inclusive	To 1/4 inclusive	3	1/8
Over 6 to 13	Over 1/4 to 1/2	5	3/16
Over 13 to 19	Over 1/2 to 3/4	6	1/4
Over 19 to 38	Over 3/4 to 1 1/2	8	5/16
Over 38 to 57	Over 1 1/2 to 2 1/4	10	3/8
Over 57 to 152	Over 2 1/4 to 6	13	1/2
Over 152	Over 6	16	5/8

Note: The weld size need not exceed the thickness of the thinner part joined.

made from one side and are usually restricted to the thinner materials; thicker materials are welded with double groove welds to minimize the amount of filler metal required. The single groove weld is used for most applications. Details of a single vee groove weld are shown in Fig. 15.12. This shows the definitions of the different aspects such as the groove angle, groove face, root face, and root opening. These factors can change dimensionally for the same type of weld and vary depending on the type of groove.

There are a number of other types of welds, but they are less popular for automation. Figure 15.13 shows the nine basic types of welds. The fillet and groove welds have already been discussed. The others are as follows.

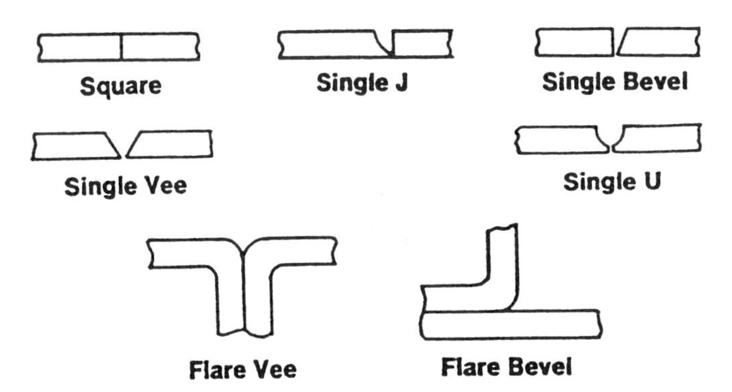

Figure 15.11 The seven basic types of groove welds.

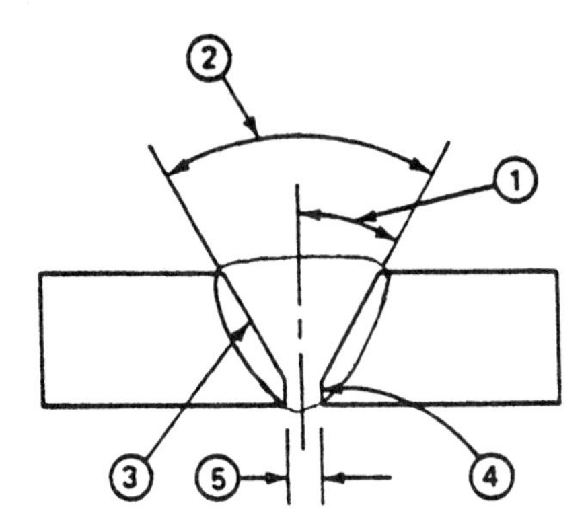

- (1) **BEVEL ANGLE** — The angle formed between the prepared edge of a member and a plane perpendicular to the surface of the member.
- (2) **GROOVE ANGLE** — The total included angle of the groove between parts to be joined by a groove weld.
- (3) **GROOVE FACE** — The surface of a member included in the groove.
- (4) **ROOT FACE** — That portion of the groove face adjacent to the root of the joint.
- (5) **ROOT OPENING** — The separation between the members to be joined at the root of the joint.

Figure 15.12 The groove weld, including definitions.

Plug or slot welds These are used with prepared holes. They are considered together because the same welding symbol is used to specify them. The important difference is the type or shape of hole prepared in one of the members being joined. If the hole is round, it is considered a plug weld; if it is elongated, it is considered a slot weld. The weld is made from the top side.

Spot or projection weld These welds are designated by the same weld symbol but can be applied by different welding processes, which make different types of welds. When the resistance welding process is used, the weld is at the interface of the members being joined; and in laser or gas metal arc welding, the weld melts through one member into the second member. They are widely used in automated spot welding.

Seam weld In cross section the seam weld looks similar to a spot weld. It is used with resistance welding, high-energy beam welding, and arc welding. With resistance welding, the weld is at the interface between the members being joined. With laser beam or arc welding the weld melts

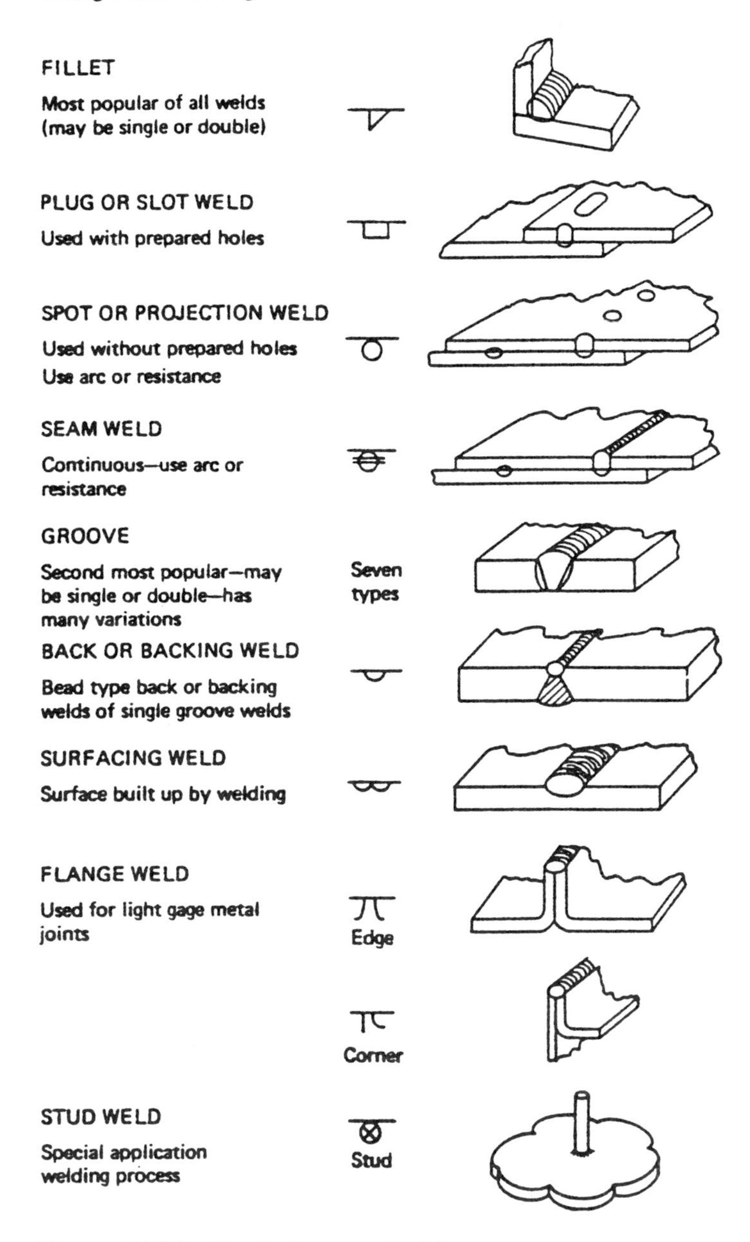

Figure 15.13 Nine basic types of welds.

through one member to join it to the second member. There are no prepared holes in either part being welded.

Back or backing weld This is a special type of weld made on the back or root side of a previously made weld. The root of the original is gouged, chipped, or ground down to base metal before the backing weld is made. This ensures complete penetration but is difficult to do with automated welding.

Surfacing weld This weld is composed of one or more beads deposited on base metal as an unbroken surface. It is used to build up surface dimensions, to provide a surface with different properties, or to protect the base metal from a hostile environment.

Flange weld The edge flange weld is used for light-gauge products and is normally made with the arc welding process. The corner flange weld is also used for light-gauge parts. In both cases, parts must be prepared for the specific joint.

Stud weld The stud weld is a special type of weld joining a stud to a flat surface. It is a separate process but can be automated.

Weld joints used to manufacture weldments often combine different types of welds. However, not all weld types can be used with each type of joint. Figure 15.14 shows what welds are applicable to each of the basic joints. Fillets and groove welds can be combined as shown by this figure. Combination weld joints are often used where full penetration is required.

Standardized joint details are included in certain specifications, both military and AWS. Individual companies have also standardized on specific joints. This means that root openings, beveled angles, etc., are standardized. Standardized dimensions of weld joints are widely used so that shop practice is consistent and a specific weld symbol will always produce the same weld detail design. A good example of standardized joints that are also prequalified is given in the AWS document *Specifications for Welding Press and Press Components* [3]. This is an aid in consistency but should not be used in every situation. If standard angles, root openings, etc., need to be changed, they should be so specified and perhaps shown in cross section on drawings. Figure 15.15 shows other possible variations of the different weld types.

Standardized welds take on added significance when we consider the economics of welding. For example, Fig. 15.2 can be used to show the economic effect of doubling the strength of a fillet weld size. It is for this reason we want to keep fillets as small as possible. Figure 15.16 illustrates root opening variations. A root opening equal to half the width of a bevel will double the amount of weld metal required. The root opening is also related to the welding process, and it can be narrower when gas metal arc welding is used. Analysis shows that bevel angles and root openings have an interrelationship that cannot be ignored. Obtaining a quality root weld depends on the included angle of the groove and

OTHER WELD TYPES	SYMBOL	THE FIVE BASIC JOINT TYPES				
		BUTT	CORNER	TEE	LAP	EDGE
PLUG OR SLOT WELD		NO				NOT APPLICABLE
SPOT OR PROJECTION (ARC OR RESISTANCE)		NO				NOT APPLICABLE
SEAM WELD (ARC OR RESISTANCE)		NO				NOT APPLICABLE
BACK OR BACKING WELD					NOT APPLICABLE	NOT APPLICABLE
SURFACING		NOT APPLICABLE	NOT APPLICABLE	NOT APPLICABLE	NOT APPLICABLE	NOT APPLICABLE
FLANGE WELD EDGE (SHEET METAL)		NO	NO	NO	NO	
FLANGE WELD CORNER (SHEET METAL)		NO		NO	NO	NO

Figure 15.14 Combination of welds for specific joints.

Weld Types		Basic Joints				
		Butt	Corner	Tee	Lap	Edge
Fillet		Special	Yes	Yes	Yes	Special
Square groove		Yes	Yes	Yes	---	Yes
V groove		Yes	Yes	Yes	---	Yes
Bevel groove		Yes	Yes	Yes	Yes	Yes
U groove		Yes	Yes	---	---	Yes
J groove		Yes	Yes	Yes	Yes	Yes
Flare V groove		Yes	Yes	---	---	---
Flare Bevel groove		Yes	Yes	Yes	Yes	---

Figure 15.15 Variations of weld types.

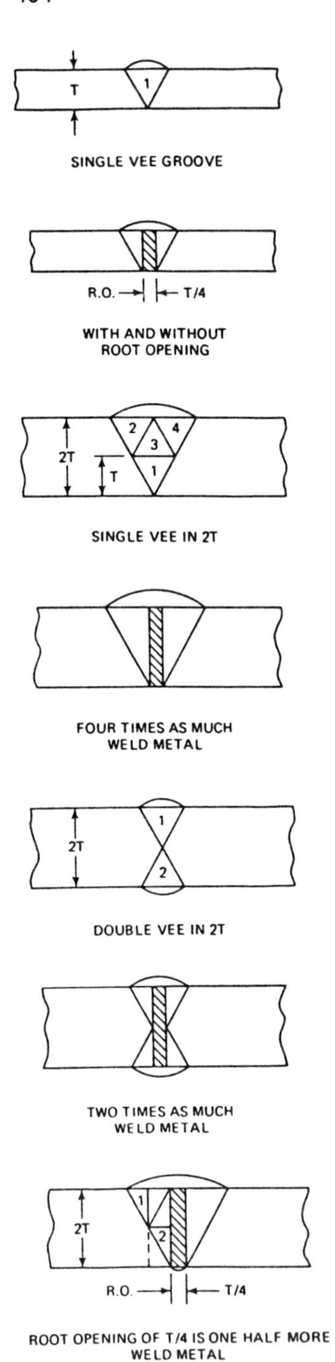

Figure 15.16 Root opening variations and extra filler metal.

the root opening. Figure 15.17 shows the relationship between single and double groove welds. The use of double groove welds for heavy material reduces the amount of filler metal required; however, the preparation is more expensive and the part must be welded from both sides. In general, backing welds made from the opposite side of the joint are not recommended for automatic welding.

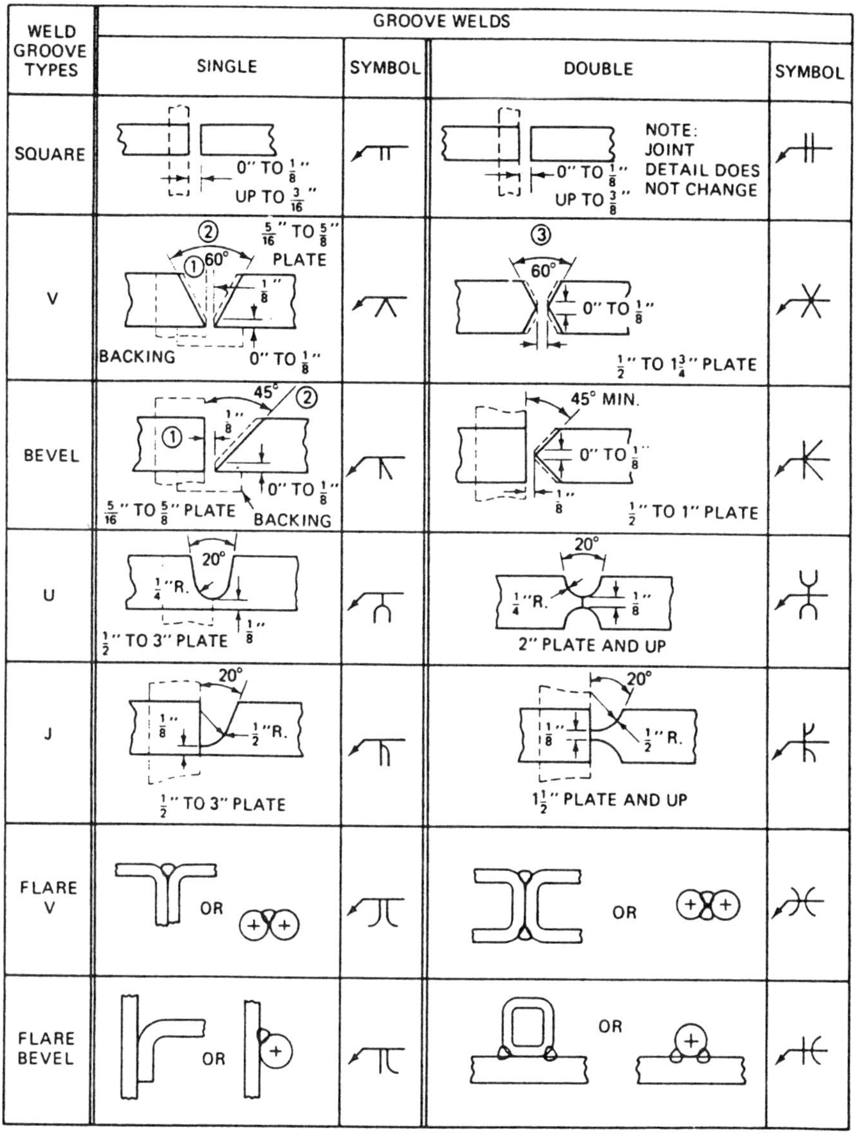

WELD GROOVE TYPES	GROOVE WELDS			
	SINGLE	SYMBOL	DOUBLE	SYMBOL
SQUARE	0" TO $\frac{1}{8}$" UP TO $\frac{3}{16}$"		0" TO $\frac{1}{8}$" UP TO $\frac{3}{8}$" NOTE: JOINT DETAIL DOES NOT CHANGE	
V	② $\frac{5}{16}$" TO $\frac{5}{8}$" PLATE ① 60° $\frac{1}{8}$" BACKING 0" TO $\frac{1}{8}$"		③ 60° 0" TO $\frac{1}{8}$" $\frac{1}{2}$" TO $1\frac{3}{4}$" PLATE	
BEVEL	45° ② ① $\frac{1}{8}$" 0" TO $\frac{1}{8}$" $\frac{5}{16}$" TO $\frac{5}{8}$" PLATE BACKING		45° MIN. 0" TO $\frac{1}{8}$" $\frac{1}{8}$" $\frac{1}{2}$" TO 1" PLATE	
U	20° $\frac{1}{4}$"R. $\frac{1}{8}$" $\frac{1}{2}$" TO 3" PLATE		20° $\frac{1}{4}$"R. $\frac{1}{8}$" 2" PLATE AND UP	
J	20° $\frac{1}{8}$" $\frac{1}{2}$"R. $\frac{1}{2}$" TO 3" PLATE		20° $\frac{1}{8}$" $\frac{1}{2}$"R. $1\frac{1}{2}$" PLATE AND UP	
FLARE V	OR		OR	
FLARE BEVEL	OR		OR	

Figure 15.17 Double versus single groove welds.

A very important factor to consider is the full-penetration joint versus a partial-penetration joint. A full-penetration weld joint has weld metal throughout its entire cross section. The partial-penetration joint is designed to have an unfused area where the weld does not penetrate the joint completely. The stress patterns are shown for both full- and partial-penetration welds in Fig. 15.18. The weld joint must have full penetration when the weldment is to be subjected to impact, fatigue, or low-temperature service because of the presence of stress risers with partial-penetration welds. A stress riser can cause stresses two to ten times greater than nominal at the unfused weld joint. Also, for fatigue service the contour of a fillet weld should be concave and rounded to aid in distributing stress. In designing for fatigue loading, internal notches, discontinuities, metallurgical notches, sharp internal corners, and similar features must be avoided.

The selection of appropriate joint details depends on several factors. In general, single groove welds are applicable to thin materials. When the material is thicker, the double groove should be specified. However, double groove welds require more time for preparation and must be welded from both sides. The J- and U-groove joint designs require machining to accomplish and are rarely used except on round members that can be easily machined. Some groove welds have composite or compound bevel angles, which again are provided only by machining.

For thin material, sheared edges or edges produced by punching are the most economical; however, there are limits to the thicknesses that can be employed, depending on available shop facilities. It is therefore essential that joint preparation be coupled to weld metal deposits to arrive at the most economical detail.

Fillet and square-groove weld designs should be used whenever possible. Other weld joint details should be used only after careful study.

An example of weld and weld joint selection is provided by the design for closing the end of a small circular tank. Figure 15.19 shows a variety of designs.

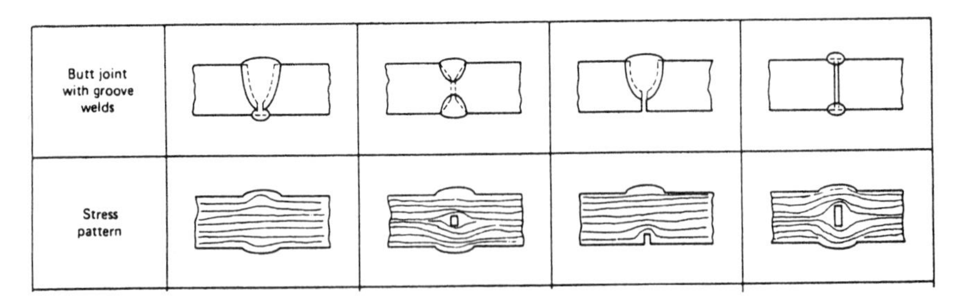

Figure 15.18 Full and partial penetration stress patterns.

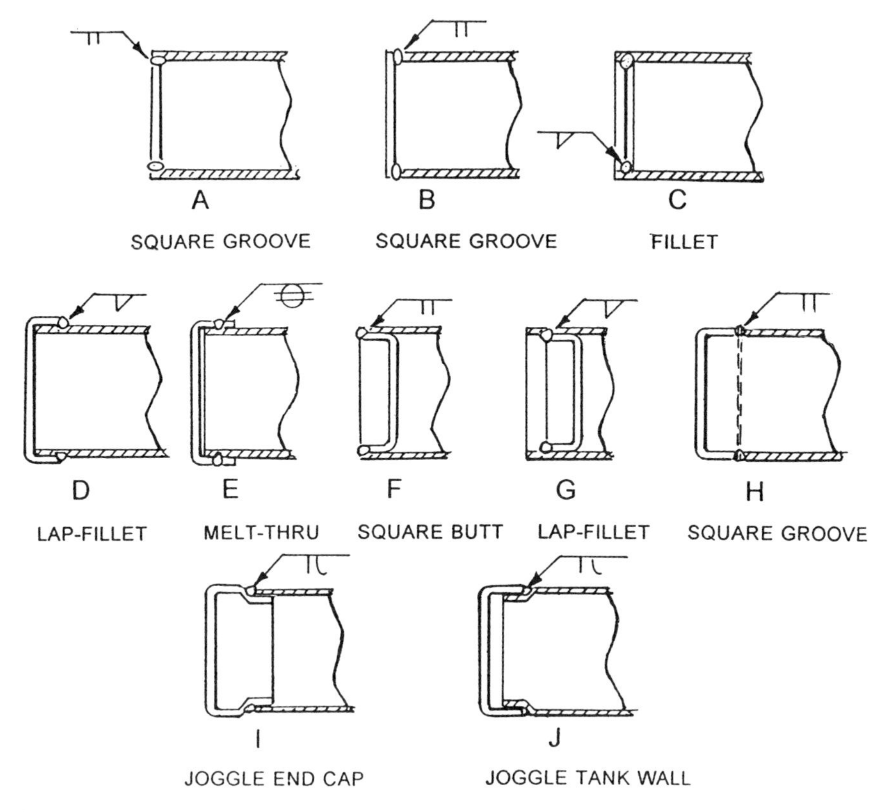

Figure 15.19 End closure of small tank.

The simplest from the point of view of piecepart preparation is to use a flat plate, shown by Figs. 15.19A–C. What welding procedure to use then becomes the question. Both B and C use the square-groove weld, but this is difficult to accomplish because fit-up must be perfect if penetration is expected to be uniform or a backing ring must be used, which increases cost. Also, a fixture must be used to hold the end plate in place prior to welding. Additionally, in Fig. 15.19B the weld is made in the horizontal position, which is more difficult. In detail C, the fillet can be used, which requires less precise fit-up. Also, both ends can be welded simultaneously, which increases productivity. The formed head is a better solution. Of course, the formed head is more expensive than the flat plate; however, a fixture is not required to hold it in place, assuming parts are to size—friction should hold it in place. D and G require fillet welds, and both ends can be welded simultaneously. Detail E can be used with thinner material, and a laser beam could be used for welding. For thinner material, GMAW would be used for a melt-through weld. Detail F has the same problem as detail A, and detail H has the same problem as detail B. Backing rings should be avoided. They add to the cost and may increase corrosion problems, depending on the contents of the tank. The joggled end design has many advantages but requires special tooling, which might adversely affect the cost. The welding procedure is the easiest.

This type of analysis can be applied to the selection of all joint designs.

15.3 Welding Symbols

Welding symbols are the shorthand of welding. They enable the engineer and drafter to convey complete "instructions for welding" to the shop on blueprints and drawings for use by set-up staff, production welders, and parts preparation personnel. Welding symbols are developed and promoted by the American Welding Society and are incorporated in AWS document A2.4 [4]. These symbols were originally developed in the early 1930s and have been frequently updated ever since. In addition, they are in general agreement with ISO standards, so that a uniform system of welding symbols is used throughout the world.

The purpose of welding symbols is to describe the desired weld accurately and completely. They can be used to transmit other information such as specifications and procedures. Welding symbols can also be combined with nondestructive examination symbols. The welding symbol consists of up to eight elements. These are

1. Reference line
2. Arrow
3. Basis weld symbol
4. Dimensions and other data
5. Supplementary symbols
6. Finish symbol
7. Tail
8. Specification, process, or reference

At least the first and second elements and either the third or seventh must be used to make an intelligible welding symbol. The other elements may or may not be incorporated, depending on the information to be transmitted.

The basic element of the welding symbol is the reference line, which is in the horizontal position and should be drawn near the weld joint that it is to identify. The other components of the symbol are constructed on this reference line. Each element must be placed in proper location on the reference line in accordance with the standard locations provided by the specification. The elements that describe the basic weld, the dimensions and other data, and the supplementary and finish symbols are always located on the reference line in the proper relationship to one another.

The secondmost important element of the welding symbol is the arrow. This is a line from one end of the reference line to the arrow side or arrow-side member of the weld joint. When the symbol is used for joints that require the preparation of one member, the arrowhead should point with a definite break to the member of the joint that is to be prepared. The other end of the reference line carries the tail of the arrow. Specific information can be placed in the tail to provide reference to a welding procedure, standard joints, specification, etc.

The basic weld symbol that describes the weld is placed under the reference line to define the weld on the arrow side of the joint or the arrow-side member, and above the reference line to describe the weld made on the other side or other-side member of the joint. Placing the basic weld symbol on both sides of the reference line indicates that the weld is to be made on both sides of the joint.

The various dimensional notations that help describe and define the weld are located with specific relationships to the reference line and weld symbol. These include the size of the weld, the root opening, and similar factors. The relationship to all dimensions and factors is illustrated by Fig. 15.20.

The use of welding symbols will result in the following advantages:

1. Controls specific design instructions to the shop regarding weld sizes and plate edge preparation; eliminates overwelding or underwelding (that can result in either increased production costs or unsafe fabrication) because of lack of definite information.
2. Eliminates the need to indicate weld sizes and specifications on drawings. Welding notes are minimized.
3. Establishes a common understanding of design intent and requirement between engineering, shop, inspection, customer, and code inspection authorities.
4. Allows standardization, not only within the company but also industry-wide. AWS welding symbols are in standard use both nationally and internationally.

The symbols shown in Fig. 15.20 are taken from AWS A2.4 [4]. Consult this standard for complete information on welding symbols.

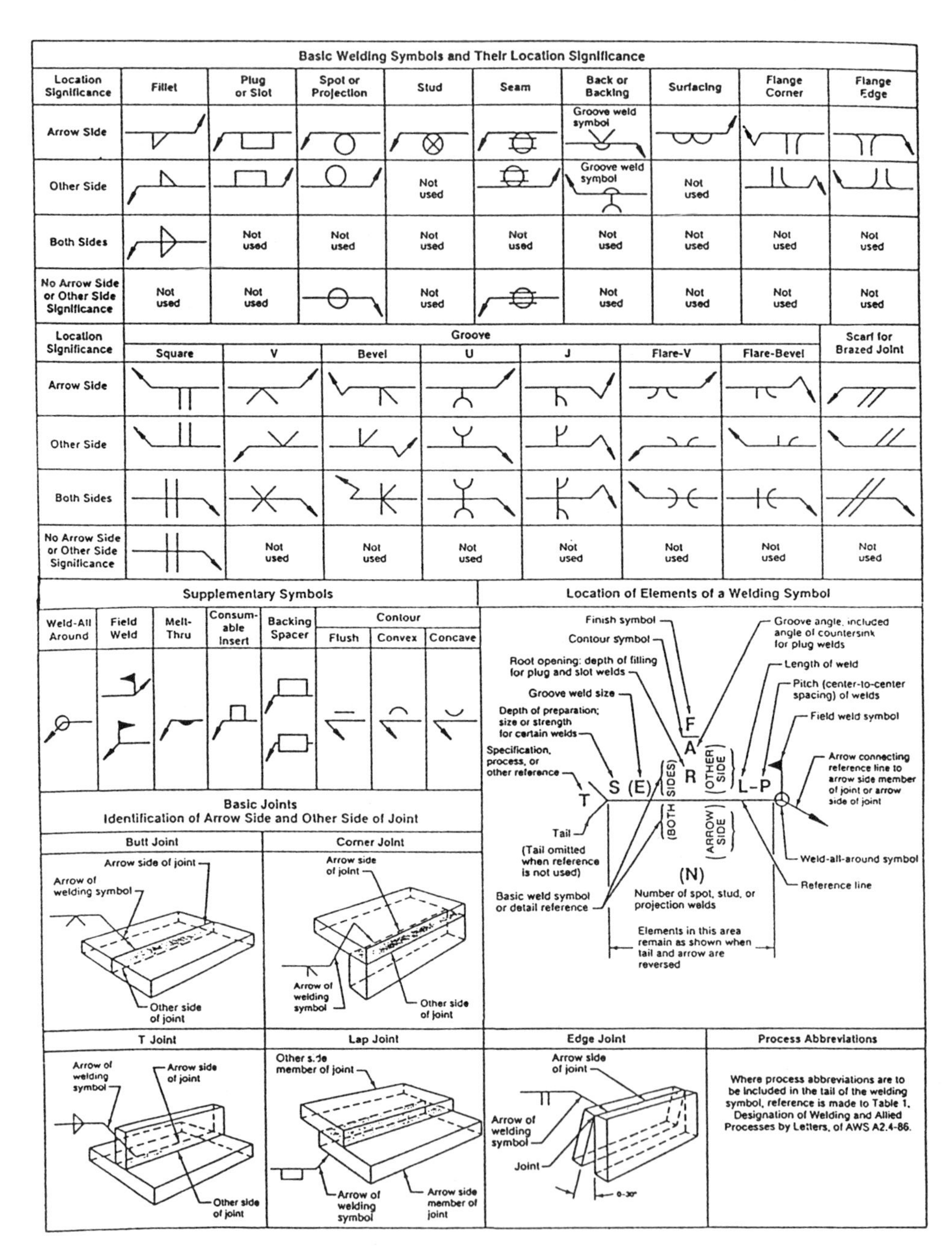

Figure 15.20 Summary of welding symbols.

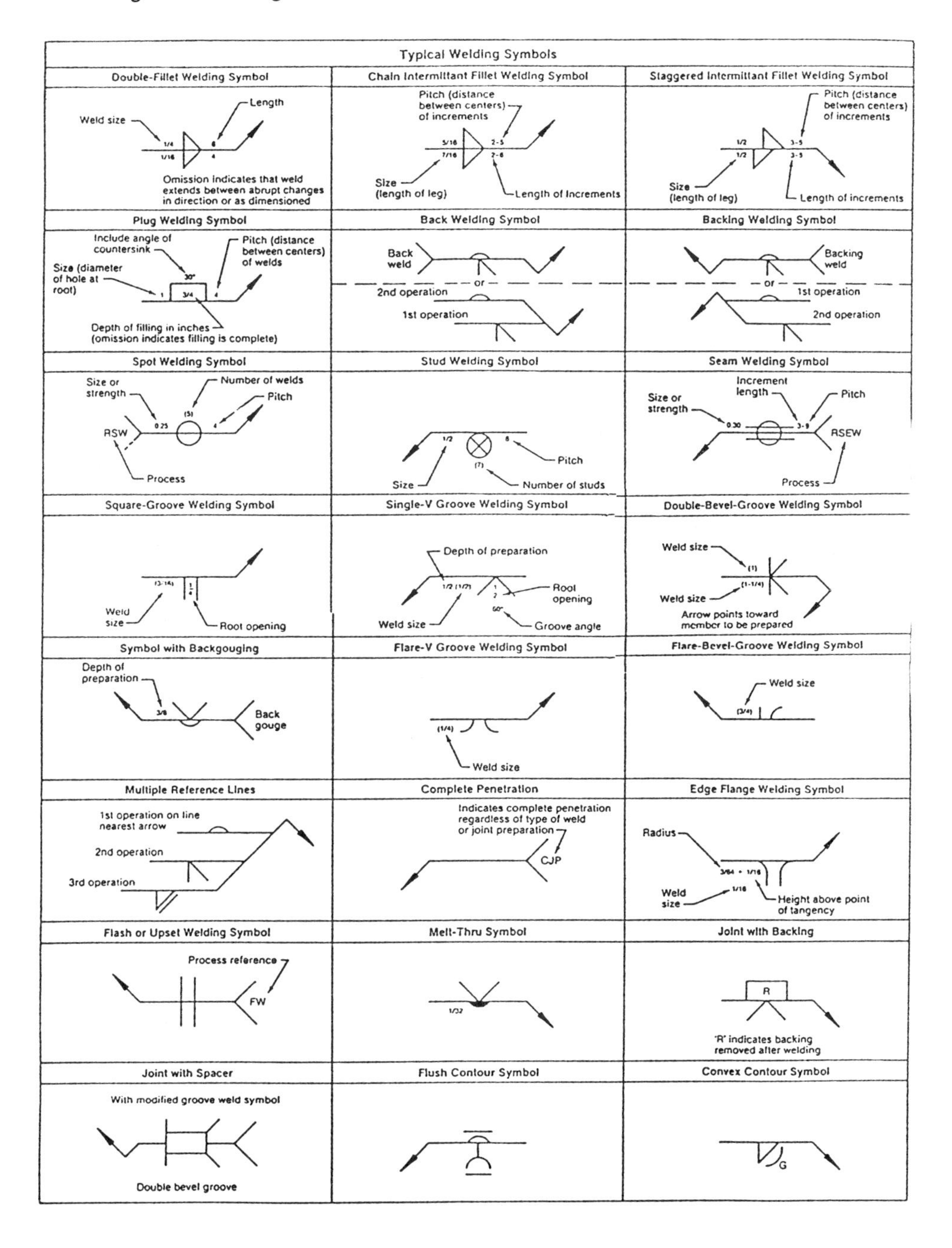
Typical Welding Symbols
Double-Fillet Welding Symbol
Length
Weld size
Omission indicates that weld extends between abrupt changes in direction or as dimensioned
Chain Intermittant Fillet Welding Symbol
Pitch (distance between centers) of increments
Size (length of leg)
Length of increments
Staggered Intermittant Fillet Welding Symbol
Pitch (distance between centers) of increments
Size (length of leg)
Length of increments
Plug Welding Symbol
Include angle of countersink
Pitch (distance between centers) of welds
Size (diameter of hole at root)
Depth of filling in inches (omission indicates filling is complete)
Back Welding Symbol
Back weld
or
2nd operation
1st operation
Backing Welding Symbol
Backing weld
or
1st operation
2nd operation
Spot Welding Symbol
Size or strength
Number of welds
Pitch
RSW
Process
Stud Welding Symbol
Pitch
Size
Number of studs
Seam Welding Symbol
Increment length
Pitch
Size or strength
RSEW
Process
Square-Groove Welding Symbol
Weld size
Root opening
Single-V Groove Welding Symbol
Depth of preparation
Root opening
Weld size
Groove angle
Double-Bevel-Groove Welding Symbol
Weld size
Weld size
Arrow points toward member to be prepared
Symbol with Backgouging
Depth of preparation
Back gouge
Flare-V Groove Welding Symbol
Weld size
Flare-Bevel-Groove Welding Symbol
Weld size
Multiple Reference Lines
1st operation on line nearest arrow
2nd operation
3rd operation
Complete Penetration
Indicates complete penetration regardless of type of weld or joint preparation
CJP
Edge Flange Welding Symbol
Radius
Weld size
Height above point of tangency
Flash or Upset Welding Symbol
Process reference
FW
Melt-Thru Symbol
Joint with Backing
R
'R' indicates backing removed after welding
Joint with Spacer
With modified groove weld symbol
Double bevel groove
Flush Contour Symbol
Convex Contour Symbol
G

15.4 Welding Fixtures

A work-holding fixture for arc welding must accurately locate and hold the component parts of the weldment in their proper location for welding. It must also locate the joints accurately and maintain the correct fit-up. It speeds up the operation and improves the dimensional accuracy of the weldment. There are many types of welding fixtures and many reasons for employing them.

Originally welding fixtures were used in conjunction with manual shielded metal arc welding (stick welding) to eliminate the costly and time-consuming hand layout and tack welding of the parts prior to welding. With manual welding, if a sufficient number of weldments were made, the cost of the fixture would be recovered quickly due to the elimination of the setup and tack welding operation. The fixture also increased the accuracy of the weldment by eliminating the errors that could occur during the setup operation. With the advent of mechanized, automated, or robotic arc welding, the advantages of fixtures became even more pronounced. Efficient fixtures allow unattended welding. Once the parts are in the fixture and it is properly located, the automated or robotic welding process can start and operate without human observation, monitoring, or supervision.

It is important to keep the fixture in operation as much of the time as possible to quickly recover its cost. In addition to the savings provided by the elimination of the layout operation, productivity was greatly increased because fixtures could be loaded and unloaded while the welding machine was making welds. Arc-on time was much higher, running as high as 90%. It also improved the safety of robotic welding because double-ended indexing positioners were normally used. Two fixtures were placed on the indexing positioner, which would rotate to position one fixture inside the welding cell and the operator remained on the outside of the cell unloading and loading the other fixture.

There are basically two types of fixtures used for robotic arc welding: (1) those used only for tack welding the parts together and (2) those that hold the weldment during the complete welding operation. The second type are sometimes called *strongbacks* and are heavier and more robust than tacking fixtures. They are used to hold the parts, maintain accurate alignment, and resist warpage of the weldment. The work-holding fixture is customized for each weldment. It is unique and must be reworked if the design of the weldment is changed.

For automated or robotic arc welding, the work-holding fixture is placed on an indexing positioner. This provides operator safety and increases productivity. Each end of the positioner may have two axes of motion such as horizontal and vertical and/or tilt motion, which can be integrated by the controller. This allows the fixture and the work to move so that the welds can be made in the flat position and also maintains accessibility for the welding torch. The work-holding device on each end of an indexing positioner need not be for the same weldment. The robot can be programmed to weld different products on the two ends of the positioner. There is an exception to this. When rotary tables are used for loading, moving the work to the welding station and moving the weldment to the unload station, the fixture must be for the same weldment.

The time required for unloading a finished weldment, loading the next weldment's pieceparts and properly locating them and clamping them must be less than the arc welding time for either weldment. This allows the operator time for inspecting, moving material, etc. The total time from beginning to load parts to unloading the weldment is the factor that determines the system's production rate. Indexing fixtures change the index position when instructed by the operator rather than by the program of the robot.

Good fit-up is required to obtain high-quality welds. The size of the root opening is related to the speed of welding. If the root opening is excessive, the root pass will burn through, resulting in the need to rework the joint.

Often a third type of fixture is used to attach the tack welded weldment to the positioner table to hold it in the proper location for the robotic program. It can also be used to facilitate quick attachment and easy removal. Clamping might not be necessary if the work table of the positioner remains horizontal.

In contrast to the previous information, keep in mind that extremely simple work-holding devices can be made quickly and will pay back after being used for a few batch runs of production. These fixtures can be built around a finished weldment. Assuming that the weldment is dimensionally accurate, the parts produced in the fixture will be accurate. Fixtures used for manual welding can be upgraded for automatic welding. They must be properly identified, stored, and called up again for the next production run of the same part. This keeps the automatic welding system running at full capacity, pays back quickly, and produces good quality weldments.

It is essential that the weldment and the fixture provide accessibility for the welding gun to make the necessary welds. It may be necessary to redesign the fixture or weldment to allow the weld gun access to the weld location.

15.5 Fixture and Tooling Design

The work-holding fixture for automated or robotic arc welding must accurately locate and hold the component parts of the weldment during tacking or welding. The fixture is successful only if the weldment produced is accurate and the welds are to size. Welding fixtures or work-holding devices are extremely complex and should be designed only by experienced people. As the weldment becomes more complex, the fixture does also, and the cost increases accordingly. However, good fixtures, when properly used, will pay for themselves quickly and can pay back their cost in a relatively short time.

Coengineering is suggested for designing the fixture and weldment. The assembly technique should be to build one piece at a time on the base or foundation piece of the weldment. We should anticipate the day when fixtures will be loaded by robots. Grippers will be used instead of torches, and in these cases fixtures and weldments should be designed so that the parts can be loaded from the top in straight-line operations. Ease of assembly, especially automated assembly, should be a paramount design consideration.

The fixture designer must have complete information concerning the weldment in question—its estimated weight and size, the materials to be used and their form, including shape and thickness. The design should indicate dimensions and which ones are critical, dimensional tolerances, how much distortion can be allowed, and how much material is allowed for machining. The preparation of pieceparts should be reviewed by the fixture designer, who should know the tolerances involved with the different processes. The specific welding process to be used must be understood. In addition, weld size, welding position, whether groove welds are to be made, whether the back side of the groove weld is to be welded, and weld fillet size should be discussed, and all weld details understood. It is worthwhile to review the design of fixtures made previously for similar projects for background.

The designer must know the make and model of the robot to be used because this provides the work envelope data. He should have specifications covering the wrist action and type of torch. The robot company may be able to provide detailed drawings of the torch or an actual torch. This will aid in the design of clamps, for example. The designer should also be told of the quantity of parts to be produced, the expected batch size, and the time allowed for loading, unloading, and welding. A design budget should be drawn up. Fixtures can be expensive and can represent up to 25% of the total cost of the robotic welding system.

The designer should also be aware of the availability of positioners and what type of positioner will be used. The fixture weight must be added to the positioner's load, which must not exceed certain limits of weight or off-center load. The design must allow for the work being welded and the fixture to be electrically connected to the positioner, which in turn is electrically connected to the welding power lead through rotary connectors. Material unloading systems have to be designed to work with other material handling or moving devices. Once the design is specified and agreed to by the weldment designer and the fixture designer, it is reduced to writing. If the fixture is to be built by an outside company, select a builder with experience in producing automatic welding equipment of the type required.

The fixture designer must analyze the weldment to determine if it can be welded on the robot in question, if there are blind welds that cannot be made, if certain welds need backing welds, and if it would be best made as subassemblies instead of a complete assembly. This is important because if at all possible all pieces needed to make the weldment should be loaded at the beginning of the welding cycle. If this cannot be done, subassemblies may be required.

The fixture for automated or robotic arc welding must be properly designed and accurately built in order to produce dimensionally accurate weldments. Weldment fixtures must be designed for maximum productivity; they must maximize the arc welding time and minimize loading and unloading time. The fixture must be designed to present the part properly to the welding gun. The gun must have access to all welds. Automatic clamps may be operated by the robot controller, which can open or close them during the robotic welding sequence. These clamps need sensors to provide feedback to the controller to prevent a "crash."

The fixture designer must determine what heat buildup or heat flow problem may be encountered and whether backing systems are required for full-penetration groove welds. Typical backing systems are shown in Fig. 15.21. If steel backing is used, it is usually left in the joint; copper backing is never left in the joint; and backing gas is not used for steel weldments. Weld backing is extremely important because it affects potential warpage. Intermediate operations such as chipping or grinding are to be avoided.

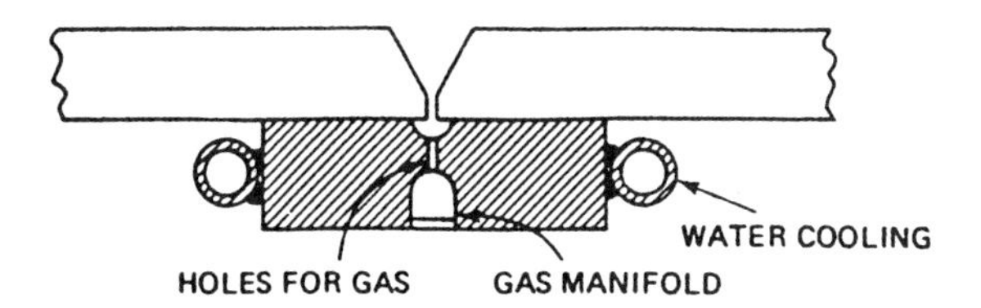

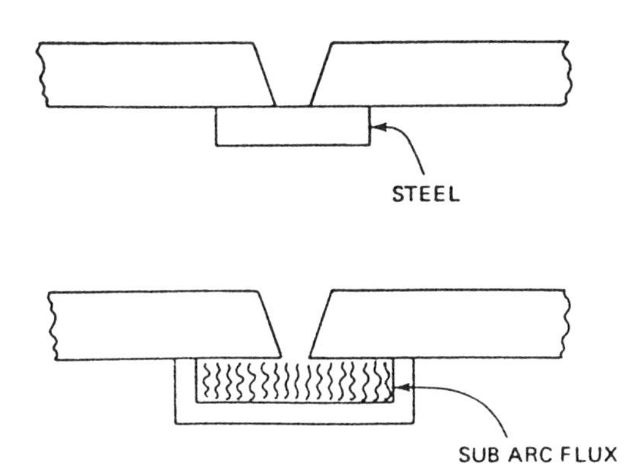

Figure 15.21 Weld backing system.

The fixture designer should also decide if the welds should be made in one position or the weldment repositioned for certain welds. The loading and unloading of the fixture is extremely important and is a major factor in its design. Heat warpage causes control distortion that affects the unloading of a fixture. Large weldments must be unloaded from the fixture by cranes, whereas small weldments are manually unloaded. Locating points in the weldment are extremely important, and, if possible, self-locating or self-jigging should be employed. Special holes, tabs, or notches may have to be designed into the weldment and fixture for locating purposes to ensure accurate positioning for the weld.

Ease of loading, locating, and clamping is essential. The fixture loading points, support points, register points, and clamping points must be located so as to not obstruct the travel of the welding torch. Locating points should be at or near surfaces where close tolerances or machine finishing are specified. They must also be kept free of weld spatter. Copper-aluminum alloys are sometimes used because spatter does not stick to them. For thin material, manual toggle clamps (Fig. 15.22) are often used. For tubular products, a different design is

Figure 15.22 Manual toggle clamps.

used; a typical example is shown in Fig. 15.23. For heavier material, clamps are often attached by screw threads using air wrenches. In some cases, automatic clamping with air or hydraulic power (use nonflammable fluid) can be used. Clamps must be positioned so that they do not introduce distortion or deflection into the parts. Clamps should withdraw from the fixture area when loading is in process and should not prevent the unloading of the weldment if distortion occurs.

In a discussion of welding fixtures, it is necessary to bring up the subject of automatic loading. As time goes by, automatic loading will become more prevalent, especially for small parts. An example will show how automatic loading can be accomplished. The parts being loaded are small pipe nipples with a plug in the end. They are fed into an automatic dial feeder as shown in Fig. 15.24. In the feeder they are arranged with the plug at the top, and they come out of the dial feeder in single file. They are then fed to a conveyor belt (see Fig. 15.25) that runs the length of the automated welding station. They are easily grasped by a gripper attached to the GMAW torch (Fig. 15.26) that is ready to pick up the nipple. The torch then takes the nipple to the weldment, where the robot inserts it into a hole and the GMAW gun tacks one side to hold it in place. This type of feed system and auxiliary gripping system can be used to insert small parts into larger weldments.

Another example of automatic loading used for resistance welding is the feeding of weld nuts from a hopper to the welding workstation. A rotary feeder positions the weld nuts in the proper orientation, as shown by Fig. 15.27. They are fed to the fixture and properly located. The resistance welder makes the projection weld to complete the part shown in the picture.

A fixture must be rugged and strong to allow rough handling yet maintain parts in the correct position. In the case of strongbacks, sufficient strength should be built into the fixture to compensate for warpage stresses in the weldment. It should eliminate, or at least greatly reduce, the weld distortion inherent in the arc welding process.

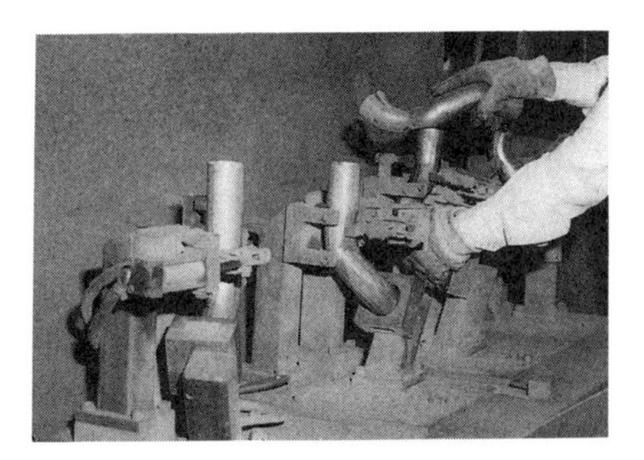

Figure 15.23 Clamping tubular parts.

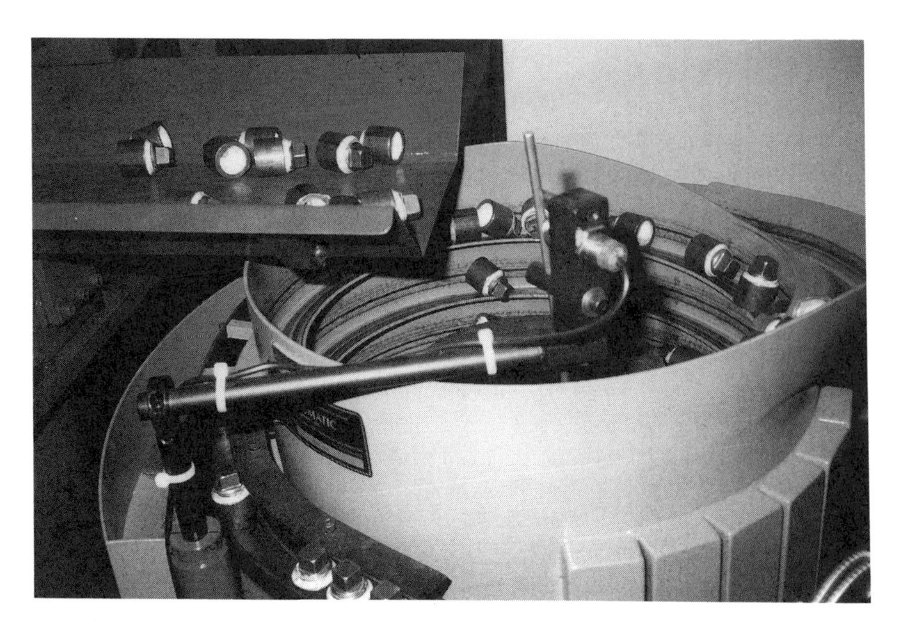

Figure 15.24 Automatic dial feeder.

Figure 15.25 Parts on conveyor belt.

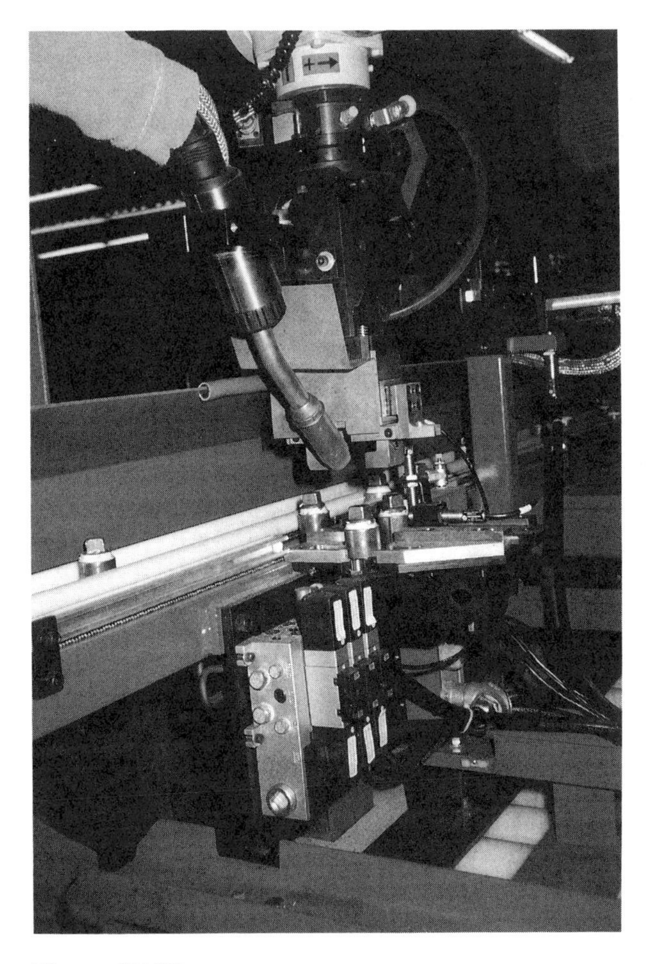

Figure 15.26 Gripper on welding gun.

The fixture design should be approved by the welding department before the fixture is built. If the fixture is to be built by a separate organization that produces work-holding, work-positioning fixtures, they must understand the specifications and should be experienced in similar fixtures. The quality of the manufactured fixture should be tested by making test weldments with the mechanized equipment to be used for the final product to ensure that the weldments will meet the agreed-upon specifications.

The resultant fixture should be easy to load, allow for unattended welding, hold the weldment in proper location and prevent distortion, and it should be unloaded easily by a person or by a crane and easily transferred to the next workstation.

Figure 15.27 Feeding resistance weld nuts.

All fixtures must be properly identified and stored so they can be called up again for the next production run of the weldment for which they were designed.

Robot Tooling Fundamentals

1. Fixture the weldment so the weld joint is accurately located. Review what this does to other dimensions that must be accurately held. Open the other dimension tolerances, if possible, so as not to move the weld joint.
2. The tolerance on weld joints should not be more than plus or minus half the electrode diameter for gas metal arc welding. For gas tungsten arc welding, the tolerance should be ±0.020 in. (±0.5 mm).
3. Protect machined areas, including shafts, slides, and rotation joints, from weld spatter. Use a boot or similar protective device.
4. Protect air lines and electrical leads from weld spatter. Use protective hoses or similar protective devices.
5. Check the design for accessibility to the welding torch. Keep an actual

welding torch handy. If the actual weldment is not available, use a cardboard mock-up.

6. If the fixture requires clamps to be repositioned during the welding operation or to complete the weld, always put sensors on the clamp so the robot knows what position the clamp is in when the torch is welding near it.
7. Program the robot to clear manual clamps when they are either open or closed. If a clamp is left open, this could prevent a collision.
8. Dedicated work-holding fixtures are usually mounted on a welding positioner. Be sure that the weight of the fixture, including the weight of the workpiece, is within the specified weight capacity of the positioner. Make sure also that it is within the off-center load capacity of the positioner.
9. Mark the weight of the fixture as well as identification numbers on the fixture for future reference.
10. Bevel the edges of the fixture parts that are close to the weld joint. This provides more torch accessibility and may prevent crashes.

Most robot manufacturers will provide a videotape of the robot welding the part. This can be used to confirm tooling design.

References

1. AWS, *Standard Welding Terms and Definitions,* AWS A3.0-89, The American Welding Society, Miami, FL.
2. AWS, *Classification and Application of Welded Joints for Machinery and Equipment,* AWS D14.4, The American Welding Society, Miami, FL.
3. AWS, *Specification for Welding of Presses and Press Components,* AWS D14.5, The American Welding Society, Miami, FL.
4. AWS, *Standard Symbols for Welding, Brazing, and Nondestructive Examination,* AWS A2.4, The American Welding Society, Miami, FL.

16

Selecting Welding Materials

16.1 Introduction

All metals used for strength are weldable; they possess *weldability.* Weldability is the capacity of a material to be welded under shop conditions into a properly designed part that will perform satisfactorily for its intended service. All metals are weldable, but some metals are more difficult to weld than others. The welding process and welding procedure must be considered when determining the weldability of a particular metal.

To produce a high-quality product you must know the product's service requirements and loading and the environment to which it will be exposed. You must understand the design of the welds so that the joint's strength will be equal to or greater than that of the material being welded. The material to be welded must be considered from all points of view—its physical properties, mechanical properties, chemical composition, metallurgical structure, and so on.

This chapter provides information on selecting the materials needed to make the weld, to match the welding procedure so that the weld will be as strong as the base metal.

The physical properties of a metal include its density, melting point, boiling point, hardness, color, and electrical conductivity. The composition or analysis of the base metal must be known; this can be determined by its specification. Other physical properties of the metal are its thermal conductivity, which is the rate at which heat is transmitted through the part; coefficient of thermal expansion, which is a measure of the linear increase per unit length per unit change in temperature; and specific heat, which is a measure of the quantity of heat required to increase the temperature of the metal. All of these properties are tabulated in engineering or metallurgical handbooks.

The mechanical properties of a metal include its tensile strength, yield strength, ductility, which is measured by the elongation or stretch of the metal, and impact resistance.

The most often welded metal is mild or low-carbon steel. However, alloy steels, stainless steels, ultrahigh-strength steels, nonferrous metals, and even cast iron are welded with automatic systems.

For most automated and robotic welding, the gas metal arc process is most commonly used. However, both flux-cored arc welding and submerged arc welding are employed in appropriate cases. The gas tungsten arc welding process is often used for welding more exotic metals; in other cases plasma arc welding is employed. Process selection is vitally important because the process helps determine what filler metals are to be employed.

16.2 Selecting the Filler Metal

In selecting the filler metal, every effort should be made to match the physical and mechanical properties and chemical composition of the base metal. If the exact chemical analysis cannot be provided, attempt to overmatch the strength of the filler material with that of the base metal. In overmatching, the yield strength and ductility of the deposit weld should exceed those of the base metal. Match these mechanical and physical properties with a filler material or electrode and provide a welding procedure to produce a quality weld deposit.

Selecting the filler metal for welding a particular base metal becomes much easier when the filler metal specifications of the American Welding Society are employed. The AWS provides over 30 different filler metal specifications, covering almost every filler metal obtainable. These are consensus standards sponsored by the society and accepted by the welding industry throughout the United States and used throughout the world. AWS specifications are based first on the type of material involved, usually an electrode, but they also cover fluxes. They then relate to a particular welding process or several processes, and finally they relate

Table 16.1 AWS Filler Metal Specifications and Applicable Welding Processes[a]

AWS specification	Specification title	GTAW	GMAW	FCAW	SAW
A5.7	Copper and Copper Alloy Bare Welding Rods and Electrodes	X			
A5.10	Bare Aluminum and Aluminum Alloy Welding Electrodes and Rods	X	X		
A5.12	Tungsten and Tungsten Alloy Electrodes for Arc Welding and Cutting	X			
A5.17	Carbon Steel Electrodes and Fluxes for Submerged Arc Welding				X
A5.18	Carbon Steel Filler Metals for Gas Shielded Arc Welding	X	X		
A5.19	Magnesium Alloy Welding Electrodes and Rods	X	X		
A5.20	Carbon Steel Electrodes for Flux Cored Arc Welding			X	
A5.22	Flux Cored Corrosion-Resistant Chromium and Chromium Nickel Steel Electrodes			X	
A5.23	Low-Alloy Steel Electrodes and Fluxes for Submerged Arc Welding				X
A5.28	Low Alloy Steel Filler Metals for Gas Shielded Arc Welding	X	X		
A5.29	Low Alloy Steel Electrodes for Flux Cored Arc Welding			X	

[a]An X signifies that the filler metal is suitable for that process.

to the metal to be welded. A list of the specifications that are available from the American Welding Society is included in their catalog. Table 16.1 is a summary of AWS specifications most commonly used for mechanized welding. The AWS specifications can be classified into four basic categories: (1) covered electrodes, (2) solid (bare) electrode wire or rod, (3) fabricated (tubular or cored) electrode wire, and (4) fluxes for welding, brazing, etc.

In general, AWS specifications are written to provide the specific chemical composition and mechanical properties of the deposited weld metal. They may also specify other properties such as toughness, and many of them include usability factors showing the welding position, welding current, quality standards, and size and packaging information.

Each specification includes a classification numbering system for delineating specific electrodes. These are peculiar to the specification and are different for the different processes. In general, they relate to the tensile strength of the deposited metal. Obtain copies of the AWS specification that applies to the process and metal you plan to use.

It is beyond the scope of this book to cover welding metallurgy, weld microstructure, or the exact filler metal recommended for high-alloy steels or nonferrous metals. Refer to the book *Modern Welding Technology* [1] for this information.

For welding steel with GMAW use the specifications AWS A5.18 and A5.28. The classification system is essentially the same for both (see Fig. 16.1). The first digit, E, indicates that it is an electrode. The next two digits stand for the minimum tensile strength in thousand pounds per square inch. For carbon steel electrodes, this number is 70; for low-alloy steel electrodes the number can be 80 or higher. The next digit, which is the letter S, indicates a solid electrode wire. The last digit (X) follows a dash and indicates the shielding gas or chemical composition. These data are summarized with the mechanical properties in the specific specification. By referring to the specification, decide which dash

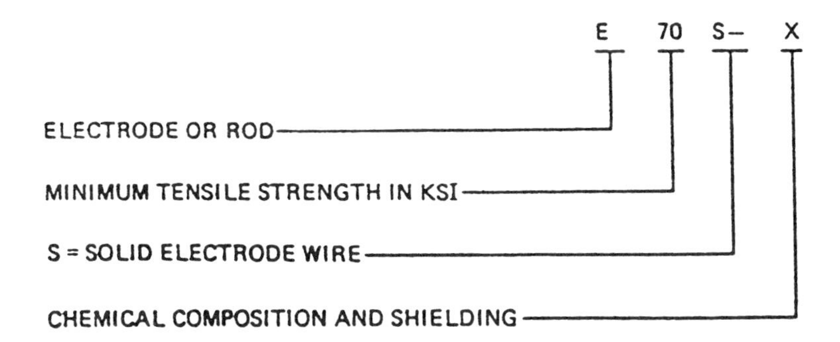

Figure 16.1 AWS designations for GMAW electrodes for steel.

number is appropriate. The specifications give the welding current and polarity, the normal shielding gas, the minimum tensile strength, the minimum yield strength, and the elongation. They also will show the impact requirements when specified.

The size of the electrode depends on the welding position, the variation of the process, the shielding gas, the size of the weld desired, and the thickness of the material being welded.

For the flux-cored arc welding process, use AWS specification A5.20 (carbon steel electrode) or A5.29 (low-alloy electrode). See Fig. 16.2. Since this process uses an electrode, the first digit in the classification is the letter E. The second digit stands for the minimum tensile strength, as welded, in 10 thousand pounds per square inch. The third digit stands for welding position—0 indicates flat or horizontal position; 1 indicates all-position welding. The next digit, T, indicates a tubular or flux-cored electrode. The digit following a dash designates the external shielding medium and welding power to be employed. There are four options.

1 indicates the use of CO_2 gas for shielding and direct current electrode positive (DCEP).
2 indicates the use of argon plus 2% oxygen for shielding and DCEP.
3 indicates no external shielding gas, i.e., self-shielding, and DCEP.
X indicates that gas shielding and polarity are not specified.

The designation is slightly different for AWS A5.29 for low-alloy flux-cored welding electrodes. In this case, the first digit following the T indicates the usability and performance capability. This is followed by a dash and other digits

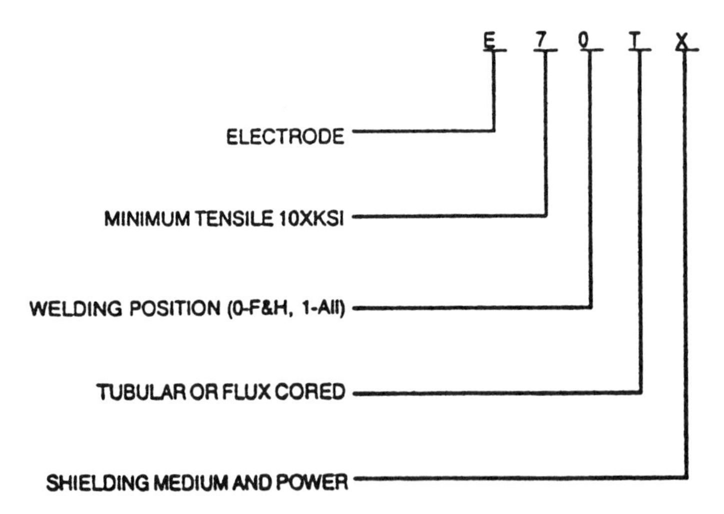

Figure 16.2 AWS designation for FCAW electrodes.

that designate the chemical composition of the deposited weld metal. Please refer to the specification for complete information.

In the case of stainless steel electrodes, the first three digits following the E indicate the AISI designation, and the digit following the T indicates the shielding medium to be employed. By using this designation and referring to the specifications it is possible to match the filler metal to the base metal.

The specification for submerged arc welding filler materials are entirely different from those just mentioned because submerged arc welding uses two consumable materials, the welding electrode and the welding flux. The specifications are AWS A5.17 for carbon steel electrodes and fluxes and AWS A5.23 for low alloy steel electrodes and fluxes.

The flux is specified by a four-digit system, shown in Fig. 16.3. Flux is specified by the letter F, which is followed by a two- or three-digit number that indicates the minimum tensile strength in increments of 10 thousand pounds per square inch. This is followed by a letter that indicates the condition of heat treatment of the weld: A for "as welded" and P for postweld heat-treated. This is followed by a one- or two-digit number that indicates the minimum temperature in degrees Fahrenheit of impact tests to provide 20 ft-lb of energy absorption, or the minimum temperature in degrees Celsius of an impact test to provide 27 J of energy absorption. A Z here indicates no requirement.

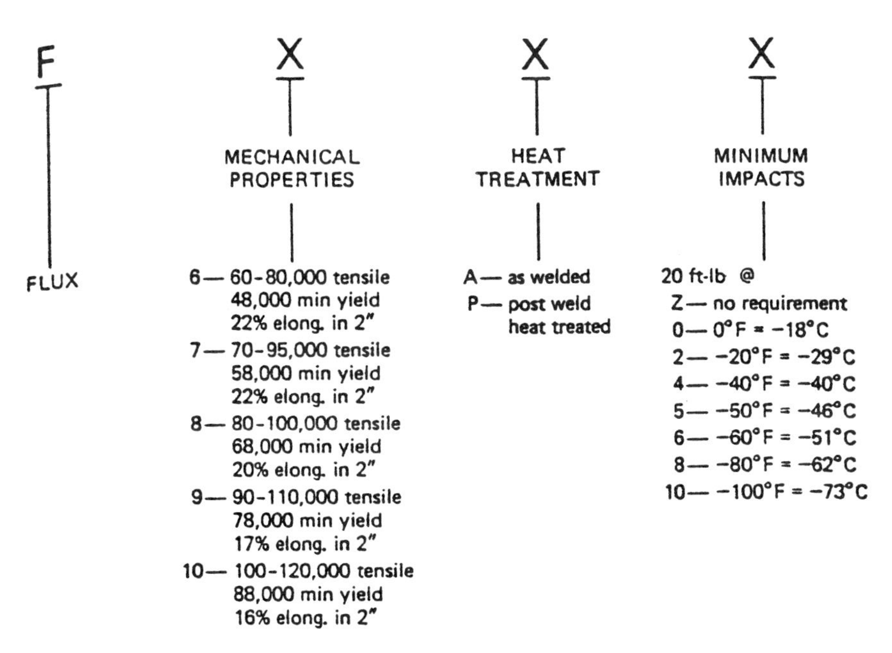

Figure 16.3 AWS designation for submerged arc flux.

The submerged arc welding electrode is specified by the letter E followed by a series of digits. However, the E can be followed by the letter C if the electrode is of composite construction, that is, flux-cored. (See Fig. 16.4.) The omission of C indicates a solid steel electrode. The next digit designates the manganese content. This is followed by a one- or two-digit number used to indicate the nominal carbon content in hundredths. These digits are sometimes followed by the letter K, which indicates that the electrode steel was silicon-killed. If the steel was of another type, the K is omitted. Two final digits may be used to indicate the alloys that are present. This system does not apply to low alloy steels. For complete information on these steels, refer to AWS A5.23.

Submerged arc welding is rarely used for robotic or fully automated systems, but it is used in automatic and mechanized systems. One of the drawbacks of submerged arc welding is that flux is abrasive and can enter mechanical joints on fixtures, etc., and create problems.

Aluminum is normally welded with the gas metal arc or gas tungsten arc welding process and in some cases with plasma arc welding. Aluminum is an active metal; it reacts with oxygen in the air, which produces a thin, hard film of aluminum oxide on its surface. The melting temperature of aluminum oxide is almost three times the melting temperature of pure aluminum, and for this reason the oxide must be removed prior to welding. Anodized coatings must also be removed from the welding area before welding. This can be done mechanically or by chemical means. With alternating current welding, there is a phenomenon known as cathodic bombardment that occurs during the half-cycle when the electrode is positive (reverse polarity DCEP). This electrical phenomenon blasts away the oxide coating to produce a clean surface. A portion of the

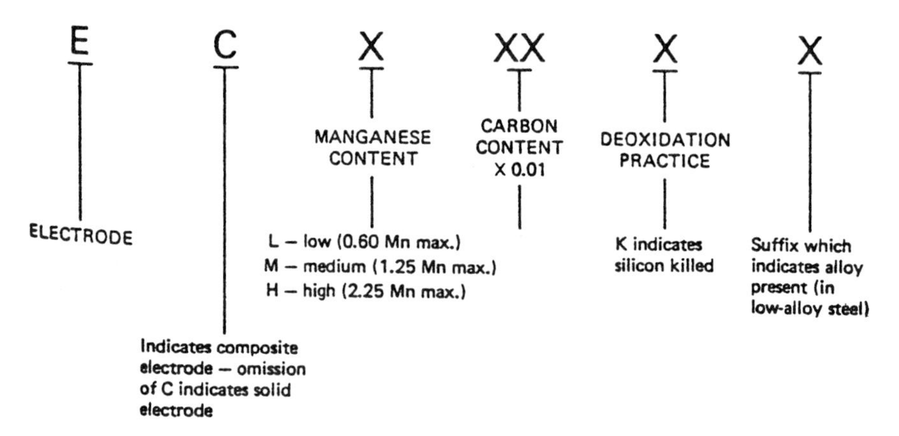

Figure 16.4 AWS designation for submerged arc electrode.

electrode-positive cycle creates the cleaning action and allows for quality welds. Alternating current is not used for gas metal arc welding.

It is important to match the analysis of the filler material with that of the base metal. A four-digit aluminum identification system is used. All aluminum can be identified by means of this system. There is also a designation system indicating the temper of the aluminum. It is important to remember that extra strength provided by heat treatment or cold working of the aluminum base metal is destroyed adjacent to the weld during the welding operation. Please refer to AWS A5.10 for designations for filler materials. This specification includes a chart showing the filler metal to be used for joining different aluminum alloys.

For welding other metals such as titanium, high-strength steels, nickel-based alloys, and copper-based alloys, the chemical composition of the weld metal must clearly match the chemical composition of the base metal. This is to provide the weld with properties that closely match those of the base metal.

For welding other materials, refer to the proper AWS filler metal specification for information on matching the filler metal to the base metal.

16.3 Selecting a Shielding Gas

Shielding is required while making any arc weld to prevent contact between the atmospheric air and the molten weld metal. In shielded metal arc welding (stick welding), gas for shielding is generated by the decomposition of chemical surrounding the electrode. In flux-cored arc welding, gas for shielding is generated by the decomposition of the chemical within the flux-cored electrode. External gas shielding for gas metal arc welding, gas tungsten arc welding, and plasma arc welding is provided by means of a gas nozzle that directs the shielding gas around the arc and the molten metal. Flux-cored arc welding with certain types of electrodes also requires external gas shielding. Gas for shielding is supplied in high-pressure cylinders and/or is piped through a factory to each workstation and from there to the nozzle that directs the shielding gas around the arc area.

Shielding gases are either inert gases or active gases. Only inert gases, argon and helium, are used for gas tungsten arc welding and plasma arc welding on aluminum, magnesium, and other metals that have an affinity for oxygen. Inert gases can be used with the gas metal arc welding process, but they are relatively expensive. Carbon dioxide, an active gas, is widely used for gas metal arc welding of steels. Carbon dioxide is an active gas because it dissociates in the arc, leaving oxygen and carbon monoxide in contact with the molten weld metal. Hydrogen, in small amounts, is sometimes used in a shielding gas mixture because it is a reducing gas. Hydrogen may be used in both mixtures of inert gases and mixtures of active gases.

Shielding gases used for arc welding must be dry. They are specified as "welding grade," which indicates that they have a dew-point temperature of –40°F or lower. Wet shielding gas is a source of porosity in weld metal. Table 16.2 lists properties of inert and active gases.

Shielding gases are mixed by gas suppliers and are provided in cylinders by local welding distributors. Different shielding gas mixtures are provided to produce specific characteristics for the weld. Some three-component gas mixtures are provided to reduce or greatly eliminate weld spatter. Other gas mixtures are produced to increase deposition rate, etc. Table 16.3 is a summary of the applications of shielding gases that are commonly used for welding.

With flux-cored arc welding, when a self-shielding electrode wire is used, an external gas shielding system is not. This eliminates the need for tubing, gas nozzles, solenoids, and controls.

For high-volume users of shielding gas, bulk gas-handling systems or cryogenic cylinders that provide gas in liquid form are used. The gas is normally piped through the factory to individual workstations. Mixing valves may be installed at each workstation to provide the exact mix of shielding gas required.

Table 16.2 Shielding Gas Properties

Property	Argon	Carbon dioxide	Helium	Hydrogen
International symbol and cylinder marking	Ar	CO_2	He	H_2
Type of gas	Inert	Active oxidizing	Inert	Active reducing
Structure	Monatomic	Diatomic	Monatomic	Diatomic
Molecular weight	39.94	44.01	4.003	2.016
Boiling point (at 1 atm)				
°F	–302.6	–109°	–452.1	–422.9
°C	–184	–78.5	–269	–252
Specific volume (ft^2 lb) at 70°F, 1 atm	9.67	8.76	96.71	192.0
Density (lb/ft^3) at 70°F and 1 atm	0.1034	0.1125	0.0103	0.0052
Specific gravity (air = 1)	1.380	1.530	0.137	0.069
Thermal conductivity (Btu/hr)	0.0093	0.0085	0.0823	0.096
Ionization potential (eV)	15.7	14.4	24.5	13.5
Maximum allowable concentration (ppm)	—[a]	5000 ppm	—[a]	—[a]

[a]Nontoxic asphyxiant.

Table 16.3 Summary of Shielding Gases and Their Applications

Shielding gas	Gas reaction	GMAW and FCAW	GTAW and PAW
Pure gases			
Argon, Ar	Inert	Nonferrous	All metals
Helium, He	Inert	Nonferrous	Al, Mg, and copper and alloys
Carbon dioxide, CO_2	Oxidizing	Mild and low-alloy steels; some stainless steels	Not used
Two-component mixtures			
Argon mixtures			
Ar + 20–50% He	Inert	Al, Mg, and Cu and alloys	Al, Mg, and Cu and alloys
Ar + 1–2% O_2	Oxidizing	Stainless and low-alloy steels	Not used
Ar + 3–5% O_2	Oxidizing	Mild, low-alloy, and stainless steels	Not used
Ar + 20–30% CO_2	Slightly oxidizing	Mild and low-alloy steels; some stainless steels	Not used
Ar + 2–4% He	Reducing	Not used	Nickel and alloys; austinetic stainless steel
Helium mixtures			
He + 25% Ar	Inert	Al and alloys; Cu and alloys	Al and alloys; Cu and alloys
CO_2 mixtures			
CO_2 + up to 20% O_2	Oxidizing	Mild and low-alloy steels (used in Japan)	Not used
CO_2 + 3–10% O_2	Oxidizing	Mild and low-alloy steels (used in Europe)	Not used
Three-component mixtures			
Helium mixtures			
He + 75% Ar + 25% CO_2	Slightly oxidizing	Stainless steel and low-alloy steels	Not used
Argon mixtures			
CO_2 + 3–10% O_2 + 15% CO_2	Oxidizing	Mild steels (used in Europe)	Not used

16.4 Other Electrodes and Materials

The gas tungsten arc and plasma arc welding processes use a tungsten electrode in the arc circuit between the torch and the workpiece. This is a non-filler-metal electrode and is considered nonconsumable even though the tungsten will gradually erode during the course of welding. Tungsten is chosen because it has the highest melting point of any metal (6170°F; 3410°C). The American Welding Society specifies seven classes of tungsten and tungsten alloys (see AWS specifications A5.12). Table 16.4 gives the classification, approximate composition, and color code for each class. This specification also gives the electrode size; electrodes are available with diameters ranging from 0.2 in. (0.5 mm) to ¼ in. (6.4 mm). They are available in lengths of 3 to 24 in. (75–610 mm).

An important consideration for tungsten electrodes is the finish of the tungsten. "Clean finish" indicates that the electrode is chemically cleaned after drawing or swedging, and "ground finish" means that the electrode has been centerless ground to a uniform outside diameter and has a bright polished surface. The extremely smooth and perfectly round electrode is better able to conduct heat to the collet that holds the electrode in the torch. The ground finish is recommended for automatic welding applications.

The tungsten electrode classification is related to the electrode's current-carrying ability, strength, and cost. Electrodes of the EWP class, pure tungsten, are the least expensive and are used for general-purpose work. The alloyed electrodes contain ceria oxide, lanthanum oxide, thorium oxide, or zirconium oxide. They have better arc-starting characteristics and higher current-carrying characteristics and are more expensive.

Electrodes of different sizes and types have different current-carrying capabilities when run using either alternating current or direct current with the electrode positive. Current ranges are listed in Table 16.5. It is advisable to choose an electrode of the size that will be working close to its maximum current-

Table 16.4 Tungsten Electrode Specifications

AWS classification	Approximate composition	Color code
EWP	Pure tungsten	Green
EWCe-2	97.3% tungsten, 2% cerium oxide	Orange
EWLa-1	98.3% tungsten, 1% lanthanum oxide	Black
EWTh-1	98.3% tungsten, 1% thorium oxide	Yellow
EWTh-2	97.3% tungsten, 2% thorium oxide	Red
EWZr-1	99.1% tungsten, 0.25% zirconium oxide	Brown
EWG	94.5% tungsten, remainder not specified	Gray

Table 16.5 Current Ranges for Tungsten Electrodes

Tungsten electrode diameter		Typical current ranges for tungsten electrodes[a]					
				Alternating current unbalanced wave		Alternating current balanced wave	
in.	mm	DCEN, EWX-X	DCEP, EWX-X	EWP	EWX-X	EWP	EWX-X
0.010	0.30	Up to 15	na	Up to 15	Up to 15	Up to 15	Up to 15
0.020	0.50	5–20	na	5–15	5–20	10–20	5–20
0.040	1.00	15–80	na	10–60	15–80	20–30	20–60
0.060	1.60	70–150	10–20	50–100	70–150	30–80	60–120
0.093	2.40	150–250	15–30	100–160	140–235	60–130	100–180
0.125	3.20	250–400	25–40	150–200	225–325	100–180	160–250
0.156	4.00	400–500	40–55	200–275	300–400	160–240	200–320
0.187	5.00	500–750	55–80	250–350	400–500	190–300	290–390
0.250	6.40	750–1000	80–125	325–450	500–630	250–400	340–525

na, Not applicable.
[a]All are values based on the use of argon gas.

carrying capacity. The electrode should remain shiny after use. If it is too large or the current is too low, the arc will wander erratically over the end of the electrode surface. If the electrode is too small or the current too high, the electrode tends to overheat and appears to have a wet surface. Changing type or size will correct this problem.

Another consumable material used in automatic welding is the submerged arc flux. This was briefly mentioned in the section on filler materials for submerged arc welding. Metal powders are sometimes added to the flux to provide increased productivity or to enrich the alloy content.

In thermal spraying applications, the electrode wires used for metallizing are normally specified by brand name because there is no AWS specification. Thermal spray applications also use metallic powders. They are used for the plasma transferred arc (PTA) welding process and sometimes for laser welding. Powder types are usually sold by brand name also, but there are ASTM specifications covering many of them. In addition, there are ASTM specifications covering the size of individual particles, which must be matched to the process and the application.

16.5 Filler Metal Packaging

Filler metals, specifically continuous electrode wires, are packaged in a variety of forms to meet the needs of the user. For automatic welding operations it is advisable to use large packages of filler materials so there is no need to shut down the operation to change or replenish the electrode wire supply. Specifications require that all wire packaged on a spool or in a container must be of one continuous length. Once the spool or coil is started in the equipment, it will continue feeding until the supply is exhausted.

Continuous filler wire, both solid and flux-cored, is available in spools that range from 1 to 60 lb (0.45–27 kg) of wire (steel) per spool. They also come as coils, which require a special spider to hold them in the wire-feeding equipment. Smaller coils come in 50–60 lb (22½–27 kg) sizes (steel). These are usually for the smaller diameter wires and are individually packed in plastic bags and corrugated cardboard cartons.

Larger packages of carbon steel, solid electrode wire, and flux-cored wires are provided on reels. Large reels require special equipment for dispensing; motorized dispensers are available and are used to reduce the load on the wire feed motor. Reels of electrode wire are available in 250, 750, and 1000 lb (113, 340, and 453 kg) sizes. For flux-cored wire there are 250, 600, and 800 lb (113, 272, and 363 kg) sizes. Reels are normally palletized and protected with shrink wrapping. Reels are normally made of wood and are nonreturnable.

Another method of providing large quantities of electrode wire is by the use of drums or payoff packs. Drums are made of heavy cardboard and will contain 250, 500, or 700 lb (113, 227, and 317 kg) of electrode wire (steel). Wire is placed in the drums to ensure a snarl-free payoff, usually while the drum is rotated. If wire is removed from a payoff pack and the pack is not revolved, there will be a twist caused by each revolution of the coil of wire. This twist can be a problem in some feeding equipment and should be avoided by using a rotational device to dispense the electrode wire. Drums are covered with shrink film for protection and are usually nonreturnable.

Electrode wire is also provided in large coils, usually 1000 lb (453 kg). Normally coils are strapped to a pallet and shrink wrapped for their protection. Large coils require special dispensing equipment.

All packaging methods, except for the smaller reels, require some type of special dispensing equipment. The heavy inertial load on a wire feed motor to start the motion of large coils or drums shortens their life and reduces positive, accurate starting. Motorized equipment is available to help reduce the inertial loads when starting wire feeding from large coils or drums. Equipment of this type is available through welding distributors. Motorized dispensing equipment will add to the life of wire feed motors and make arc starting more reliable.

In addition to reducing the time needed for changing electrode wires, the larger drums provide electrode wire at a lower cost. This is an economic advantage.

There are two factors, cast and helix, related to the type of electrode wire packaging. AWS specifications require that the cast and helix of electrode wires must not interfere with feeding when automatic equipment is used. The *cast* of an electrode wire is measured by removing a loop or ring of wire from the spool. When cut from the spool and laid on a flat surface, the wire should form an unrestrained circle of not less than a specified minimum diameter based on the diameter of the coil and the wire specification.

The *helix* of a coiled wire is also measured on a loop or ring of wire, which is placed on a flat surface without restraint. The maximum distance of any portion of the loop above the flat surface must not be greater than a specified dimension.

Cast and helix can both cause wavering or uneven feeding of electrode wire by the wire feed system.

Reference

1. H. B. Cary, *Modern Welding Technology,* 3rd ed., Prentice-Hall, Englewood Cliffs, NJ, 1994.

17

Safety of Automated Welding Systems

17.1 Arc Welding Safety

Welding is an established manufacturing process with known potential hazards. However, it is no more hazardous than any other metalworking occupation, provided proper precautionary measures are followed. Welding is also a very fatiguing occupation. The introduction of the industrial robot seemed like a natural solution to remove the welder from a dangerous environment and at the same time eliminate the fatigue factor. Robotic welding largely eliminated eye and muscular fatigue but unfortunately introduced a new potential hazard. For this reason, potential welding hazards and robot safety hazards must be reviewed together.

Even though the human welder is further from the arc in mechanized welding, welding safety rules must still be followed. Every area where welding is performed must have a welding safety program. Management must insist that adequate safety precautions be taken to protect all employees. All welders and operators, as well as employees working around welding operations, must follow documented precautions and procedures. The source for this information is the ANSI standard *Safety in Welding and Cutting* [1].

Workplace safety is a moral and humanitarian issue. In addition, the costs of workplace accidents and illness are continuing to increase. Also, people are more productive and efficient when they are confident and comfortable in their environment. For numerous reasons, then, it is important that management provide a safe working environment.

Potential safety hazards associated with arc welding include arc radiation, air contamination, electrical shock, fire and explosion, compressed gases, and other hazards.

Arc Radiation

The welding arc generates intense radiation in the ultraviolet (UV), visible, and infrared (IR) ranges of the electromagnetic spectrum. The intensity of arc radiation depends on the welding current, the welding process, the size of the electrode, and the shielding medium. In many mechanized welding operations, the arc can be shielded from the operator and others by such means as curtains, doors, and shields, because the arc travel pattern is permanently established. In the case of robotic arc welding, the arc travel pattern routinely changes, and the welding operator and others may need to be shielded from the entire operation. Skin exposed to the arc can suffer ultraviolet and infrared burns similar to sunburn. Welding can produce a painful burn in a very short time. Welders and operators should wear protective clothing suitable for the welding to be done.

The eyes should be shielded from the radiation to eliminate arc *burn,* which will result if the eyes are not protected. Exposure of the eyes to an ultraviolet arc light results in a form of conjunctivitis that is painful for about 24–48 hr. Usually it does not cause permanent injury to the eyes. Infrared arc rays can cause fatigue of the retina of the eye. The effects of infrared radiation are not nearly as noticeable or as immediate as the effect of ultraviolet rays.

No one should be close to a welding arc without protection. The best protection for the eyes and face is provided by a head shield that has a window in it with a filter lens. Head shields are generally made of lightweight fiberglass. The filter lens is made of dark glass that is capable of absorbing infrared rays, ultraviolet rays, and most of the visible light radiated by the arc. It should be dark enough that the arc can be viewed without discomfort but not so dark that the welder cannot see the arc. Table 17.1 shows the lens shade recommended for use in arc welding [1,2]. The higher the lens number, the darker the lens. A clear glass cover plate should be put on the outside of the welding lens to protect it. Lenses are available in various sizes.

The closer the arc, the more intense the radiation. However, even people at a distance from the arc should have some type of eye protection if they can see

Table 17.1 Lens Shades Recommended for Viewing the Welding Arc

Welding or cutting operation	Electrode size (mm) or metal thickness	Welding current (A)	Minimum protective shade	Suggested shade no.[a] (comfort)
Shielded metal arc welding	< 3 (2.5)	< 60	7	—
	3–5 (2.5–4)	60–160	8	10
	5–8 (4–6.4)	160–250	10	12
	> 8 (6.4)	250–550	11	14
Gas metal arc welding and flux cored arc welding		< 60	7	—
		60–160	10	11
		160–250	10	12
		250–500	10	14
Gas tungsten arc welding		< 50	8	10
		50–150	8	12
		150–500	10	14
Air carbon	Light	< 500	10	12
Arc cutting	Heavy	500–1000	11	14
Plasma arc welding		< 20	6	6 to 8
		20–100	8	10
		100–400	10	12
		400–800	11	14
Plasma arc cutting	Light[b]	< 300	8	9
	Medium[b]	300–400	9	12
	Heavy[b]	400–800	10	14
Torch brazing		—	—	3 or 4
Torch soldering		—	—	2
Carbon arc welding		—	—	14
	Plate thickness			
	in.	mm		
Gas welding				
Light	< ⅛	< 3.2		4 or 5
Medium	⅛–½	3.2–12.7		5 or 6
Heavy	> ½	> 12.7		6 or 8
Oxygen cutting				
Light	< 1	< 25		3 or 4
Medium	1–6	25–150		4 or 5
Heavy	> 6	> 150		5 or 6

[a]As a rule of thumb, start with a shade that is too dark to see the weld zone. Then go to a lighter shade which gives sufficient view of the weld zone without going below the minimum. In oxyfuel gas welding or cutting where the torch produces a high yellow light, it is desirable to use a filter lens that absorbs the yellow or sodium line in the visible light of the (spectrum) operation.

[b]These values apply where the actual arc is clearly seen. Experience has shown that lighter filters may be used when the arc is hidden by the workpiece.

the arc. At a great distance, tinted safety glasses with side shades of dark material are probably sufficient. Welding areas should be surrounded by radiation screens to protect adjacent workers.

Air Contamination

Welding smoke and fumes are generated in the arc by the melting of the electrode material and the decomposition of fluxing ingredients. The welding fumes contain two types of air contamination, particulate matter and gases. Warning labels are attached to all packages of electrode and filler wire. They mention that the welder should keep his or her head out of the fumes and/or use enough ventilation exhaust at the arc to keep fumes and gases from the breathing zone and the general area. This is a major advantage of mechanized welding. The welder doing manual welding cannot get very far from the arc.

The welding work area must be adequately ventilated. Covered electrode welding, which is not done by mechanized procedures, and flux-cored arc welding produce the most contaminants. Gas metal arc welding, gas tungsten arc welding, plasma arc welding, and submerged arc welding produce the least. Ventilating systems that collect the contaminants at the arc are most efficient. Two types of systems are used; one collects fumes near the arc, and the other uses special nozzles that collect the fumes. Both types of local ventilation are shown in Fig. 17.1. They can also be used for mechanized and robotic welding applications. General ventilation is also required to keep the contamination level, or the

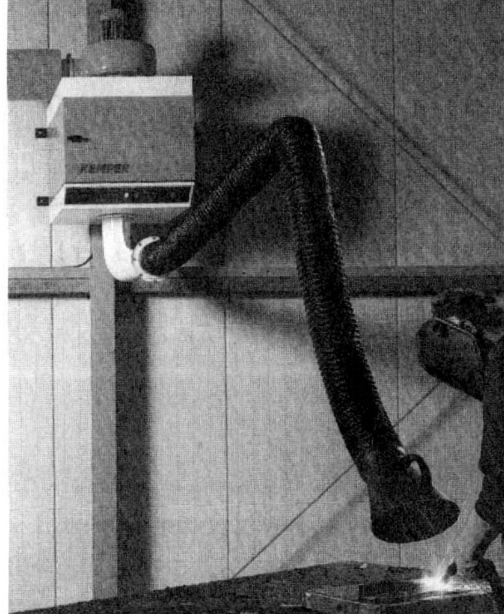

Figure 17.1 Local ventilation equipment.

smoke level, down throughout the entire manufacturing area. Specific requirements are presented in Ref. 1.

Smoke and fumes can also be created during welding on coated materials. Drawing compounds, oil, protective finishes, and so on will generate fumes when exposed to the welding arc. The coated metal should be thoroughly analyzed from a ventilation point of view. Welding on galvanized or zinc-coated material, on cadmium-coated material, on lead-coated steel, or on chromium- or nickel-plated steel can cause hazardous fumes. Extraspecial precautions must be taken when welding on these coated metals.

Certain welding processes and procedures will cause a concentration of gases. Ozone and nitrogen oxides are severe respiratory irritants. Carbon monoxide is generated in the arc when carbon dioxide shielding gas is used; however, away from the arc the CO recombines with oxygen. Carbon dioxide, argon, and helium are nontoxic, but they can displace oxygen when welding is done in a confined, unventilated space for a long period of time.

Welding should not be done near degreasing operations. Arc rays will cause the disintegration of certain degreasing fumes and create lethal gases.

Electrical Shock

Precautions should be taken to reduce the hazard of electrical shock to the welding machine operator. It is important to make sure that the arc welding equipment, mechanized equipment, or robot is installed according to code and properly grounded to earth. Electrical equipment should be maintained in accordance with applicable codes and regulations. The cabinet of welding power sources, electrical controls, and operator pendants must be electrically grounded. Welding cables, electrical connections, and terminals must be tight and insulated. Cables with frayed or cracked insulation and faulty or badly worn connections can cause electric short circuits and shocks. Welding cable must be of the proper size. This is particularly important in automatic applications because higher duty cycles and currents are normally used. Overloading of welding cables creates excessive heat buildup in the cables and destroys the insulation. Improperly insulated welding cable is both an electrical shock hazard and a fire hazard. The welding area should be dry and free of any moisture such as standing water that will increase the chance for electrical shock.

Fires and Explosions

Welding produces hot metal and molten slag that can cause severe burns. Welding sparks, hot metal, and spatter can start a fire or cause an explosion. The welding area must be kept free of flammable, volatile, or explosive materials. Welding on tanks or containers that may have held flammable liquids can be dangerous because explosive atmospheres may exist in such "empty" tanks. Fuel gases often used in welding operations may collect in low areas and cause explosions. Fires can also be started by electrical shorts or by overheated, worn cables. Halon 1211 fire extinguishers must be available in all welding operation areas around electronic equipment.

Compressed Gases

Compressed gases such as shielding gases, fuel gases, and oxygen are used in welding departments. All compressed gas cylinders should be handled with extreme care. The cylinder caps should be on the cylinders when they are not in use. All gas cylinders should be secured to the wall or other structural support. They should never become a part of an electric circuit, and they should be protected from mechanical shocks. When compressed gas cylinders are empty, the valves should be closed and the cylinder marked empty. Compressed gas cylinders should be stored in a safe place with good ventilation. Fuel gas cylinders should be stored separately from oxygen cylinders. Acetylene cylinders should be stored and used in the vertical position only. Gas hoses and lines should be checked periodically for leaks. Gas apparatus such as regulators, flow meters, and solenoid valves should also be checked frequently for proper operation and for leaks. Hose and apparatus should be repaired as necessary.

Other Welding Hazards

Welders and welding machine operators are exposed to cumulative trauma disorders. Work at automatic welding stations requires much bending, stooping, and stretching during loading and unloading. These motions can put tremendous strains on muscles, especially in the lower back. Welders and machine operators should minimize stooping or crouching, always get help when lifting heavy loads, and use proper lifting techniques or equipment.

Most welding processes are not excessively noisy; however, noisy work such as grinding, polishing, and cutting are often done in the welding department. Noisy operations should be isolated as much as possible from the actual welding. These include slag shipping and weld grinding. Appropriate protective clothing and equipment should be worn.

17.2 Hazards of Mechanized and Robotic Welding Systems

Robots were originally designed to duplicate the job functions of a human. They were designed to relieve humans of the drudgery of unpleasant, fatiguing, or repetitive tasks and also to remove humans from a potentially hazardous environment. In this regard, robots can replace humans in the performance of dangerous jobs and are considered beneficial for preventing industrial accidents. On the other hand, robots have caused fatal accidents.

The potential hazards arising from the use of mechanized or robotic welding equipment are the inherent dangers of moving machinery. For example, a person may be injured by moving machinery as a result of being hit with it or of being trapped between the moving machine and fixed object. A person can also be injured by becoming entangled in the machinery or by being struck by material ejected from the machinery. These are mechanical hazards. Other potential hazards can include electrical hazards, air contamination from leaking gases, burns from hot metal or spatter, and, of course, arc welding hazards. Mechanized welding equipment poses these threats, but robotic arc welding equipment can be even more dangerous. In general, mechanized welding machines have a fixed location and operate within a particular base area. They usually have only one or two axes of motion, and they move at relatively slow speeds. Guards are easily attached. Automatic welding machines can be made safe and practically "idiot-proof" because most of them have only one or two operating stations.

Robots work beyond their base area and have large work envelopes that may overlap with those of adjacent machinery. The travel speed is fast, and robots are multidirectional, operating with as many as six or more axes. Additionally, they start up suddenly and change direction abruptly during motion.

Due to the variable nature of robot applications, specific safety hazards for each installation must be studied on an individual basis. For a robot to be truly effective, it must maintain a high degree of flexibility. This implies that the working envelope must be unrestricted to allow for program and path changes. Robots work best when they stand where a person once stood next to other people. Unfortunately, a robot performing the same function as a human will occupy more space than the human. The primary safety rule is that the human and the robot should not occupy the same working space at the same time. The nature of arc welding requires a person to be close to the arc while programming or analyzing a program. To remove the human from the arc area limits the flexibility and accuracy of the robot. Hence, one of the major problems associated with robotic arc welding is the presence of the human programmer in close proximity to the welding torch held by the robot. For small weldments this is not a major problem. For very large weldments, the parts are heavy and the completed weldment will require cranes for loading and unloading. These introduce the normal problems of safety with respect to materials handling. However, they also introduce the problem of bringing heavy pieces of material into the robot's work envelope.

One of the best solutions for robot safety is to purchase a complete welding cell from a robot vendor. A complete cell includes barriers, all necessary safety devices, and a method of loading and unloading the workstation. It is best used for the production of smaller weldments. In general, a turntable, turnover, or shuttle device is used for loading parts outside the robot's work envelope. The parts are then presented to the robot inside the barrier where the welding is performed. After the welding is completed, the parts are transferred to outside the barrier, where they are unloaded and sent to their next destination. Figure 17.2

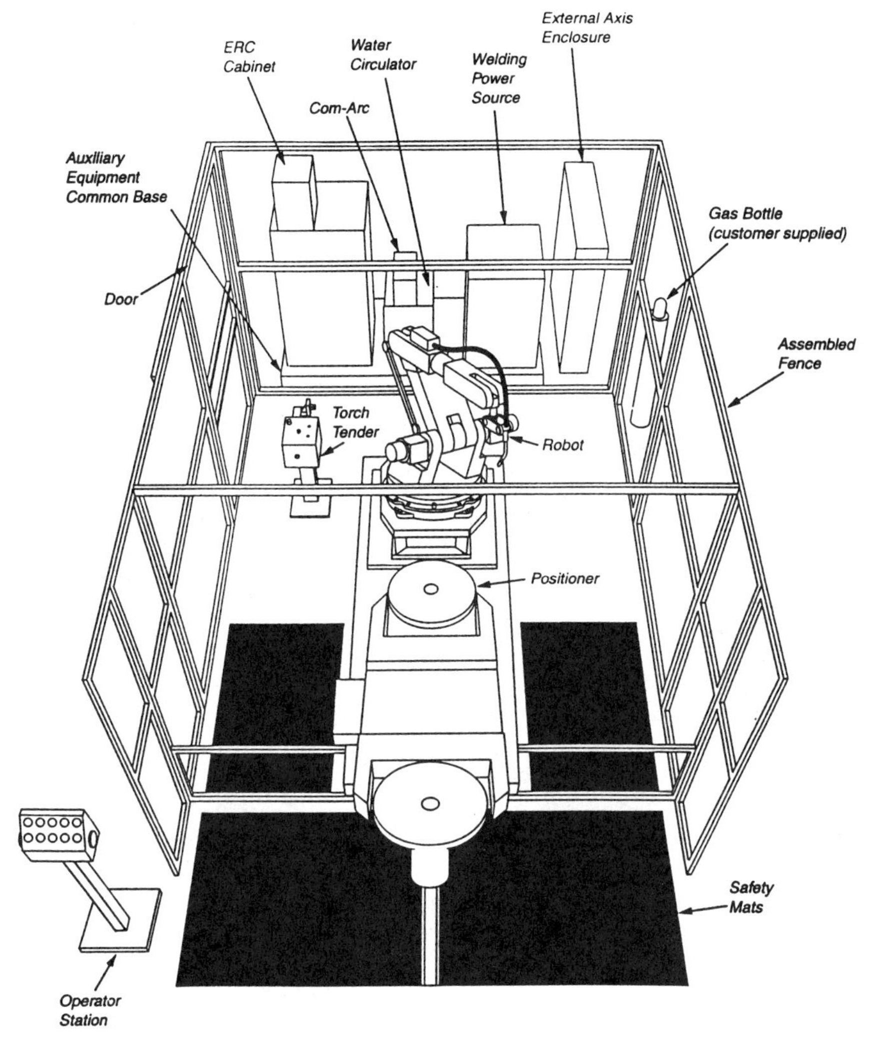

Figure 17.2 Layout of a robot cell with 180° index positioner.

shows a pictorial view of a robot cell with a rotary 180° indexing table at the workstation. A photograph of a robot cell similarly equipped is shown by Fig. 17.3. This cell is for small weldments that can be loaded and unloaded outside the barrier. Figure 17.4 shows a turnover index positioner. A robotic welding cell layout with two pneumatic indexing shuttle positioners and two robots is shown in Fig. 17.5. Work-holding devices can also be loaded and unloaded automatically, in which case they must work with a material motion device that presents additional safety hazards. This then becomes a special engineered project.

The robot industry has adopted special safety graphics (Fig. 17.6). These have not yet been accepted in the National Standard R15.06 but are used by many robot suppliers. They should be posted at appropriate locations. Finally, one warning sign that is agreed upon by all is shown in Fig. 17.7: DANGER—DO NOT ENTER—MACHINE MAY START AT ANY TIME. This should be posted at the cell access door.

Figure 17.3 Photo of a robot cell with 180° index positioner.

Figure 17.4 Robot cell with turnover index positioner.

17.3 Safe Design and Layout

Management is responsible for providing a safe workplace for its employees. Each robot installation must be carefully planned from a safety viewpoint to eliminate hazards [3]. When the robot is in operation it is necessary that people remain outside the work envelope. Barriers or fences should be in place around the robot. This is difficult to accomplish when robots are welding on large weldments or when two or more multiaxis robots are working together. Restricted space within a barrier must be kept at a minimum. Bringing parts to the robot and removing weldments from the work area must be continuous. Barriers must be engineered for each specific installation. Barriers themselves can create pinch points; therefore, there must be sufficient space between the robot's work envelope and the barrier to avoid crushing a human. Openings

Figure 17.5 Robot cell with two shuttle index positioners.

in barriers must have interlocks that immediately remove power from the robot when a door is opened.

Barriers must be designed to completely surround the robot and eliminate the possibility of people climbing over or under to get inside the barrier. Signal lights must be arranged on the robot or in the robot area, either flashing or rotating to indicate that the robot is powered.

Emergency stop buttons must be placed on the exterior of the barrier to enable personnel to stop the robot if an accident occurs or is perceived. Restart

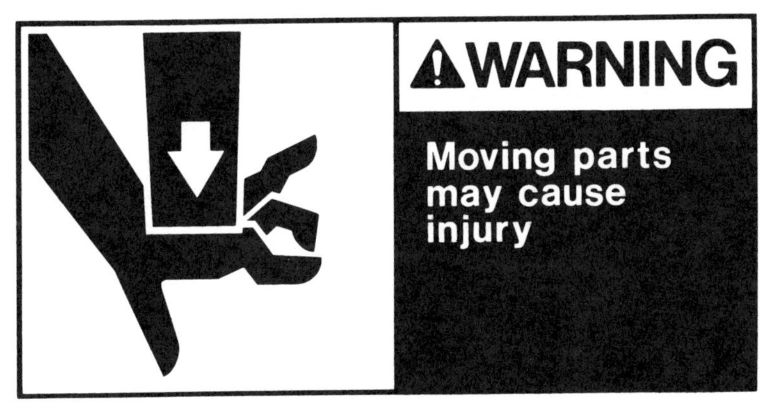

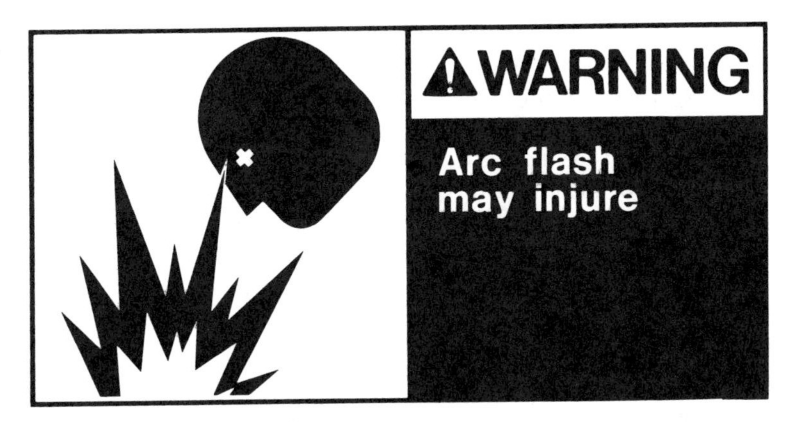

Figure 17.6 Safety graphics.

Figure 17.7 Danger warning sign.

buttons should be placed at a control station so that the robot can be restarted after it is powered down for any reason.

Special rules must be set out with respect to robots. They should prohibit anyone from being inside the barrier when the robot is in its production cycle. Sometimes it is allowable to provide positive stops for the robot for certain motions or provide certain restrictions to the work envelope. It is understood that such stops provide pinch points that are dangerous.

The robot's computer controller software must disallow specific motions or locations that would interfere with adjacent machinery or other accommodations. Software should not be relied upon to provide a fail-safe stop mechanism.

Pneumatic or hydraulic robots are rarely used for welding. However, they may be used for fixture motions and for powering auxiliary equipment. Provisions should be made to deenergize such systems in the case of an emergency stop situation. Hydraulic fluid should be nonflammable.

Loading and unloading work is potentially hazardous. One of the most widely used safety devices also increases robot productivity. This is the double-ended fixture, that has one end inside the barrier where the robot makes welds while the other end is outside the barrier where the operator loads the parts to be welded and removes weldments. There are several types of two-position indexing positioners. The most common is the 180° indexing table, which is restricted to small weldments. Another type rotates about a horizontal axis and is designed for long slender weldments. These were shown in the previous section. When indexing positioners are used in conjunction with fixed barriers, they prevent the operator from entering the robot's work envelope.

Robot manufacturers provide basic safety guidelines for their robots, but these guidelines are not included in the design of the robot. It is the responsibility of the owner to provide a safe layout and safe interrelationship with other mechanized units or robots.

Once the robot is installed, it should be carefully inspected from the point of view of risk avoidance. It should be proof-tested and every attempt made to circumvent the safety features. If any problems are discovered, they should be rectified and the proof test repeated. If the robot passes the safety evaluation, it should be turned over to the production department.

17.4 Safeguarding the Welding System

Safeguarding the arc welding robot requires very close attention. It requires the services of people with experience. Robot vendors and robot integrators can be very helpful in providing assistance in safeguarding the robots. For additional information, refer to the ANSI safety standard referenced previously [3]. This document, sponsored by the Robots Industries Association, is the most complete source of robot safety information. The section on installation covers earth grounding of equipment, location of controls, robot system clearances, power disconnects, mechanical limiting devices, environmental conditions, precautionary labels, restricted envelopes (space identification), dynamic restricted envelopes, and robot emergency stops. It suggests that safeguarding may include the presence of safeguard devices, awareness barriers, and awareness signals. It also brings out the fact that the owner of the robot system is ultimately responsible for its safe operation.

Many commercial devices are available to ensure the safety of those working with robots. These include at least the following.

1. Barriers needed to prevent personnel from entering the restricted envelope or workspace. Required access must be through interlock doors that will stop the robot. To restore automatic operation, it must be necessary to exit the restricted envelope, restore the safeguards, and initiate a deliberate start-up procedure. This procedure should be designed for fail-safe operation but should not prevent the normal stopping action of the robot. It should not allow the robot to operate until the component failure has been corrected.
2. Lockout, tag-out systems for power are required to safeguard maintenance and repair personnel. It is understood that maintenance personnel will be inside the restricted barrier. If during maintenance it becomes necessary to bypass any safeguards, alternative safeguards must be provided and the bypass method identified. Personnel servicing the robot must have total control of the robot during the maintenance period.
3. The problem of fixed barriers has been fully explored, and the fact that the barrier can contribute to a pinch point suggests that 18 in. (457 mm) of clearance should be provided. Interlocks for doors in the fixed

barrier must be provided and cannot be bypassed. Portable barriers can be used but are not recommended.

4. Other types of barriers such as light curtains and ultrasonic curtains have been widely used in the past for press operators. When the barrier is broken, the emergency stop action takes place. Other safety devices include two types of pressure-sensitive mats. One type requires the operator to be in position for the robot to function; the other can immediately initiate the emergency stop function if the person or operator comes within its area. These are awareness barriers that activate an awareness signal.
5. Limiting devices are mechanical stops that limit the motion of the robot to prevent it from moving into certain areas. Limiting devices create pinch points. There are also nonmechanical limiting devices that are associated with the control system and are capable of stopping robot motion. They must be designed to be fail-safe. Presence sensors are of many types and are used to detect the presence of humans within restricted areas. They must be connected to the emergency stop circuit so that if a person is in the wrong location, a sensor will immediately stop the robot operation.
6. Floor markings are used to indicate the work envelope of the robot and associated equipment. They cannot be relied upon as a positive safety feature. Indicator lights of various types are useful and make people aware that the robot is powered and is in the operating mode. The major problem is an unexpected or unintended robot movement. Many robots have built-in safety features. This includes such items as collision protection, which causes the robot to stop when the torch hits an object. Breakaway torch holders are also available. Another important feature is the slow-speed lock mode, which provides the teach speed limitation; this is usually 20% of the normal program speed. There are also machine locks, control cabinet door interlocks, twin palm button interlocks, which require the operator to push two buttons to energize the robot cycles, and internal diagnostic signals.

Safety Equipment Glossary

Alarm signals: bells, horns, sirens, whistles, light, etc., to alert, warn, caution, or prevent in an emergency situation.

Awareness signals: visual, audible, or both. To alert, warn, caution, etc., in an emergency situation.

Barrier, barricades: chain, rope, railing, fencing, traffic cones, etc. An area protection structure to prevent, caution, warn, or stop personnel from entering a potentially hazardous work area. Can be set up to automatically shut down the operating system when unauthorized entry occurs.

Card or key access system: A safety alarm lock system requiring a key or plastic card with a magnetic strip to gain entry to a potentially hazardous area. System provides for quick exit in case of emergency.

Double safety lockout devices: A lockout system requiring two people to operate before entry. Examples: two locks, two-key lock, two release buttons, two plastic cards, combination lock with two people required to enter numbers to open the lock. See also *Lockout device*.

Electric switch floor mats: Heavy-duty nonskid floor mats; when used to guard the working envelope, there is a switch wire to shut down or place operating equipment in a hold position.

Emergency lighting systems: Battery-powered lamps that automatically provide instant illumination in the event of failure of the central power supply.

Fencing: Steel woven wire fencing can provide low-cost and custom-designed panels, partitions, cages, or security rooms around work envelopes.

Fiber-optic photoelectric control system: Light curtain perimeter guard system provided by fiber-optic modulated light-emitting diodes.

Flashing light rope/chain: Flashing red or yellow lights built into a sturdy rope or chain; can be erected as a barrier to identify the work envelope and to warn of existence of a potentially dangerous area.

Floor/aisle area markers: Vinyl safety markers or tapes marking equipment for identifying hazardous areas, to caution personnel and direct traffic flow.

Floor cable guards/covers: Means of protecting and covering cables that lie on the floor. Prevent tripping and injury and withstand heavy traffic, reducing replacement and lost time.

Fume exhaust systems: High-volume industrial blower system that quickly removes fumes, smoke, and vapors from work area.

Handrailing: Steel, aluminum, or wood; makes an excellent barrier or guard for robots where the work area has limited access and the need is to prevent personnel from accidentally entering the work area.

Infrared light curtains: Invisible band or curtain of infrared light erected around an area to detect entry or penetration of a dangerous zone. System will automatically shut down or place in a hold condition all operating equipment during unauthorized entry.

Interlock gates or barriers: A protection system designed so that control of entry and machine operation coordination are interdependent in order to prevent unauthorized or accidental entry.

Laser sensing systems: A protection/detection system to prevent entry into a dangerous area or to warn of presence of a dangerous area using laser light sensors.

Light-shielding curtains: A curtain of PVC or similar material that combines see-through welding protection and temporary walls or partitions to isolate hazardous areas.

Lockout device: A special double interlocking device permitting a maximum of six separate locks to reduce accidents and injury resulting from the operation of dangerous equipment; Made of case-hardened steel with rustproof plating and tamperproof. See also *Double safety lockout device.*

Lock-out tag: An identity tag used in conjunction with the safety lock-out device to quickly identify the party or parties having their locks on the device. Tag also warns all other personnel of the existing danger.

Mezzanine/balcony partitions: Create additional space in almost any area providing safer working environment without incurring large costs.

Nonslip floor coating: Protects personnel against slips and skids on wet, oily, or smooth surfaces.

Photoelectric controls: Electronic control devices used with protection systems such as illumination curtains, reflective scanners, entry control, and equipment shutdown devices.

Pressure-sensitive floor mats: Heavy-duty vinyl floor mats with built-in pressure-sensing switches. Can be placed around a work envelope as a protection or warning barrier and interlocked to shut down or place operating equipment in a hold condition when the area is penetrated accidentally.

Proximity sensors: A device that senses that an object is only a short distance away and/or measures how far away it is. Proximity sensors work on principles such as triangulation of reflected light or elapsed time for reflected sound. Various types and models allow them to be used for many types of protection and prevention.

Retro-reflective sensing systems: Uses a single photoelectric sensor with a retro-target that is highly reflective and not critical as to angular alignment. Retro-reflective systems are a compromise between operation range and simplicity of installation.

Safety tape: Durable, high-visibility, fluorescent vinyl tape for marking and/or dramatizing dangerous or hazardous areas.

Tamper-resistant sensors/switches: Devices designed and housed to make it difficult to bypass, alter, remove, and/or affect or change the safety system they control.

Transmitted beam sensing systems: These systems offer the longest operation ranges but require more installation effort. Unit requires separate sensor heads for the source and detector. This system allows the greatest flexibility for special applications requiring precision detection.

Two palm button operator control stations: Workstation designed to place the operator outside robot working range. Start/stop buttons that must be operated with both hands at the same time allow the operator to control the system.

Voice-actuated warning system: Safety prevention system interfaced with pretaped voice warning or caution messages. Employees like system and will follow suggested messages. System allows messages to be changed frequently to meet specific problems.

Warning signs: Eye-catching phosphorescent signs to indicate or warn of dangerous areas, equipment, operation, protrusion, etc.

Wire mesh (steel): Inexpensive construction material for safety walls, partitions, cages, doors, rooms, etc.

To ensure a safe robot operating system, solicit help from your safety engineer, from public officials who provide such services, from insurance company consultants, and from the robot supplier, the robot integrator, or the supplier of safety equipment.

17.5 Safety Training

From the standpoint of safety considerations, there are three phases of robotic welding: During the programming of the robot, personnel must stand within the robot's work envelope; during the actual production operation of the robot, personnel are not permitted to enter the work envelope; and during maintenance or servicing of the robot, personnel must be close to the robot within its work envelope.

Statistics indicate that 90% of robot accidents occur during either the programming or maintenance and servicing phases.

As discussed earlier, a robot's owner and company managers are responsible for its safe operation. It is their responsibility to ensure that the robot is correctly installed, that the workplace is safe for personnel who work with the robot. Unfortunately, employees who work near and around robots often take them for granted; they do not take the hazards of working near robots as seriously as they should. They may even bypass safety devices. Continuous safety training is required. This should include frequent refresher courses emphasizing safety for programmers, operators, maintenance people, and others working with and near robots.

Close supervision is necessary for everyone, especially for newly trained employees. Supervisors must stress that operators should never be in the work area while a robot is operational. Supervisors must strictly enforce safety rules with disciplinary action for anyone who enters the work area while a robot is operational.

The following is a list of safety items that should be stressed repeatedly.

1. Always keep your eye on the robot during operation.
2. When programming, always plan an escape route from the robot work envelope. Know where the safe zones are within the barrier area.

3. When changing from one type of robot to another, provide special training for the new unit.
4. Keep the teach pendant with you at all times while working inside the barrier.
5. Test safety equipment at the start of every shift to verify its proper operation.
6. Keep unauthorized people out of the robot's work barrier area. The operator is responsible for the safe operation of the robot.
7. Know and understand the robot program. Recognize the motions and speeds so you can immediately detect any unexpected movement. Familiarize yourself with any changes that have been made.
8. Shut down the system if you suspect that the robot is not operating correctly.
9. Use lock-outs and lock-out tagging when performing any type of service or maintenance.
10. Inform your partner and the operator on the next shift of any changes in program and any problems that you suspect.

The ANSI standard for robot safety [3] suggests the following training program. Training appropriate to each assigned task should include but not be limited to the following:

1. Review applicable industry safety procedures and standards.
2. Review applicable robot vendor safety recommendations.
3. Provide an explanation of the purpose of the robot system and its operation.
4. Provide an explanation of the specific tasks and responsibilities of each person.
5. Identify the person or persons (by name, location, and phone number) to contact when the actions required are beyond the responsibility and training of the person being trained.
6. Identify the hazards associated with each task.
7. Identify an appropriate response to unusual operating conditions.
8. Provide an explanation of the functions and limitations of all safeguards and their design characteristics.
9. Provide an explanation for function testing or otherwise ensuring the proper functioning of safeguarding devices.

Personnel should be retrained whenever the operating system changes or when the supervisor has reason to believe that additional training is required.

Refresher courses that reemphasize safety and discuss new technological developments should be provided for experienced programmers, operators, and maintenance personnel. Unfortunately, no amount of safeguarding can prevent all accidents. It is the worker attitude that will keep the environment safe. Safety must become a habit. People assume that a robot will always move the same way at the same speed. This is not a correct assumption. Since robots are normally consistent in their operation, it is easy for people to be lulled into a false sense of security. It is this complacency that creates problems. Complacency encourages people who have knowledge of the robot system and the repeatability of the robot motion to justify breaking the rules.

Supervisors should ensure that no one is allowed to pass the barrier to the robot's operational area without first locking out the robot, putting it on hold, eliminating power, or reducing the operating speed. Supervisors must insist that people working with the robot not become complacent, overconfident, or inattentive to the hazards inherent to robot operation. The very thing that makes the robot a valuable tool to industry also makes it dangerous.

References

1. ANSI *Safety in Welding and Cutting,* ANSI.ASC Z49.1, American Welding Society, Miami, FL.
2. *Occupational Safety and Health Standards,* Code of Federal Regulations Title 29, Labor, Part 1910, Subpart Q, Superintendent of Documents, U.S. Govt. Printing Office, Washington, DC.; also *Federal Register,* April 1990.
3. ANSI, *American National Standard for Industrial Robots and Robot Systems—Safety Requirements,* ANSI/RIA R15.06-1986, American National Standards Institute, New York.

18

Justification, Selection, and Introduction of Automated Welding

18.1 Justification

A welding robot or automated welding equipment properly applied to a manufacturing situation can be a real profit producer. On the other hand, if not properly applied, it can be a drain on the profitability of the company. There have been many horror stories of the operation of robotic welding systems or automatic welding machines that did not turn out as expected. These failures have caused huge financial losses and organizational disruption and have set back the acceptance of automated welding immeasurably. Sometimes it is worthwhile to find out what went wrong.

Before introducing any new automated welding system, it is wise to review the situation from several points of view. Can we justify this new machine? Can we learn enough about it before selection and purchase? Can we review the cost picture and payback? How do we evaluate the vendor and determine what would be the best machine for the job? And how should we introduce the concept, prepare for its coming, and implement it to obtain the expected goals?

Introducing the first automated or robotic welding machine into a plant is difficult. However, if done correctly it paves the way for smooth acceptance of the machines that follow. It is essential to establish a plan for introducing automated equipment.

There should be a good solid reason for introducing automated welding or robotic welding into the shop. It is not enough that our competitor has a robot, our neighbor has one, and we should therefore have one. Adopt robotics only if it will improve your operation.

A good approach is to conduct a preliminary investigation. A small group should be appointed by management to evaluate the use of automatic arc welding machines and arc welding robots for the type of work the company does. They should learn all they can from manufacturers' literature, textbooks on welding and automation, technical society conferences and seminars, industry associations, plant visits, visits with automated equipment producers and robot suppliers. This group must digest all this material and decide whether the adoption of automated welding equipment or welding robots makes sense and would improve the operation. If the decision is positive, the group should then report to top management or ownership with sufficient information to convince them of the advantage of adopting automated welding.

Top management, if in agreement, should then appoint a committee to thoroughly investigate the possibility of adopting automated or robotic welding. They should be sufficiently convinced of the merits of automated welding before making a commitment to install automated welding in the plant. This should go so far as to state that if economically feasible, the management will provide total approval and financial support for an automation welding program.

The group or committee should investigate the possibilities of automated welding for the company's product line. It should look into the ways in which automated welding would benefit the company. Finally, it should formulate a plan for introducing not just the first robot but a robot system, and perhaps even a robotic manufacturing operation. The personnel in this group should include welders, welding supervisors, welding inspectors, process and methods engineers, product designers, company sales representatives, and accountants. It is essential to have a responsible person chair this group who is not directly related with the production but who has sufficient time to properly lead the group and initiate the plan. The group should be instructed to develop specific objectives such as Improve weld quality, Improve welding productivity, Eliminate hazardous occupations, Reduce or eliminate boring work and hostile environments. Top management must become involved with the investigation and help develop specific management objectives or targets.

The investigative team must thoroughly investigate many facets of the shop welding operation. They should seek recommendations from welding shop personnel. These people are aware of welding problems such as inaccurate piecepart preparation, problems with existing tools and fixtures. The team should check the attitudes of the welders, welding supervisors, and seek their support. They should check other sources such as the labor relations department to determine weld shop hazard areas and sources of shop grievances. They should determine who is using automated welding in their city, the types of applications, whether automated welding is paying off, and if it improves the operation. They should recognize the fact that improved quality is essential—good quality is cheap but poor quality is expensive. Quality improvements may be a greater savings than direct cost savings. With respect to productivity, they should engage the experience of robotic and automated welding equipment vendors.

The team should recognize that automation will change the mode of operation of the welding shop in that much of the control for welding will pass from the welder to a machine. They should review shop routines and welding procedures to determine whether higher weld travel speeds and higher currents will improve productivity without sacrificing quality. They should sell the concept of automation to shop personnel, realizing that the welding supervisors may be the most reluctant to change. They should also let the shop welders realize that they will be used to set up, program, and supervise the robots. Welders are aware that the repetitious and fatiguing aspects of their job, which is normally hot and unpleasant, can be ameliorated by adopting automatic systems, that robots can eliminate their exposure to the bright arc and welding fumes. And finally, it should be brought out that the welders will not be replaced by robots, that lower production costs usually increase business and increase employment. Automation means that more skilled jobs would be required to teach, monitor, and maintain robot cells. The team should investigate the manufacturability of the company's products and seek the assistance of design and process engineers to make the product easier and more accessible for welding. They should seek the cooperation of all who will make welding automation work.

The investigation should result in a specific plan of action. This program must include financial data to show the cost savings and payback. Top management should thoroughly review the report, discuss it with the investigative team, and issue their statement that they are solidly behind the new technology. They must establish specific objectives such as the need to improve welding productivity, the need to improve weld quality, and that the use of robots would relieve the welders from boring, fatiguing, and potentially hazardous jobs.

Following this, management would appoint a team, most likely the investigative team, to follow through, select vendors, select equipment, and supervise the implementation of the new equipment in the factory.

18.2 Cost and Payback

Every company has slightly different accounting and financial record keeping methods. In general, the cost of new capital equipment, such as robots, is based on its payback time—the savings generated by reducing manufacturing cost. Accountants and top management will not approve a project until they are satisfied with the savings that will be realized. In some cases, the purchase of one robot may be approved to gain more knowledge of its potential cost savings. It is important that the investigative team gather sufficient data and information to (1) establish the total cost of the proposed new system and (2) establish the cost savings that will result from the installation of the new system.

Experience indicates that a robotic system will pay for itself in less than 1 year if it is replacing manual shielded metal arc (stick) welding operations. Experience also indicates that if the robot is replacing semiautomatic gas metal arc welding, it can pay for itself in 2 years or less. The same payback applies to dedicated automated welding equipment. The payback period, or return on investment (ROI), is usually calculated as follows:

$$\frac{\text{Initial investment}}{\text{Yearly savings}} = \text{payback period (years)}$$

The difficult part is determining the total cost of the investment and the expected yearly savings.

The initial investment figure must include the cost of everything involved. The estimate would include at least the cost of following:

The robot manipulator and controller
The welding power source, wire feeder, electrode wire dispensing system, torch, and auxiliary equipment such as wire straighteners, wire trimmers, and shielding gas supply system
Safety equipment: barriers, interlock systems, alarm systems, safety mats, etc.
Welding positioners including indexing and material motion devices
Work-holding fixtures
Installation expense of the robot cell
Plant relocation expense
Meetings and training of all personnel in aspects of programming, maintenance, etc.
Material moving equipment if required
Disruptions of factory routines during installation

Undoubtedly other expenses can be charged to this project and should be for conservative estimates. These data must be accepted by the accountants.

The yearly savings expected to result from the installation are more difficult to determine. It is beyond the scope of this book to provide information for calculating welding costs, which are available from other sources. However, it is necessary to make calculations showing the increased productivity of the projected system. Labor savings account for the major part of the cost savings. This amount is best obtained by determining the cycle time under the proposed new robotic welding system. One way to establish this savings is to make projective studies of the weldments that will be robotically welded. Prospective vendors can provide cost studies on specific weldments. This service is normally provided at a cost to prospective customers. It is the most accurate way to estimate the potential savings. The more weldments studied in this manner the higher the accuracy of the total yearly savings. After sufficient laboratory work, worksheets must be developed showing the current manufacturing cost versus the cost of manufacture with the projected robot cell. Savings determined in this manner should be quite accurate.

The major savings is in the cost of labor. With the robot or complex automated equipment, two work-holding fixtures are employed and the arc-on time is much higher than in manual welding since the operator is loading and unloading one fixture while the robot is welding on the other. The robot arc-on time, or duty cycle, can be as high as 85% of the total time with the robot. With semiautomatic welding the duty cycle will average 30–35%. For manual shielded arc welding, the duty cycle will average 10–30%. This is due to the fatigue factor of humans and shows a major cost savings advantage for the robot cell.

The automated welding machine or robot can use higher welding current, which increases the deposition rate. It can use a faster travel speed, which reduces the welding cycle time. Many robotic arc welding applications use single-pass fillet welds. Cycle time is reduced due to the increased travel speed of the robot and faster cycle time per weldment. This is easily calculated and entered into the cost savings between current procedures and expected robotic welding. In general, the labor cost improvement with robots can provide sufficient savings to justify automated robotic arc welding systems. If this does not happen, more sophisticated cost analysis methods should be used. Your accounting department should be consulted with reference to calculating ROI.

One course of action is for the company to purchase one welding robot, place it in a laboratory, and study its performance. This involves taking typical weldments from the shop and running them through the laboratory to compare procedures and costs. This will provide accurate cost information and should be used if at all possible.

Once the cost of a particular weldment is estimated for the robotic system, it is compared to the cost of the same weldment under present conditions. The cost savings per item times the number of units normally produced in a year should justify the cost of a robot. In addition, the greater output of the robot system increases the number of units welded per year, which further reduces the cost per unit and increases the expected savings per year.

Once the data are collected, work sheets can be developed to show the economic advantage. These data become the financial information in the project team's report to management. It is best to be conservative in cost savings estimates so that the actual resulting cost savings will equal or exceed those projected.

There are many other cost savings that result from robotic or automated arc welding. These are summarized as follows:

Reduction in the use of shielding gas. The machine starts and stops the gas flow and allows gas flow only while welding.

Welding robots use only the filler metal needed to make the welds, less electrode wire is wasted.

Robots and automated systems produce consistent welds; this results in less rework and improves quality.

An automated system will produce a constant number of units on a time base period; this aids in scheduling, determining shipping dates, etc.

Fume removal systems can be less expensive because the fumes are collected at the arc and much less air needs to be moved.

Finally, robots produce more product in the same factory space during the same period of time. This increases factory capacity and provides more product for sale.

18.3 Evaluation and Selection

Evaluation and selection of a robot go hand in hand. Once a decision has been made to invest in an arc welding robot, it is wise to investigate the purchase of a robotic welding cell and consider the entire welding system. It is also wise to select the vendor who will supply the arc welding robot. The vendor could be an arc welding distributor, a systems house, a welding systems integrator, the robot manufacturer, or other supplier. It is important to evaluate the vendor because continuing support is required. Arc welding is a complex process; adding the robot makes it even more complex. Make sure the vendor takes total responsibility for the entire system. Deal with a vendor who has experience with arc welding robots. The vendor should be able to troubleshoot a welding problem from the electrode wire point of view, the welding equipment point of view, or the robotic point of view. The vendor should have qualified sales personnel as well as trained and experienced service personnel backed up by a spare parts inventory. The vendor should provide presale service such as live demonstrations on your parts, feasibility studies, application engineering, time-cost studies, tooling recommendations, and other technical consulting advisory services. Check out the vendor's reliability by asking his customers their opinions and reactions.

It is then necessary to select the most suitable robot for your particular application. This is best done in consultation with the vendor's robot sales engineer to establish the specifications for the desired robot. Every robot data sheet contains complete specifications. These include a listing of welding and cutting process capabilities, the number of axes of motion, the payload or weight-carrying capacity, and a diagram showing the working range, the maximum and minimum reach. The specifications also include wrist motions, the method of providing motion, and the speed of operation. The control system is described, including the processor, the language used, the controller (teach pendant), the memory system, backup ability, method of readout, and so on. The welding equipment should also be specified, the arc welding power supply capacity, the wire feeder and its ability to feed different size electrode wires, the torch, the gas control system, etc. Auxiliary equipment, if used, should be specified; this includes positioners, indexing devices, electrode wire dispensing systems, fume removal system, and so forth.

Consult with the vendor's sales engineer to determine if the weldment you propose to weld fits within the work envelope of the selected robot. Many arc welding robot vendors will, at a price, actually weld your weldment in their laboratory to show that it can be welded with the robot in question. At the same time it is possible to get cost data based on robotic welding of the sample weldment. The vendor application laboratory can also make recommendations for positioning equipment and other auxiliary devices.

Once a robot has been selected, purchase a single unit and examine it, weld your product on it, etc. It is smart to set up a robotic laboratory and learn how best to use the robot. Determine which of your weldments can be welded with it. Some type of fixturing will be required. This can be a locating device or a parts-holding fixture for the weldments to be processed. For laboratory work, start with temporary fixturing that merely locates the weldment parts. The weldment should be tack welded together ready for robotic welding. In this way, the robot can be programmed. Your weldments can be welded on the robot, and cost studies developed to determine the welding procedure and times involved. This will provide cost information for justifying the purchase of the robot.

In evaluating a robot system, a number of selection criteria should be considered. These can be classified into three categories: (1) critical factors, (2) objective factors, and (3) subjective factors. Critical factors include the return on investment and the budget ceiling for the entire project. Objective factors include the purchase price of the system and the installation cost plus all other costs including education, training, software, and tooling costs. There are many subjective factors that must be considered such as purchase and installation assistance, quality and type of training provided by the vendor, application engineering assistance, and, of course, increased throughput resulting from the robot installation. These must all be determined and totaled for each potential vendor.

Finally the decision has been made. The robot and assorted equipment is installed and people are trained. Everyone is waiting for the robot to be placed in production and to determine if it will produce the predicted savings.

18.4 Introducing the Robot into Service

It must be understood that the introduction of a robot into a manufacturing operation will change many things. The groundwork has been laid for introducing the robot through discussion groups and training programs for all employees. Management has explained why the decision was made to introduce robots. It is a matter of making the company more efficient, better able to compete on the local, national, and international level. Everybody must know what is being done and why.

The normal reaction from the production worker is that the use of robots will reduce employment. This must be faced and the long-range viewpoint sold. Supervisors will be afraid that automation will add to their workload. This may be true until the robot is working smoothly. Manufacturing management would prefer the status quo, believing that the robot is a risky investment. They too must be sold on the company's long-range program. And, of course, executives, having made the initial outlay, want proof that it was a wise decision. The payoff can be determined quickly once the robot is put into production.

Once the robot cell has been installed, it is time to put it to work. The investigative committee has studied the question of which weldments it is possible to weld with the robot. Shop management should review the data and if possible the laboratory results of preliminary studies. They should then select a simple weldment that indicates a quick payoff or one that the welders dislike working on, feeling that the work is disagreeable or boring. It is assumed that by now several welders have been trained to program the robot at the vendor's training school. They should use the locating fixtures and program the robot system. Trial runs should be made on these sample weldments. After completing several weldments, the program should be revisited and adjusted for higher productivity and higher quality. This should be done with the assistance of vendor technicians who have experience programming arc welding robots. Improvements should include increasing travel speed and welding current where possible, reducing "air cut" time, and avoiding unnecessary motions or waiting periods. Additional products should be programmed and refined. Fine tuning should be considered to further improve quality and decrease welding time. A cost analysis should be made and the robot operation evaluated to determine if the anticipated savings were achieved. By this time, the system should be in normal operation. Measure the system's output and compare it with the anticipated output. The data should be broadcast and posted; obviously, they will be favorable.

It is expected that shop practice will be modified on the basis of these initial tests. The result should be the anticipated increase in production. Personnel problems will gradually disappear, and the robot operation will become routine. Once this happens, it is time to plan for more robots.

19

Installation, Maintenance, and Repair of Equipment

19.1 Installation

Before beginning to install a robotic welding cell, you need to have an engineering drawing or layout of it. This drawing will show the location of each component in the cell. One purpose of the layout is to ensure that the cell was designed for safe and productive operation. The design must conform with all safety regulations as discussed in Chapter 17. A second purpose is to show the robot cell with respect to work flow in the shop. There must be access for incoming material and space for racks and bins, etc., of unwelded material and for completed weldments. There should be room for loading and unloading the fixtures with sufficient space for worker safety. In the event that the new robot cell is part of an assembly line or if any type of conveyor is used, the loading and unloading areas must be laid out for safety. This is particularly important if the robot fixture is to be mechanically loaded or unloaded. Third, the layout must show the location of utilities needed to supply the robot cell—electric power, compressed air, water (and the necessary drains), and shielding gas supply.

Finally, the new robot cell must be installed in a dry area in the shop. Some robot vendors provide typical layouts for their robots or robot system packages.

The layout for the new robot cell must be approved by the proper authorities. This layout must be perfectly understood before installation is begun. Robotic welding and associated equipment must be installed in accordance with the manufacturer's instructions, the company's own standard practices, and all local, state, and national regulations.

The robot cell must be installed in compliance with quality electrical procedures. Make sure that the layout is in agreement with the AWS *Standard for Components of Robotic and Automatic Welding Installations* (ANSI/NEMA/AWS D16.2). This standard is particularly informative concerning the electric circuitry and grounding. Most welding machines, robots, and heavy equipment are manufactured so that different incoming line voltages can be used. The voltage is quickly changed by moving links between studs to properly connect the input voltage to match the incoming voltage. A wiring diagram showing this information is usually posted on the inside of the equipment. Disconnect switches and fuses of the correct size must be used. The metal cases of welding machines and of the robot controller, the robot manipulator, etc., must be grounded to earth. All cables for the welding circuit must be properly sized for the current to be employed.

If gas tungsten arc welding or plasma arc welding is to be used, and if high-frequency current is used for arc starting, special precautions are required. In general, special radio-frequency shielding is required for equipment using high-frequency arc stabilizers. This is particularly important with respect to computer controllers for robots. Robot controllers for GTAW should be hardened against high-frequency stabilized current. See NEMA Standard EW-1, Section 10, "Arc Welding Power Sources with High Frequency Arc Starting and/or Stabilizing," for recommended installation and test procedures. The robot must be installed away from sources of high-frequency radiation.

The installation, maintenance, and adjustment of robotic welding machines and associated equipment are usually done by different people. In many plants, the original installation is done by construction people, riggers, and electricians. Their work is complete once the equipment is operating. Maintenance is an ongoing operation and is usually performed by plant maintenance technicians. Minor adjustments of a routine nature such as changing torch tips and nozzles, gas cylinders, electrode wire coils, meter charts, etc., are made by the welding operators, who must be trained to perform these duties.

Finally, the installation of the equipment must be inspected to make sure that it is in accordance with the layout and all applicable standards and codes and approved.

19.2 Maintenance

Preventive maintenance is the routine scheduled maintenance performed on equipment so that it does not deteriorate rapidly. Preventive maintenance for a robotic welding cell must cover the safety equipment as well as all equipment in the cell. Many robot suppliers or welding cell integrators provide a checklist of preventive measures for the equipment in a cell. If such a list has not been provided, it is recommended that you prepare your own.

Maintenance manuals for each piece of equipment in the work cell are available from the manufacturers. Suppliers should provide these manuals when the installation is complete, the run off successful, and the equipment accepted. Each manual provides a maintenance schedule that must be followed. The checklist is usually based on calendar days from the time the cell was installed. The information from the different manuals should be correlated so that you can establish a routine maintenance schedule and checklist. This work is usually scheduled by the maintenance department. How these schedules should be managed and who performs the different operations depends on how the company is organized.

To be sure the robot cell is operating properly, the maintenance department should provide at least the following services related to motion of the robot and coordinated fixtures. Lubrication of gear boxes, drive motors, etc., is particularly important for the robot manipulator and coordinated and indexing positioners. Electrical slip rings and electrical pick-up brushes on positioners should be checked frequently for wear. Constant motion creates wear that affects the conduction of electric current. The electric welding current must be steady. Electrically conductive grease should be used. Another important point often overlooked is the dryness of the compressed air used to control fixture clamps and some types of index tables. All items of this nature should be included in the maintenance department schedule.

Specific daily maintenance tasks should, however, be the responsibility of the welding operator. These relate more to the welding operation, which is the welding operator's responsibility and include adjustments, changes of consumable supplies, and calibrations. This type of work is best scheduled according to the use of electrode wire. The welding operator normally replaces coils or spools of electrode wire when the supply is exhausted. At the same time, the conduit leading from the electrode wire supply coil to the wire feeder should be blown out with compressed air and inspected. The conduit from the wire feeder to the gun or torch should also be blown out and inspected. Accumulated dirt at the wire feed rolls should be removed. Wire feed rolls and guide tubes should be inspected, adjusted as required, aligned and centered, and replaced if necessary. The electrode torch pick-up tip should be inspected and replaced if necessary. The hole through which the electrode passes is subject to rapid wear. If it becomes enlarged, current pick-up will be erratic. The torch nozzle should be cleaned of spatter and inspected. This is because it is exposed to the heat of the arc and deteriorates rapidly. If a gas tungsten arc or plasma arc is used, the torch should be inspected for centering of the electrode, cleanliness of the electrode tip, and possible deterioration of the orifice and nozzle.

At the same time, the welding operator should check the shielding gas delivery system. Shielding gas supply hoses flex continually, deteriorate, and may leak. Gas leaks affect the quality of the weld, and a serious leak reduces the effectiveness of the shielding gas. The cylinders supplying the shielding gas should be checked routinely to make sure there is sufficient shielding gas available. Cooling water hoses to the torch should be checked for damage caused by the heat and continuous flexing. Water leaks can drastically affect the quality of the weld. Other items that need to be checked are breakaway torch holders, torch changers, nozzle cleaners, the supply of antispatter compound, wire clippers, etc. If they are employed, supplies of meter charts or printout paper should be checked to be sure they are sufficient to finish the shift. Sensors or information pick-up devices should also be checked. At the same time, automatic clamps for welding fixtures, backup bars, gas or water supply to back-up bars, and similar items should be inspected.

In the event that the large supply of electrode wire lasts for more than one shift, the above comments apply to the shift change time. Cooperation between operators on different shifts is essential for smooth operation of the welding cell.

At least once a day, the safety interlocks of access doors, pressure-sensitive mats, safety screens, warning lights and signals, and everything relating to the safety of the cell should be checked.

Trash, debris, and combustible material must not be allowed to accumulate in the welding cell. This can create fire hazards and deteriorate the efficiency of the robot work cell. It should be removed immediately.

19.3 Troubleshooting

Troubleshooting is required at a robotic welding cell when the welds being produced are not up to acceptable quality level. The main difficulty in troubleshooting robotic welding is to determine if a problem is a welding problem or an equipment problem. Visual inspection is most commonly used and is a quick method to help diagnose the problem. Refer to Chapter 7, Section 7.4, for help in diagnosing problems. If the problem is related to piecepart preparation and cleanliness and resulting poor fit-up, this could be determined by viewing the resulting weld. On the other hand, if the problem is that no arc is produced, a power or equipment problem is indicated.

If it is determined that the problem is a welding problem, the welding operator or programmer should check the parameters of the weld being made. This is done by referencing the voltmeter and ammeter, which are usually on the welding equipment, and checking travel speed and wire feed speed. If these values do not agree with those of the established procedure, adjustments should be made to bring them into compliance. Shielding gas should be checked for proper flow and dryness. If there is no gas flow, it could be the result of an empty supply tank, broken or pinched hose, inoperative solenoids or valves, or a draft from a fan, or another problem related to the shielding gas flow at the nozzle. Usually the loss of shielding gas results in surface porosity or just bad welds. Similar problems occur when the weld is made through puddles of water or oil, which create hydrogen in the arc and overcome the beneficial effect of the shielding gas. Any source of hydrogen in the arc atmosphere can cause this problem.

The electrode wire feed should be checked for supply availability and uniform feed. One problem is weld wander, which can be caused by excessive cast or helix in the electrode wire. It is advisable to check the wire straightener every time the electrode wire supply is replaced. The cast and helix should be measured. Problems with them can usually be overcome by adjusting the wire straightener. If the electrode does not feed when all conditions are correct, it could be that there is an obstruction in the wire feed system such as an accumulation of debris in the conduits or a bird's nest in the wire feed rolls. In this case, clean out the conduit and properly adjust the wire drive rolls and guide tubes. It is wise to check the surface of the electrode wire, either solid or cored, for cleanliness. Surface contamination will interfere with current pick-up to the electrode wire. This may cause the wire to stop or cause variations in the weld appearance.

If the welding parameters are correct but the weld appearance is erratic, it is wise to check the welding programmer. Problems are often found in the welding power supply with respect to preflow time, crater fill time, burnback time, postflow time, etc. One or more of these factors could be out of adjustment, which would affect the quality of weld starts or finishes. The welding parameters sometimes change during the total weld path. If the change does not occur when scheduled, the parameters will be out of line for a particular portion of the weld.

In the case of gas tungsten arc welding, if the arc does not start, the problem is in the high-frequency or capacitor start circuits. These systems are in operation for a very short period of time, and the arc may not start immediately. If the arc does not start, the cause may be a dirty tungsten electrode, an out-of-adjustment spark gap, or a loose connection.

With gas tungsten arc welding, if an automatic voltage control circuit is involved, it may be out of adjustment and cause the arc length to vary, which in turn will cause the weld appearance to change.

If pulsed GMAW is being used, the power source program with respect to the pulsing frequency and pulse wave shape should be checked to ensure quality and balance between welding current, pulsing, pulsing wave shape, and pulsing frequency.

If it is determined that the parameters are in-line and that the shielding gas and electrode wire are feeding properly, check the equipment—both the mechanical robot equipment and the computer controller. Visual inspection can determine whether the weld follows the joint or wanders. If the weld missed the joint, there may be a sensor problem or a controller problem, or pieceparts may be out of dimensional tolerance. It is also possible that the controller memory is affected or lost, which could result in missing the joint path.

It has been found that the welding equipment is more often at fault than the robot controller or manipulator. The robot manipulator is rarely a problem.

Diagnosing welding problems usually means that the operator or maintenance technician is in the welding cell with the robot while it is in operation. Precautions should be taken, and the robot should be operated at lower than standard speed while diagnostic work is being done. Lockouts showing that personnel are inside the cell should be prominently displayed so that others will not speed up the robot to normal speed. It is a fact that more injuries are incurred around robots during the maintenance operation than at any other time.

The above information should help determine and fix any welding problem. This information is rarely covered in the instruction manuals for robots, controllers, arc welding power sources, or wire feeders.

19.4 Diagnostic Procedures

The robot vendor or cell integrator will provide a manual for each piece of equipment in the welding cell. Each manual will include troubleshooting, diagnostic tables, charts, checklists, and maintenance schedules. You should coordinate this material for your particular welding cell. It is beyond the scope of this book to become specific with respect to different pieces of equipment in the robotic welding cell.

You should receive a manual for each of the following parts of equipment that is included in your cell:

Computer controller
Robot manipulator
Coordinated positioner
Arc welding power source
Wire feeder
Gun or torch and cable assembly
Smoke exhaust system
Safety equipment
GTAW arc starting system

GTAW torch
Cooling water circulator
Automatic fixture clamps

Most manuals use standard flow chart symbols for diagnostic charts. An overview of flow chart symbols that must be understood by all maintenance personnel is provided by Table 19.1 and by Fig. 19.1.

The controller is a very complex piece of equipment, and many safeguards and checks are built into it to ensure safe, reliable operation. Because of its complexity, a sophisticated diagnostic system is included. This, with the maintenance manual, should make it possible for a qualified technician to do much of the maintenance work. This equipment has one problem not common to ordinary electrical equipment: Static electricity, which can arise from numerous sources, can severely damage its logic boards. It should be the rule to always use antistatic wrist bands, grounding mats, and other static reduction techniques when handling

Table 19.1 Basic Flow Chart Symbols

Activity Symbol

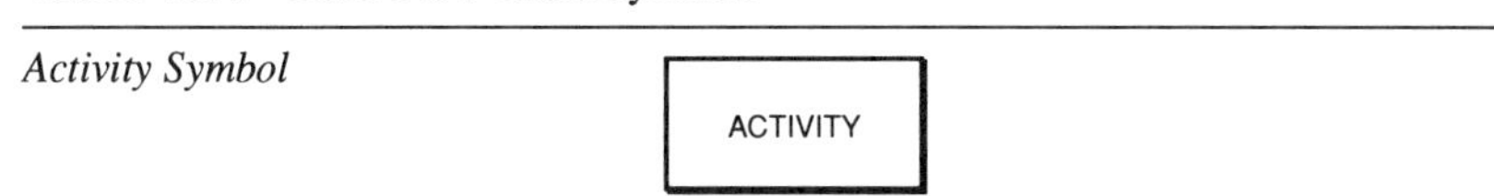

A rectangle is used to indicate that an activity occurs at this point in the process. A brief explanation or identifier of the activity is placed inside the symbol.

Flow Line Symbol

A flow line indicates the direction of the path that connects the elements of the flow chart.

Connector Symbol

An off page connector is used when you run out of space on a page but haven't yet reached the end of the process. A letter or number should be placed inside the symbol on each page to indicate which ones match.

Decision Symbol

A diamond or a hexagon represents a *decision point* in process. A question is written inside the symbol. The path of the next step depends on the answer to the question (see Fig. 19.1).

At any decision point, the path should show two or more possible ways to proceed, depending on the answer (see Fig. 19.1).

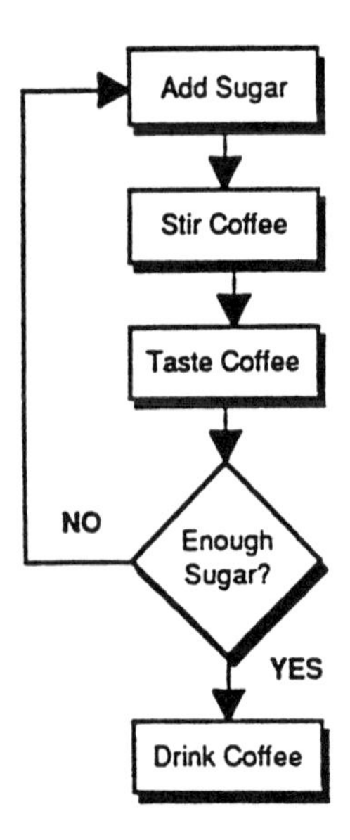

Figure 19.1 A simple flow chart.

or transporting printed circuit boards. Wiring diagrams, pin connector identifications, and other information are included in the maintenance manual.

The robot manipulator contains motors, gear boxes, gearing systems, and bearings. These are of high precision because of the need for accurate positioning. Fortunately, robot manipulators provide very good service in welding applications. Proper lubrication will keep the manipulator operating for many years. Diagnostic alarm codes generated by the controller in connection with the manipulator form a self-diagnostic troubleshooting system, which can be utilized after thoroughly studying the maintenance manual. Fault diagnostics are classified into two groups—faults that can be rectified by your own maintenance personnel and those that must be left to the manufacturer's service personnel.

Positioners can be indexed or coordinated. Indexed positioners move from one position to another using electric, hydraulic, or air power. The important thing is that they come to a stop at a specified location. Locating pins are sometimes used on an indexed positioner to lock it into an exact location. Coordinated positioners are much more complex and include highly sophisticated motor drive systems that are controlled by the controller. They require the same attention as the motion systems of the robot. In some cases, the controller for the coordinated positioner will provide a diagnostic system with alarm codes. Consult the manual for the coordinated positioner.

The arc welding power source is either a transformer rectifier type or an inverter type. Many machines contain programmers used, in the case of pulsed MIG welding, to control the pulsing wave shape and all aspects of pulsing. In many cases these can be adjusted by the user. Other machines include welding programmers that actuate gas valves, contactors, water valves, and other controls for such purposes as weld starting and stopping and providing post- and pregas flow. The instruction manual for the power source will include details for diagnosing problems with these controllers and programmers. The power source may also include a high-frequency stabilization arc starting system. Such a system requires continual attention and must be adjusted periodically in accordance with the instructions.

The wire feeder and wire feeding system, including the torch, is the most troublesome part of the entire welding cell. This is because the torch is exposed to the high temperature of the welding arc and the wire feeder contains the moving parts that drive the wire through the conduit and the torch. Constant attention, adjustment, and cleaning are required. This is spelled out in the maintenance manual.

The gun or torch and cable assembly, mentioned above, requires continual attention because of wire flaking contaminants on wire and other surfaces. In some cases, lubricants or lubricating devices are placed in the system to provide easier passage of the wire through the conduit. The instruction manual will provide schedules.

The smoke exhaust system must be kept in good operating condition to perform properly. Cleanouts, disposal of accumulated particles, and similar tasks should be done on a routine basis. Filters, if included, must be changed periodically, and accumulated material must be cleaned from the conduits.

The safety equipment that is used to protect the workers relates primarily to the welding cell enclosure—the interlock switches, pressure mats, alarms, and similar items. These should be checked at least once a day by maintenance personnel and repaired whenever problems occur.

Cells include such equipment as the GTAW arc starting system, GTAW torch, cooling water circulator, and automatic fixture clamps. There should be maintenance manuals available for all pieces of equipment, and they should be referred to as often as required. Automotive antifreeze coolants should not be used for the water circulators. Automotive antifreeze has stop-leak additives that will clog the small openings in GTAW torches and valves. Ethylene glycol with fungus inhibitors and emulsion lubricants should be used.

If you need additional information, check with your robot vendor or the robot cell integrator.

20

The Future of Automated Welding

20.1 The Factory of the Future

In this book we have attempted to make the transition from manual and semiautomatic welding to automated welding. Automated welding is subdivided into mechanized welding, which is under operator control; automatic welding without operator control; adaptive welding, which uses sensors and complex control systems; and robotic welding, which can be either open- or closed-loop depending on the use of sensors, feedback systems, and controllers. In the 1970s the majority of industries used the manual and semiautomatic methods of welding. Mechanized and automatic welding were used to a limited degree, but only in the high-volume industries.

In the mid-1980s, robotic welding entered the picture and promised to remove the human from the welding operation. This did not happen; largely, robotics have been used to weld parts for the high-volume industries. However, manufacturers decided that robots should also be used for other than high-volume production. With the addition of specific sensors and ingenious holding fixtures, robots could be used for low-volume production where the pieceparts were not perfectly prepared.

The future is especially bright for robotic welding because the welding industry is still growing, however slowly, because of the inherent economic advantage of using welding for joining metals. Robotics is a high-growth industry due to its promise of improving productivity. Robotic welding is still in its infancy.

Automatic welding ("hard automation") with adaptive control and robotic welding with adaptive control make the factory of the future possible. The factory of the future can be defined as an automated highly flexible manufacturing complex that can be operated for an eight-hour shift without human production workers. Some call it the untended shop, or the lights-out factory, because it will make use of intelligent manufacturing systems (IMSs). Some predict that the factory of the future will manufacture products at the lowest possible cost and the highest possible quality. To be practical, the untended shop cannot rely on fixed automation and long production runs. Consumers want the new, different, and better product; this dictates shorter life cycles for the products. The operation must accommodate rapid design changes and customizing of products. The key is flexibility. Individual operations must be able to adjust to parts of different sizes and designs.

The robot, which is considered a stand-alone machine, must be integrated into the total production operation. This means automatic loading and unloading and automatic material transport, which involves mobile robots. It requires factorywide computer networks and total integration.

The total lights-out factory may not be necessary to remain competitive in a world economy. A total lights-out factory would undoubtedly be very expensive. The high expense may not be justified, and hence the partially manned factory may be the practical solution. In other words, total integration may not be necessary. The task of the robot remains that of improving the quality of life of welders, relieving them of monotonous, hazardous, and uncomfortable work. Robots will also improve the quality of the product by providing more consistent processing and improve productivity with the higher operator factors made possible by overcoming normal human welder fatigue. To accomplish all of this, the product to be manufactured must be designed for robotic welding and manufacture. Robots must become more agile and more flexible. The robot controller must be more powerful and faster and adapt to higher hierarchy instructions. It must use multiple sensors and artificial intelligence and fuzzy logic. The robot controller must store massive databanks and provide sufficient data to ensure real-time quality control.

20.2 Welding That Can Be Automated

Products are designed to meet specific service requirements. Loads and stresses are calculated. Material types and thicknesses are specified, size and shape are considered, the working environment is analyzed, prototypes are made and tested, and the final design evolves and is approved. Many factors must be considered in the design of a new product. Many criteria must all be satisfied, but today the design must emphasize weldability and manufacturability. This is especially true if a product is to meet its economic targets, if manufacturing is to be automated, and if automated arc welding is involved.

It is clear that a systematic engineering approach must be taken at all stages of weldment product design and fabrication if robotic mechanization of the welding operation is to be successful. Designers must have a thorough understanding of the welding processes. They must select gas metal arc or flux-cored welding for flexibility, they must have an understanding of the manufacturing processes both leading up to the welding operation and following it. A welding operation is an assembly operation with the joining done by welding. Designers must have welding application information and must understand the function of the controllers, sensors, parts feeding, and work-holding devices to design a product that can be manufactured by welding using automated techniques. They must understand the integrating system of the total product. This will benefit the automation-oriented elements of the total manufacturing operation.

For a truly unmanned manufacturing facility, the pieceparts must be engineered and processed in such a way as to allow mechanization. The assembly of the pieceparts to make the total welded product must also be mechanized. Individual tasks may not always be cost-effective. The production of sufficiently accurate pieceparts for automated welding must be studied. At this juncture the unattended factory concept may need to be reevaluated. This becomes increasingly difficult when there are a variety of sizes or design modifications of the product. It is agreed that if the design can be accomplished with the flexibility that is demanded, there will be a tremendous economic advantage to untended production. However, it must be tempered with total capital cost.

Robots are becoming more powerful with more powerful processors. Integrated motion between the robot and positioners is becoming more common, and automatic loading of work-holding fixtures and unloading of weldments will soon be commonplace. Other processing operations are being integrated with welding cells to make up a total manufacturing cell.

The greatest advances will come in the industries that do not have high-volume production runs. Small specialized portable robots will make an appearance in shipbuilding and in building construction. These robots, designed to weld specific joints, will be carried to the welding site and placed at the joint where welding is specified, where they will find the joint and make the specified weld. This will greatly improve productivity. Small portable controllers will be required. It is expected that the laser will be coupled to portable robots for making welds that require filler material. These will be developed with a high degree of safety for use in shipbuilding and construction.

It is expected that in normal factory operation the robots will perform welding and material-moving functions with combination gripper/torch tools.

20.3 Computer Controllers

The rate of development of robotic welding has largely been paced by the development of increasingly powerful integrated computer controllers. Fortunately, at the time this has been accomplished, the cost of computer power has decreased. With the rapid advances in computer technology, it is important to increase the capability of robot controllers, operating systems, and "smart" computer-based peripheral components. We now have the fully computer-integrated robotic work cell. It is important that we have compatibility of work cell controllers with computer-aided design (CAD) and computer-aided manufacturing (CAM) systems. Broad integration capabilities will provide the means for direct transfer of welding designs and procedural databases to the welding work cell. This will be done without human interaction. It is important to have computer-integrated manufacturing to provide off-line programming and procedure development. This lack of integration is one of the major inefficiencies in the utilization of arc welding robots in unmanned manufacturing.

The necessary hardware and software are already available. A powerful controller to control an arc welding cell can control any other manufacturing process from a single console. It is built around the latest microprocessor chip with the ability to process vast amounts of data at extremely high speed. It includes a large amount of storage with a hard disk for storing programs and historical data plus floppy disk drives for making copies of welding procedures or collecting current data. It controls up to 48 axes of integrated motion and will accommodate the most complex welding procedures, including multiple processes. The controller console has a color graphics monitor and a keyboard with integrated accessibility. It can also be remotely controlled with a teach pendant. Security is provided because operators can call up only those programs that can be operated at the time. A supervisory system assigns privilege levels. If incorrect procedural information is entered, the controller will suggest proper procedures. The controller has a standard VME chassis with many open slots to allow expansion. Backup is provided by cassette tape units, and the system includes a printer for hard copy records. The software uses user-friendly pulldown menus. Data are displayed in either conventional or metric units, and an optional modem is available for remote communications and diagnostics. It is a user-friendly system that is quickly learned.

20.4 Sensors and Adaptive Control

Adaptive control systems are required for any automatic welding system to adjust for deviations in joint location or joint design. These variations require adjustments in welding parameters to produce the high-quality weld required. This is a problem related to tolerances and fixturing equipment in the robot system and most important in part preparation. The cost to eliminate dimensional variations would be prohibitive, so sensors and adaptive controllers that adjust weld location and parameters are required.

Sensors developed for arc welding have proliferated in the last few years. It was mentioned previously that sensors are used to sense variations in the actual weld conditions. Sensors provide information that can be fed back to the controller to compensate for differences between actual weld conditions encountered in use and those normally anticipated to make the weld as designed. The welding variables that require sensory feedback and control include joint tracking, joint geometry, weld penetration, cooling rates and part temperature, electrode and part geometry, impurities in or on the electrode or filler metal, impurities of the base metal, impurities in the shielding gas, the shielding gas coverage efficiency, and others to produce the weld with the desired or specified geometrical shape and metallurgical characteristics.

No single sensor can provide all the data necessary in the manner of the human welder's brain. It is obvious that separate sensors cannot be used to collect each type of data. It therefore becomes necessary to determine what types of sensors are needed for each particular welding situation or station. Experience indicates that only a limited number of sensors can be attached to a welding torch at any one time. It is necessary that sensors that operate in the arc area pick up signals and send them to remote detectors. Sensors must be made stronger, more rugged, and adapted to operate in the hostile arc area. This work is even more important for the control of arc welding in an unmanned factory.

20.5 Real-Time Quality Control

Real-time quality control means that when the arc is extinguished and the weldment is completed, it is instantly known if the quality requirements have been met. The computer mentioned above has open slots available for the addition of numerous sensing devices, and software can be written to accept this feedback information and modify the welding parameters and travel path to make the changes necessary to make the desired high-quality weld. The computer also has the capabilities of printing out weld parameters on a continuous basis. It can also be programmed to shut down the machine if any welding parameter is outside specific limits.

We now have sensing devices that can analyze the composition of the shielding gas and other gases in the arc area. They can, for example, determine if hydrogen is present in the arc atmosphere. Hydrogen is objectionable and will contribute to a defective weld. On the other hand, it is not known whether the presence of hydrogen is due to dampness in the gas, oil on the workpiece, or dirt on the electrode wire. If we want the truly unmanned factory we must be able to determine the cause of the defect so that the proper action can be taken to eliminate it.

Considerably more effort is required to pinpoint defects and make corrections. Up until now, humans have been required to correct many of the problems discovered. This is why the semi-unmanned factory seems more possible in the near future than the total lights-out factory.

Appendixes

Appendix 1

Definitions of Robotic Terms

The terms and definitions of the Robotic Industries Association* are used in this book. The more important ones are as follows:

Accuracy: (1) Quality, state, or degree of conformance to a recognized standard or specification; (2) degree to which actual position corresponds to desired or commanded position.

Active accommodation: Integration of sensors, control, and robot motion to achieve alteration of a robot's preprogrammed motions in response to sensed forces. Used to stop a robot when forces reach set levels or to perform force-feedback tasks like insertions, door opening, and edge tracing.

Actuator: A motor or transducer that converts electrical, hydraulic, or pneumatic energy to effect motion of the robot.

Adaptive control: A control algorithm or technique where the controller changes its control parameters and performance characteristics in response to its environment and experience.

**Robotics Glossary,* Robotic Industries Association, P.O. Box 3724, Ann Arbor, MI 48106.

Algorithm: A prescribed set of well-defined rules, processes, or mathematical equations for the solution of a problem in a finite number of steps.

Alphanumeric: Consisting of alphabetic, numeric, and usually other characters such as punctuation marks.

Ambient temperature: A temperature within a given volume such as a room or building.

Analog control: Control involving analog signal processing devices (electronic, hydraulic, pneumatic, etc.).

Anthropomorphic robot: A robot with all rotary joints and motions similar to a human's arm. (Also called *jointed-arm robot.*)

Architecture: Physical and logical structure of a computer or manufacturing process.

Arm: An interconnected set of links and powered joints comprising a manipulator that supports or moves a wrist and hand or end-effector.

Artificial intelligence: The capability of a machine to perform human-like intelligence functions such as learning, adapting, reasoning, and self-correction. (See also *Adaptive control.*)

Automatic operation: That time when the robot is performing its programmed tasks through continuous program execution.

Automation: Automatically controlled operation of an apparatus, process, or system by mechanical or electronic devices that take the place of human observation, effort, and decision.

Axis: A travel path in space, usually referred to as a linear direction of travel.

Backlash: Free play in a power transmission system, such as a gear train, resulting in a characteristic form of hysteresis.

Barrier: A physical means of separating persons from the robot's restricted work envelope.

Base: The platform or structure to which the shoulder of a robot arm is attached; the end of a kinematic chain of arm links and joints opposite to that which grasps or processes external objects.

Batch manufacturing: The production of parts or material in discrete runs, or batches, interspersed with other production operations or runs of other parts or materials.

Binary systems: The basis for calculations in all digital computers. This two-digit numbering system consists of the digits 0 and 1, which can be simply represented by on/off switches.

Calibrate: To determine the deviation from a standard so as to ascertain the proper corrections.

Cartesian coordinates: All robot motions travel in right angle lines to each other. There are no radial motions.

Cartesian coordinate robot: A robot whose manipulator arm degrees of freedom are defined primarily by Cartesian coordinates.

Cartesian coordinate system: A coordinate system whose axes or dimensions are three intersecting perpendicular straight lines, usually designated *x, y, z* (or *X, Y, Z*), and whose origin is their intersection. *X* and *Y* refer to directional lines that are parallel to earth and perpendicular to each other; *Z* refers to a directional line that is vertical, perpendicular to the earth's surface and to both *X* and *Y*.

Cell: A manufacturing unit consisting of two or more cells and the materials transport and storage buffers that interconnect them.

Central processing unit (CPU): (1) Another term for processor. It includes the circuits controlling the interpretation and execution of the user-inserted program instructions stored in the computer or robot memory. (2) The hardware part of a computer that directs the sequence of operations, interprets the coded instructions, performs arithmetic and logical operations, and initiates the proper commands to the computer circuits for execution. The arithmetic-logic unit and the control unit of a digital computer controls the computer operation as directed by the program it is executing.

Circular interpolation: A function automatically performed in the control of defining the continuum of points in a radius based on a minimum of three taught coordinate positions.

Closed-loop control: A method of control in which feedback is used to link a controlled process back to the original command signal. Measures the degree to which actual system response conforms to desired system response and uses the difference to drive the system into conformance.

Computer-aided design (CAD): The use of a computer to assist in the creation or modification of a design.

Computer-aided manufacture (CAM): The use of computer technology in the management, control, and operation of manufacturing.

Computer control: Control involving one or more electronic digital computers.

Computer numerical control (CNC): The use of a dedicated mini- or microcomputer to implement the numerical control function. Uses local data input from devices such as paper tape, magnetic tape cassette, or floppy disk.

Contact sensor: A device capable of sensing mechanical contact.

Continuous-path control: A control scheme whereby the inputs or commands specify every point along a desired path of motion.

Control: (1) The process of making a variable or system of variables conform to what is desired. (2) A device to achieve such conformance automatically. (3) A device by which a person may communicate commands to a machine.

Control, adaptive: See *Adaptive control.*

Control hierarchy: A relationship of sensory processing elements whereby the results of lower level elements are used as inputs by higher level elements.

Control system: Sensors, manual input and mode selection elements, interlocking and decision-making circuitry, and output elements to the operating mechanism.

Controller: (1) An information processing device whose inputs are both desired and measured position, velocity, or other pertinent variables in a process and whose outputs are drive signals to a controlling motor or actuator. (2) A communication device through which a person introduces commands to a control system. (3) A person who performs the control functions.

Coordinated axis control: (1) Control wherein the axes of the robot arrive at their respective endpoints simultaneously, giving a smooth appearance to the motion. (2) Control wherein the motions of the axes are such that the endpoint moves along a prespecified type of path (line, circle, etc.). Also called *endpoint control.*

Cycle time: The period of time between the start of one machine operation and the start of another (a repeat, in a pattern of continuous repetition).

Cylindrical coordinate robot: A robot whose manipulator arm degrees of freedom are defined primarily by cylindrical coordinates.

Cylindrical coordinate system: A coordinate system that defines the position of any point in terms of an angular dimension, a radial dimension, and a height from a reference plane. These three dimensions specify a point on a cylinder.

Database: A large collection of records stored on a computer system from which specialized data may be extracted, organized, and manipulated by a program. Any organized and structured collection of data in memory.

Degree of freedom: One of a limited number of ways in which a point or a body may move or in which a dynamic system may change, each way being expressed by an independent variable and all required to be specified if the physical state of the body or system is to be completely defined.

Diagnostic program: A user-inserted test program to help isolate hardware malfunctions in the computer or robot and the application equipment.

Documentation: An orderly collection of recorded hardware and software data such as tables, lists, and diagrams to provide reference information for any application operation and maintenance.

Downtime: The time when a system is not available for production.

Duty cycle: The fraction of time during which a device or system will be active or at full power.

Edit: To modify the form or format of data; for example, to insert or delete characters.

Elbow: The joint that connects the upper arm and forearm.

Emergency stop: A method using hardware-based components that overrides all other robot controls, removes drive power from the robot actuators, and brings all moving parts to a stop.

Encoder: (1) A rotary feedback device that transmits a specific code for each position. (2) A device that transmits a fixed number of pulses for each revolution.

End-effector: An actuator, gripper, or mechanical device attached to the wrist of a manipulator by which objects can be grasped or otherwise acted upon.

Error signal: The difference between desired response and actual response.

External sensor: A feedback device that is outside the inherent makeup of a robot system or a device used to effect the actions of a robot system that are used to source a signal independent of the robot's internal design.

Factory: A manufacturing unit consisting of two or more centers and the materials transport, storage buffers, and communications that interconnect them.

Fail-safe: Failure of a device without danger to personnel or damage to product or plant facilities.

Fault: Any malfunction that interferes with normal application operation.

Feedback: The signal or data sent to the control system from a controlled machine or process to denote its response to the command signal.

Fiber optics: A communication technique where information is transmitted in the form of light over a transparent material (fiber) such as a strand of glass. Advantages are noise-free communication not susceptible to electromagnetic interference.

Flexible manufacturing: Production with machines capable of making a different product without retooling or any similar changeover. Flexible manufacturing is usually carried out with numerically controlled machine tools, robots, and conveyors under the control of a central computer.

Forearm: That portion of a jointed arm that is connected to the wrist.

Gantry: A bridgelike frame along which a suspended robot moves. A gantry creates a much larger work envelope than the robot would have if it were pedestal-mounted.

Ground: A conducting connection, intentional or accidental, between an electric circuit or equipment chassis and the earth ground.

Ground potential: Zero voltage potential with respect to earth ground.

Guard: A physical means of separating persons from danger.

Hazard: A condition or changing set of circumstances that presents a potential for injury, illness, or property damage. The potential or inherent characteristics of an activity, condition, or circumstance that can produce adverse or harmful consequences.

Hierarchy: A relationship of elements in a structure divided into levels, with those at higher levels having priority or precedence over those at lower levels. (See also *Control hierarchy.*)

Industrial robot: A reprogrammable, multifunctional manipulator designed to move material, parts, tools, or specialized devices through variable programmed motions for the performance of a variety of tasks.

Inertia: The tendency of a mass at rest to remain at rest, and of a mass in motion to remain in motion.

Input devices: Devices such as limit switches, pressure switches, and push buttons that supply data to a robot controller. These discrete inputs are two types: those with common return and those with individual returns (referred to as isolated inputs). Other inputs include analog devices and digital encoders.

Intelligent robot: A robot that can be programmed to make performance choices contingent on sensory inputs.

Interface: A shared boundary. An interface might be a mechanical or electrical connection between two devices, a portion of computer storage accessed by two or more programs, a device for communication to or from a human operator.

Interlock: To arrange the control of machines or devices so that their operation is interdependent to ensure their proper coordination.

Job shop: A discrete parts manufacturing facility characterized by a mix of products of relatively low volume production in batch lots.

Joint: As applied to a robot, a rotational or translational mechanism providing a degree of freedom in a manipulator system.

Language: A set of symbols and rules for representing and communicating information (data) among people or between people and machines.

Lead-through: Programming or teaching by physically guiding the robot through the desired actions.

Level of automation: The degree to which a process has been made automatic. Relevant to the level of automation are questions of automatic failure recovery, the variety of situations which will be automatically handled, and the conditions under which manual intervention or action by humans is required.

Limit switch: A switch that is actuated by some part or motion of a machine or equipment to alter the electric circuit associated with it.

Limited-degree-of-freedom robot: A robot able to position and orient its end effector with fewer than six degrees of freedom.

Load deflection: The difference in position of some point in a body between a nonloaded and an externally loaded condition.

Logic: A means of solving complex problems through the repeated use of simple functions that define basic concepts. Three basic logic functions are AND, OR, and NOT.

Loop: A sequence of instructions that is executed repeatedly until some specified condition is met.

Maintenance: The act of keeping the robot system in its proper operating condition.

Major axes (of motion): The number of independent directions in which the arm can move the attached wrist and end-effector relative to a point of origin of the manipulator such as the base. The number of robot arm axes required to reach world coordinate points is dependent on the design of robot arm configuration.

Malfunction: Any incorrect functioning within electronic, electrical, or mechanical hardware. (See also *Fault.*)

Manipulation: (1) The process of controlling and monitoring data table bits or words to vary application functions according to the user's program. (2) The movement or reorientation of objects such as parts or tools.

Manipulator: A mechanism, usually consisting of a series of segments jointed or sliding relative to one another, for the purpose of grasping and moving objects, usually with several degrees of freedom. It may be remotely controlled by a computer or a human.

Mass production: The large-scale production of parts or material in a continuous process uninterrupted by the production of other parts or materials.

Maximum speed: (1) The greatest rate at which an operation can be accomplished according to some criterion of satisfaction. (2) The greatest velocity of movement of a tool or end-effector that can be achieved in producing a satisfactory result.

Memory: A device into which data can be entered, in which data can be stored, and from which data can be retrieved at a later time.

Microprocessor: An electronic computer processor implemented in relatively few IC chips (typically LSI) that contain arithmetic, logic, register, control, and memory functions. The microprocessor is characterized by having instructions that reference micro operations; functional equivalent of minicomputer instructions accomplished by programming a series of instructions.

Minicomputer: A small multiuser computer, the minicomputer is typically characterized by higher performance, a richer instruction set, higher price, a variety of high-level languages, several operating systems, and networking software.

Mobile robot: A robot mounted on a movable platform.

Modular: Made up of subunits that can be combined in various ways. (1) A modular robot is one constructed from a number of interchangeable subunits each of which can be one of a range of sizes or have one of several possible motion styles (rectangular, cylindrical, etc.) and number of axes. (2) "Modular design" permits assembly of products, software, or hardware from standardized components. (See also *Module*.)

Module: (1) An interchangeable "plug-in" item containing electronic components that may be combined with other interchangeable items to form a complete unit. (2) A mechanical component having a single degree of freedom that can be combined with other components to form a multiaxis manipulator or robot.

Monitor: (1) CRT display package. (2) To observe an operation.

Motor controller: A device or group of devices that serves to govern, in a predetermined manner, the electric power delivered to a motor.

Noise: Extraneous signals; any disturbance that causes interference with the desired signal or operation.

Numerical control (NC): A technique that provides for the automatic control of a machine tool from information prerecorded in symbolic form representing every detail of the machining sequence.

Off-line: Describes equipment or devices that are not connected to the communications line.

Off-line programming: Defining the sequences and conditions of actions on a computer system that is independent of the robot's "on board" control. The prepackaged program is loaded into the robot's controller for subsequent automatic action of the manipulator.

On-line: Describes equipment or devices that are connected to the communications line.

Open loop: Without feedback.

Open-loop control: (1) A method of control in which there is no self-correcting action for an error in the desired operational condition. (2) Control achieved by driving control actuators with a sequence of preprogrammed signals without measuring actual system response and closing the feedback loop.

Open-loop robot: A robot that incorporates no feedback, that is, no means of comparing actual output to command input of position or rate.

Optic sensor: A device or system that converts light into an electrical signal.

Pattern recognition: Description or classification of pictures or other data structures into a set of classes or categories; a subset of artificial intelligence.

Pick-and-place robot: A simple robot, often with only two or three degrees of freedom, that transfers items from place to place by means of point-to-point moves. Little or no trajectory control is available. Often referred to as a "bang-bang" robot.

Pinch point: Any point where it is possible for a part of the body to be injured between the moving or stationary parts of a robot and the moving or stationary parts of associated equipment or between the material and moving parts of the robot or associated equipment.

Pitch: The angular rotation of a moving body about an axis perpendicular to its direction of motion and in the same plane as its top side.

Playback accuracy: (1) Difference between a position command recorded in an automatic control system and that actually produced at a later time when the recorded position is used to execute control. (2) Difference between actual position response of an automatic control system during a programming or teaching run and the corresponding response in a subsequent run.

Point-to-point control: A control scheme whereby the inputs or commands specify only a limited number of points along a desired path of motion. The control system determines the intervening path segments.

Position control: Control by a system in which the input command is the desired position of a body.

Power supply: (1) In general a device that converts ac line voltage to one or more dc voltages. A robot power supply provides only the dc voltages required by the electronic circuits internal to the robot controller. (2) A separate power supply, installed by the user, to provide any dc voltages required by the application input and output devices.

Precision: The standard deviation, or root-mean-square deviation, of values around their mean.

Presence-sensing device: A device designed, constructed, and installed to create a sensing field or area around one or more robots and that will detect an intrusion into that field or area by a person, robot, or other object.

Procedure: The course of action taken for the solution of a problem.

Process: (1) A continuous and regular production executed in a definite uninterrupted manner. (2) A computer application that primarily requires data comparison and manipulation. The CPU monitors the input parameters in order to vary the output values. (As generally contrasted with a machine, a CPU process does not cause mechanical motion.)

Program: (1) (n) A sequence of instructions to be executed by the computer or robot controller to control a machine or process. (2) (v) To furnish (a computer) with a code of instructions. (3) (v) To teach a robot system a specific set of movements and instructions to accomplish a task.

Programmable: Capable of being instructed to operate in a specified manner or of accepting setpoints or other commands from a remote source.

Programmable controller (PC): A solid-state control system that has a user-programmable memory for storage of instructions to implement specific functions such as I/O control logic, timing, counting, arithmetic, and data manipulation. A PC consists of a central processor, input/output interface, memory, and programming device that typically uses relay-equivalent symbols. The PC is purposely designed as an industrial control system that can perform functions equivalent to a relay panel or a wired solid-state logic control system.

Protocol: A defined means of establishing criteria for receiving and transmitting data through communication channels.

Proximity sensor: A device that senses that an object is only a short distance (e.g., a few inches or feet) away and/or measures how far away it is.

Real-time: Pertaining to computation performed in synchronization with the related physical process.

Rectangular coordinate system: Same as Cartesian coordinate system but applied to points in a plane (only two axes used).

Reliability: The probability that a device will function without failure over a specified time period or amount of usage.

Repair: To restore to operating condition after damage or malfunction.

Repeatability: Closeness of agreement of repeated position movements, under the same conditions, to the same location.

Resolver: A transducer that converts rotary or linear mechanical position into an analog electrical signal by means of the interaction of electromagnetic fields between the movable and stationary parts of the transducer.

Robot: A mechanical device that can be programmed to perform some task of manipulation or locomotion under automatic control.

Robot system: A robot's hardware and software—the manipulator, power supply, and controller; the end-effector(s); any equipment, devices, and sensors the robot directly interfaces with; any equipment, devices, and sensors required for the robot to perform its task; and any communications interface that is operating and monitoring the robot, equipment, and sensors. (This definition excludes the rest of the operating system hardware and software.)

Robotic: Pertaining to robots.

Robotics: The science of designing, building, and applying robots.

Roll: The angular displacement of a moving body around its principal axis of motion.

Safeguard: A guard, device, or procedure designed to protect persons from danger.

Search function: A computer or robot controller programming equipment feature that allows the user to quickly display and/or edit any part of the software program.

Servo-controlled robot: A robot driven by servomechanisms, motors whose driving signal is a function of the difference between commanded position and/or rate and measured actual position and/or rate. Such a robot is capable of stopping at or moving through a practically unlimited number of points in executing a programmed trajectory.

Servomechanism: An automatic control mechanism consisting of a motor driven by a signal that is a function of the difference between commanded position and/or rate and measured actual position and/or rate.

Shoulder: The joint, or pair of joints, that connect the arm to the base.

Single-point control of motion: A safeguarding method for certain maintenance operations used to restrict entry into the work envelope of the robot.

Smart sensor: A sensing device whose output signal is contingent upon mathematical or logical operations based upon internal data or additional sensing devices.

Software: The program that controls the operation of a computer or robot controller.

Spherical coordinate robot: A robot whose manipulator arm degrees of freedom are defined primarily by spherical coordinates.

Spherical coordinate system: A coordinate system, two of whose dimensions are angles, the third being a linear distance from the point of origin. These three coordinates specify a point on a sphere.

Station: (1) (See Workstation.) (2) Any PC, computer, or data terminal corrected to, and communicating by means of, a data network.

Stepping motor: A bidirectional permanent magnet motor that turns through one angular increment for each pulse applied to it.

Subroutine: A portion of a computer program that performs a secondary or repeated function such as printing or sorting. A subroutine is executed repeatedly as required by the main program.

Supervisory control: A control scheme whereby a person or computer monitors and intermittently reprograms, sets subgoals, or adjusts control parameters of a lower level automatic controller, while the lower level controller performs the control task continuously in real time.

System: A collection of units combined to work as a larger integrated unit having the capabilities of all the separate units.

Tactile: Perceived by touch, or having the sense of touch.

Tactile sensor: A transducer that is sensitive to touch.

Teach: To move a robot through the series of points that define the robot's intended task. (See also *Program.*)

Teach control: See *Teach pendant.*

Teach pendant: A hand-held control unit, usually connected by a cable to the control system, with which a robot can be programmed or moved.

Timer: In relay-panel hardware, an electromechanical device that can be wired and preset to control the operating interval of other devices.

Tolerance: A specified allowance for error from a desired or measured quantity.

Upper arm: The portion of a jointed arm that is connected to the shoulder.

Velocity: Rate of motion; distance per unit of time.

Vision: A sensory capability involving the image of an object or scene.

Visual-optical system: A device, such as a camera, that is designed, constructed, and installed to detect intrusion by a person(s) into the robot's restricted work envelope and that could also serve to restrict that work envelope.

Warning device: An audible or visual device such as a bell, horn, or flasher used to alert persons to expect robot movement.

Work cell: A manufacturing unit consisting of one or more workstations. (See also *Workstation.*)

Work envelope: The set of points representing the maximum extent or reach of the robot hand or working tool in all directions.

Workstation: A manufacturing unit consisting of one robot and the machine tools, conveyors, and other equipment with which it interacts.

Working space or volume: The physical space bounded by the work envelope in physical space.

Wrist: A set of rotary joints between the arm and end-effector that allow the end-effector to be oriented to the workpiece.

Yaw: The angular displacement of a moving body about an axis perpendicular to the line of motion and to the top side of the body.

Appendix 2

Definitions of Welding Terms*

Adaptive control: Pertaining to process control that automatically determines changes in process conditions and directs the equipment to take appropriate action.

Adaptive control welding: Welding with a process control system that automatically determines changes in welding conditions and directs the equipment to take appropriate action.

Arc blow: The deflection of an arc from its normal path because of magnetic forces.

Arc length: The distance from the tip of the welding electrode to the adjacent surface of the weld pool.

Arc spot weld: A spot weld made by an arc welding process.

Arc stud welding (SW): An arc welding process that uses an arc between a metal stud, or similar part, and the other workpiece. The process is used without

---

*Taken from American Welding Society's publication ANSI/AWS A3.0-9X, *Standard Welding Terms and Definitions,* Draft 4, dated November 1993.

filler metal, with or without shielding gas or flux, with or without partial shielding from a ceramic or graphite ferrule surrounding the stud, and with the application of pressure after the faying surfaces are sufficiently heated.

Arc time: The time during which an arc is maintained in making an arc weld.

Arc voltage: The voltage across the welding arc.

Arc welding (AW): A group of welding processes that produces coalescence of workpieces by heating them with an arc. The processes are used with or without the application of pressure and with or without filler metal.

Arc welding deposition efficiency: The ratio of the weight of filler metal deposited in the weld metal to the weight of filler metal melted, expressed in percent.

Arc welding electrode: A component of the welding circuit through which current is conducted and that terminates at the arc.

Arc welding gun: A device used to transfer current to a continuously fed consumable electrode, guide the electrode, and direct the shielding gas.

Arc welding torch: A device used to transfer current to a fixed welding electrode, position the electrode, and direct the flow of shielding gas.

As-welded: Pertaining to the condition of weld metal, welded joints, and weldments after welding but prior to any subsequent thermal, mechanical, or chemical treatments.

Automatic: Pertaining to the control of a process with equipment that requires only occasional or no observation of the welding and no manual adjustment of the equipment controls.

Automatic welding: Welding with equipment that requires only occasional or no observation of the welding and no manual adjustment of the equipment controls. Also called automatic brazing, automatic soldering, automatic thermal cutting, and automatic thermal spraying.

Backing: A material or device placed against the back side of the joint or at both sides of a weld in electroslag and electrogas welding to support and retain molten weld metal. The material may be partially fused or remain unfused during welding and may be either metal or nonmetal.

Base metal: The metal or alloy that is welded, brazed, soldered, or cut.

Braze: A weld produced by heating an assembly to the brazing temperature using a filler metal having a liquidus above 450°C (840°F) and below the solidus of the base metal. The filler metal is distributed between the closely fitted faying surfaces of the joint by capillary action.

Butt joint: A joint between two members aligned in approximately the same plane.

Carbon arc welding (CAW): An arc welding process that uses an arc between a carbon electrode and the weld pool. The process is used with or without shielding and without the application of pressure.

Constant-current power source: An arc welding power source with a volt–

ampere relationship yielding a small welding current change from a large arc voltage change.

Constant-voltage power source: An arc welding power source with a volt–ampere relationship yielding a large welding current change from a small arc voltage change.

Consumable electrode: An electrode that provides filler metal.

Covered electrode: A composite filler metal electrode consisting of a core of a bare electrode or metal-cored electrode to which a covering sufficient to provide a slag layer on the weld metal has been applied. The covering may contain materials providing such functions as shielding from the atmosphere, deoxidation, and arc stabilization and can serve as a source of metallic additions to the weld.

Crater: A depression in the weld face at the termination of a weld bead.

Defect: A discontinuity or discontinuities that by nature or accumulated effect (e.g., total crack length) render a part or product unable to meet minimum applicable acceptance standards or specifications. The term designates rejectability.

Deposited metal: Filler metal that has been added during welding, brazing, or soldering.

Deposition rate: The weight of material deposited in a unit of time.

Depth of fusion: The distance that fusion extends into the base metal or previous bead from the surface melted during welding.

Direct current electrode negative (DCEN): The arrangement of direct current arc welding leads in which the electrode is the negative pole and the workpiece is the positive pole of the welding arc.

Direct current electrode positive (DCEP): The arrangement of direct current arc welding leads in which the electrode is the positive pole and the workpiece is the negative pole of the welding arc.

Duty cycle: The percentage of time during an arbitrary test period that a power source or its accessories can be operated at rated output without overheating.

Edge preparation: The preparation of the edges of the joint members by cutting, cleaning, plating, or other means.

Electrode extension: In flux-cored arc welding, electrogas welding, gas metal arc welding, and submerged arc welding, the length of electrode extending beyond the end of the contact tube; in gas tungsten arc welding and plasma arc welding, the length of tungsten electrode extending beyond the end of the collet.

Electrode lead: The electrical conductor between the source of arc welding current and the electrode holder.

Electron beam welding (EBW): A welding process that produces coalescence with a concentrated beam, composed primarily of high-velocity electrons, impinging on the joint. The process is used without shielding gas and without the application of pressure.

Exhaust booth: A mechanically ventilated, semienclosed area in which an air flow across the work area is used to remove fumes, gases, and solid particles.

Field weld: A weld made at a location other than a shop or the place of initial construction.

Filler metal: The metal or alloy to be added in making a welded, brazed, or soldered joint.

Fillet weld: A weld of approximately triangular cross section joining two surfaces approximately at right angles to each other in a lap joint, T-joint, or corner joint.

Fixture: A device designed to hold and maintain parts in proper relation to each other.

Flat welding position: The welding position used to weld from the upper side of the joint at a point where the weld axis is approximately horizontal and the weld face lies in an approximately horizontal plane.

Flux: A material used to hinder or prevent the formation of oxides and other undesirable substances in molten metal and on solid metal surfaces and to dissolve or otherwise facilitate the removal of such substances.

Flux-cored arc welding (FCAW): An arc welding process that uses an arc between a continuous filler metal electrode and the weld pool. The process is used with shielding gas from a flux contained within the tubular electrode, with or without additional shielding from an eternally supplied gas and without the application of pressure.

Flux-cored electrode: A composite tubular filler metal electrode consisting of a metal sheath and a core of various powdered materials, producing an extensive slag cover on the face of a weld bead. External shielding may be required.

Fuel gas: A gas such as acetylene, natural gas, hydrogen, propane, stabilized methylacetylene propadiene, or other fuel normally used with oxygen in one of the oxyfuel processes and for heating.

Fusion welding: Any welding process that uses fusion of the base metal to make the weld.

Gas metal arc welding (GMAW): An arc welding process that uses an arc between a continuous filler metal electrode and the weld pool. The process is used with shielding from an externally supplied gas and without the application of pressure.

Gas nozzle: A device at the exit end of the torch or gun that directs shielding gas.

Gas-shielded flux-cored arc welding (FCAW-G): A flux-cored arc welding process variation in which shielding gas is supplied through the gas nozzle in addition to that obtained from the flux within the electrode.

Gas tungsten arc welding (GTAW): An arc welding process that uses an arc between a tungsten electrode (nonconsumable) and the weld pool. The process is used with shielding gas and without the application of pressure.

Gas tungsten arc welding torch: A device used to transfer current to a fixed welding electrode, position the electrode, and direct the flow of shielding gas.

Globular transfer: In arc welding, the transfer of molten metal in large drops from a consumable electrode across the arc.

Groove weld: A weld made in a groove between the workpieces.

Ground connection: An electrical connection of the welding machine frame to the earth for safety.

Hand shield: A protective device used in arc welding, arc cutting, and thermal spraying for shielding the welder's eyes, face, and neck. It is equipped with a filter plate and is designed to be held by hand.

Heat-affected zone: The portion of the base metal whose mechanical properties or microstructure have been altered by the heat of welding, brazing, soldering, or thermal cutting.

Horizontal welding position: For a fillet weld, the welding position in which the weld is on the upper side of an approximately horizontal surface and against an approximately vertical surface.

Horizontal welding position: For a groove weld, the welding position in which the weld face lies in an approximately vertical plane and the weld axis at the point of welding is approximately horizontal.

Inert gas: A gas that normally does not combine chemically with another material.

Joint: The junction of members or the edges of members that are to be joined or have been joined.

Joint design: The shape, dimensions, and configuration of the welding joint.

Joint recognition: A function of an adaptive control process that determines changes in the joint geometry during welding and directs the welding equipment to take appropriate action.

Joint root: That portion of a joint to be welded where the members approach closest to each other. In cross section, the joint root may be either a point, a line, or an area.

Joint tracking: A function of an adaptive control process that determines changes in joint location during welding and directs the welding machine to take appropriate action.

Joint type: A weld joint classification based on five basic joint configurations: butt joint, corner joint, edge joint, lap joint, and T joint.

Keyhole welding: A technique in which a concentrated heat source penetrates partially or completely through a workpiece, forming a hole (keyhole) at the leading edge of the weld pool. As the heat source progresses, the molten metal fills in behind the hole to form the weld bead.

Laser beam welding (LBW): A welding process that produces coalescence with the heat from a laser beam impinging on the joint. The process is used without a shielding gas and without the application of pressure.

Level-wound: Pertaining to spooled or coiled filler metal that has been wound in distinct layers such that adjacent turns touch.

Manual: Pertaining to the control of a process with the torch, gun, or electrode holder held and manipulated by hand.

Manual welding: Welding with the torch, gun, or electrode holder held and manipulated by hand. Accessory equipment such as part motion devices and manually controlled filler material feeders may be used.

Mechanized: Pertaining to the control of a process with equipment that requires manual adjustment of the equipment controls in response to visual observation of the operation, with the torch, gun, wire guide assembly, or electrode holder held by a mechanical device.

Mechanized welding: Welding with equipment that requires manual adjustment of the equipment controls in response to visual observation of the welding, with the torch, gun, or electrode holder held by a mechanical device.

Metal-cored electrode: A composite tubular filler metal electrode consisting of a metal sheath and a core of various powdered materials, producing no more than slag islands on the face of a weld bead. External shielding may be required.

MIG welding: A nonstandard term for gas metal arc welding and flux-cored arc welding.

Nondestructive examination (NDE): The act of determining the suitability of some material or component for its intended purpose using techniques that do not affect its serviceability.

Nontransferred arc: An arc established between the electrode and the constricting nozzle of the plasma arc torch or thermal spraying gun. The workpiece is not in the electric circuit.

Overhead welding position: The welding position in which welding is performed from the underside of the joint.

Oxygen cutting (OC): A group of thermal cutting processes that severs or removes metal by means of the chemical reaction between oxygen and the base metal at elevated temperature. The necessary temperature is maintained by the heat from an arc, an oxyfuel gas flame, or other source.

Pilot arc: A low-current arc between the electrode and the constricting nozzle of the plasma arc torch to ionize the gas and facilitate the start of the welding arc.

Plasma arc welding (PAW): An arc welding process that uses a constricted arc between a nonconsumable electrode and the weld pool (transferred arc) or between the electrode and the constricting nozzle (nontransferred arc). Shielding is obtained from the ionized gas issuing from the torch, which may be supplemented by an auxiliary source of shielding gas. The process is used without the application of pressure.

Plug weld: A weld made in a circular hole in one member of a joint fusing that member to another member. A fillet-welded hole is not to be construed as conforming to this definition.

Polarity: See *Direct current electrode negative* and *Direct current electrode positive.*

Power source: An apparatus for supplying current and voltage suitable for welding, thermal cutting, or thermal spraying.

Procedure: The detailed elements of a process or method used to produce a specific result.

Procedure qualification: The demonstration that welds made by a specific procedure can meet prescribed standards.

Process: A grouping of basic operational elements used in welding, thermal cutting, or thermal spraying.

Protective atmosphere: A gas or vacuum envelope surrounding the workpieces, used to prevent or reduce the formation of oxides and other detrimental surface substances and to facilitate their removal.

Pulsed power welding: An arc welding process variation in which the power is cyclically programmed to pulse so that effective but short-duration values of power can be utilized. Such short-duration values are significantly different from the average value of power. Equivalent terms are *pulsed voltage welding* and *pulsed current welding.*

Pulsed spray welding: An arc welding process variation in which the current is pulsed to utilize the advantages of the spray mode of metal transfer at average currents equal to or less than the globular to spray transition current.

Random-wound: Pertaining to spooled or coiled filler metal that has not been wound in distinct layers.

Reducing atmosphere: A chemically active protective atmosphere that will reduce metal oxides to their metallic state at elevated temperature.

Resistance welding (RW): A group of welding processes that produce coalescence of the faying surfaces with the heat obtained from resistance of the workpiece to the flow of the welding current in a circuit of which the workpiece is a part, and by the application of pressure.

Reverse polarity: A nonstandard term for *direct current electrode positive.*

Robotic: Pertaining to process control by robotic equipment.

Robotic welding: Welding that is performed and controlled by robotic equipment. Variations are robotic brazing, robotic soldering, robotic thermal cutting, and robotic thermal spraying.

Root opening: A separation at the joint root between the workpieces.

Seam weld: A continuous weld made between or upon overlapping members, in which coalescence may start and occur on the faying surfaces or may proceed from the outer surface of one member. The continuous weld may consist of a single weld bead or a series of overlapping spot welds.

Self-shielded flux-cored arc welding (FCAW-S): A flux-cored arc welding process variation in which shielding gas is obtained exclusively from the flux within the electrode.

Semiautomatic: Pertaining to the manual control of a process with equipment that automatically controls one or more of the process conditions.

Semiautomatic welding: Manual welding with equipment that automatically controls one or more of the welding conditions.

Shielded metal arc welding (SMAW): An arc welding process with an arc between a covered electrode and the weld pool. The process is used with shielding from the decomposition of the electrode covering without the application of pressure and with filler metal from the electrode.

Shielding gas: Protective gas used to prevent or reduce atmospheric contamination.

Short-circuit gas metal arc welding (GMAW-S): A gas metal arc welding process variation in which the consumable electrode is deposited during repeated short circuits.

Short-circuiting transfer: In arc welding, metal transfer in which molten metal from a consumable electrode is deposited during repeated short circuits.

Solid-state welding (SSW): A group of welding processes that produce coalescence by the application of pressure at a welding temperature below the melting temperatures of the base metal and the filler metal.

Spatter: The metal particles expelled during fusion welding that do not form a part of the weld.

Spool: A filler metal package consisting of a continuous length of welding wire in coil form wound on a cylinder (called a barrel) that is flanged at both ends. The flange contains a spindle hole of smaller diameter than the inside diameter of the barrel.

Spot weld: A weld made between or upon overlapping members in which coalescence may start and occur on the faying surfaces or may proceed from the outer surface of one member. The weld cross section (plan view) is approximately circular.

Standoff distance: The distance between the welding nozzle and the workpiece.

Starting weld tab: Additional material that extends beyond the beginning of the joint, on which the weld is started.

Stick electrode welding: A nonstandard term for *shielded metal arc welding.*

Straight polarity: A nonstandard term for *direct current electrode negative.*

Stub: The short length of filler metal electrode, welding rod, or brazing rod that remains after its use for welding or brazing.

Stud welding: A general term for joining a metal stud or similar part to a workpiece. Welding may be accomplished by arc, resistance, friction, or other process with or without external gas shielding.

Submerged arc welding (SAW): An arc welding process that uses an arc or arcs between a bare metal electrode or electrodes and the weld pool. The arc and molten metal are shielded by a blanket of granular flux on the workpieces. The process is used without pressure and with filler metal from the electrode sometimes from a supplemental source (welding rod, flux, or metal granules).

Surface preparation: The operations necessary to produce a desired or specified surface condition.

Tack weld: A weld made to hold the parts of a weldment in proper alignment until the final welds are made.

TIG welding: A nonstandard term for *gas tungsten arc welding.*

T joint: A joint between two members located approximately at right angles to each other in the form of a T.

Transferred arc: A plasma arc established between the electrode of the plasma arc torch and the workpiece.

Travel angle: The angle less than 90° between the electrode axis and a line perpendicular to the weld axis, in a plane determined by the electrode axis and the weld axis. This angle can also be used to partially define the position of guns, torches, rods, and beams.

Undercut: A groove melted into the base metal adjacent to the weld toe or weld root and left unfilled by weld metal.

Vertical welding position: The welding position in which the weld axis, at the point of welding, is approximately vertical and the weld face lies in an approximately vertical plane.

Wave soldering (WS): An automatic soldering process where workpieces are passed through a wave of molten solder.

Weld: A localized coalescence of metals or nonmetals produced either by heating the materials to the welding temperature, with or without the application of pressure, or by the application of pressure alone, with or without the use of filler material.

Weldability: The capacity of a material to be welded under the imposed fabrication conditions into a specific, suitably designed structure and to perform satisfactorily in the intended service.

Weld axis: A line through the length of the weld, perpendicular to and at the geometric center of its cross section.

Welder: One who performs manual or semiautomatic welding.

Welder certification: Written verification that a welder has produced welds meeting a prescribed standard of welder performance.

Welder performance qualification: The demonstration of a welder's ability to produce welds meeting prescribed standards.

Weld feed speed: The rate at which wire is consumed in arc cutting, thermal spraying, or welding.

Weld gauge: A device designed for measuring the shape and size of welds.

Weld groove: A channel in the surface of a workpiece or an opening between two joint members that provides space to contain a weld.

Welding: A joining process that produces coalescence of materials by heating them to the welding temperature, with or without the application of pressure or by the application of pressure alone, and with or without the use of filler metal.

Welding arc: A controlled electrical discharge between the electrode and the workpiece that is formed and sustained by the establishment of a gaseous conductive medium called an arc plasma.

Welding electrode: A component of the welding circuit through which current is conducted and that terminates at the arc, molten conductive slag, or base metal.

Welding filler metal: The metal or alloy to be added in making a weld joint that alloys with the base metal to form weld metal in a fusion welded joint.

Welding generator: A generator used for supplying current for welding.

Welding head: The part of a welding machine in which a welding gun or torch is incorporated.

Welding helmet: A device equipped with a filter plate and designed to be worn on the head to protect eyes, face, and neck from arc radiation, radiated heat, and spatter or other harmful matter expelled during some welding and cutting processes.

Welding leads: The workpiece lead and electrode lead of an arc welding circuit.

Welding machine: Equipment used to perform the welding operation—for example, spot welding machine, arc welding machine, and seam welding machine.

Welding operator: One who operates adaptive control, automatic, mechanized, or robotic welding equipment

Welding position: The relationship between the weld pool, joint, joint members, and welding heat source during welding.

Welding power source: An apparatus for supplying current and voltage suitable for welding.

Welding procedure: The detailed methods and practices involved in the production of a weldment.

Welding procedure qualification record (WPQR): A record of welding variables used to produce an acceptable test weldment and the results of tests conducted on the weldment to qualify a welding procedure specification.

Welding procedure specification (WPS): A document providing in detail the required variables for specific application to ensure repeatability by properly trained welders and welding operators.

Welding rectifier: A device in a welding power source for converting alternating current to direct current.

Welding schedule: A written statement, usually in tabular form, specifying values of parameters and the welding sequence for performing a specific welding operation.

Welding sequence: The order of making welds in a weldment.

Welding symbol: A graphical representation of a weld.

Welding transformer: A transformer used for supplying current for welding.

Welding wire: A form of welding filler metal, normally packaged as coils or spools, that may or may not conduct electric current depending upon the welding process with which it is used.

Weldment: An assembly whose component parts are joined by welding.

Weld metal: The portion of a fusion weld that has been completely melted during welding.

Weld pass sequence: The order in which the weld passes are made.

Weld pool: The localized volume of molten metal in a weld prior to its solidification as weld metal.

Weld recognition: A function of an adaptive control that determines changes in the shape of the weld pool or the weld metal during welding and directs the welding machine to take appropriate action.

Weld reinforcement: Weld metal in excess of the quantity required to fill a joint.

Weld root: The points, shown in a cross section, at which the root surface intersects the base metal surfaces.

Weld symbol: A graphical character connected to the welding symbol indicating the type of weld.

Weld tab: Additional material that extends beyond either end of the joint, on which the weld is started or terminated.

Wire straightener: A device used for controlling the cast and helix of coiled wire to enable it to be easily fed through the wire feed system.

Work angle: The angle less than 90° between a line perpendicular to the major workpiece surface and a plane determined by the electrode axis and the weld axis. In a T joint or a corner joint, the line is perpendicular to the nonbutting member. This angle can also be used to partially define the position of guns, torches, rods, and beams.

Workpiece: The part that is welded, brazed, soldered, thermal cut, or thermal sprayed.

Workpiece connection: The connection of the workpiece lead to the workpiece.

Workpiece lead: The electrical conductor between the arc welding current source and workpiece connection.

Appendix 3

Metric Units, Conversion Tables, and Geometric Formulas

The International System of Units (SI) is being adopted throughout the world. The purpose of this system is to standardize the units used by all countries. The units and conversions in Tables A.1–A.11 are those that are the most widely used in the welding industry. The conversions used in this section are SI units or AWS preferred units. For more complete information on the SI system and other welding conversions, consult the *Metric Practice Guide for the Welding Industry.*

Table A.1 Metric Prefixes

Exponential expression	Multiplication factor	Prefix	Symbol
10^{12}	1 000 000 000 000	tera	T
10^{9}	1 000 000 000	giga	G
10^{6}	1 000 000	mega	M
10^{3}	1 000	kilo	k
10^{2}	100	hecto[a]	h
10	10	deka[a]	da
10^{-1}	0.1	deci[a]	d
10^{-2}	0.01	centi[a]	c
10^{-3}	0.001	milli	m
10^{-6}	0.000 001	micro	μ
10^{-9}	0.000 000 001	nano	n
10^{-12}	0.000 000 000 001	pico	p
10^{-15}	0.000 000 000 000 001	femto	f
10^{-18}	0.000 000 000 000 000 001	atto	a

[a]Rarely used. In scientific notation, the number before the exponential expression is between 1 and 10 (e.g., 1 kg = 1×10^3 g; 5 mL = 5×10^{-3} L; 0.001 mm = 1 μm).

Table A.2 Units Pertaining to Welding

Property	Unit	Symbol and formula
Area	Square millimeter	mm^2
Density	Kilograms per cubic meter	kg/m^3
Deposition rate	Kilograms per hour	kg/hr
Flow rate	Liters per minute	L/min
Frequency	Hertz	1 Hz = 1/s
Heat (energy)	Joule	1 J = N·m
Force	Newton	1 N = $kg{\cdot}m/s^2$
Impact strength (energy)	Joule	1 J = N·m
Length	Meter	m
Power	Watt	1 W = J/s
Pressure (gas, liquid)	Kilopascal	1 kPa = 1000 N/m^2
Temperature	Degree Celsius	°C
Stress	Megapascal	1 MPa = 1,000,000 N/m^2
Time	Hour	hr
	Minute	min
	Second	s
Travel speed	Millimeters per second	mm/s
Volume (liquid or gas)	Liter	1 L = 1000 cm^3
Volume (solid)	Cubic millimeters	mm^3
Wire feed speed	Millimeters per second	mm/s

Table A.3 Conversion Table for Welding Units

Quantity and preferred units	To convert from	Multiplication factor (or formula)	Conversion product
Area (mm^2)	$in.^2$	645.1	mm^2
	mm^2	0.001550	$in.^2$
	ft^2	0.09289	m^2
	m^2	10.76	ft^2
Deposition rate (kg/hr)	lb/hr	0.45	kg/hr
	kg/hr	2.2	lb/hr
Flow rate (L/min)	ft^3/hr	0.472	L/min
	L/min	2.12	ft^3/hr
	gal/min	0.0631	L/hr
	L/hr	15.85	gal/min
	gal/min	3.785	L/min
	L/min	0.2642	gal/min
Heat input (J/m)	J/in.	39.37	J/m
	J/m	0.0254	J/in.
Impact energy (J)	ft-lb	1.356	J
	J	0.7376	ft-lb
Length (mm)	in.	25.40	mm
	mm	0.03937	in.
	ft	0.3048	m
	m	3.281	ft
	ft	304.8	mm
	mm	0.003281	ft
Mass (weight) (kg)	lb	0.45	kg
	kg	2.2	lb
	ton (short)	0.907	tonne (metric)
	tonne (metric)	1.10	ton (short)
Pressure, gas and liquid (kPa)	psi	0.006895	kPa
	kPa	0.1450	psi
Stress, pressure (MPa)	psi	0.006895	MPa
	ksi	6.895	MPa
	MPa	0.1450	ksi
	MPa	145.0	psi
Temperature	°F	$°C = 5/9(F - 32)$	°C
	°C	$°F = 9/5(C + 32)$	°F
	°C	$K = °C + 273$	K
Travel speed (mm/s)	in./min	0.423	mm/sec
	mm/sec	2.36	in./min

Table A.3 Continued

Quantity and preferred units	To convert from	Multiplication factor (or formula)	Conversion product
Volume, solid (mm^3)	in^3	16,400	mm^3
	mm^3	0.0000610	in.^{3}
	ft^3	0.02832	m^3
	m^3	35.31	ft^3
Volume, liquid and gas (L)	gal	3.785	L
	L	0.2642	gal
	in.^{3}	0.0164	L
	L	61.0	in.^{3}
	ft^3	28.3	L
	L	0.0353	ft^3

Table A.4 Length, Speed, and Distance—Approximate Conversions

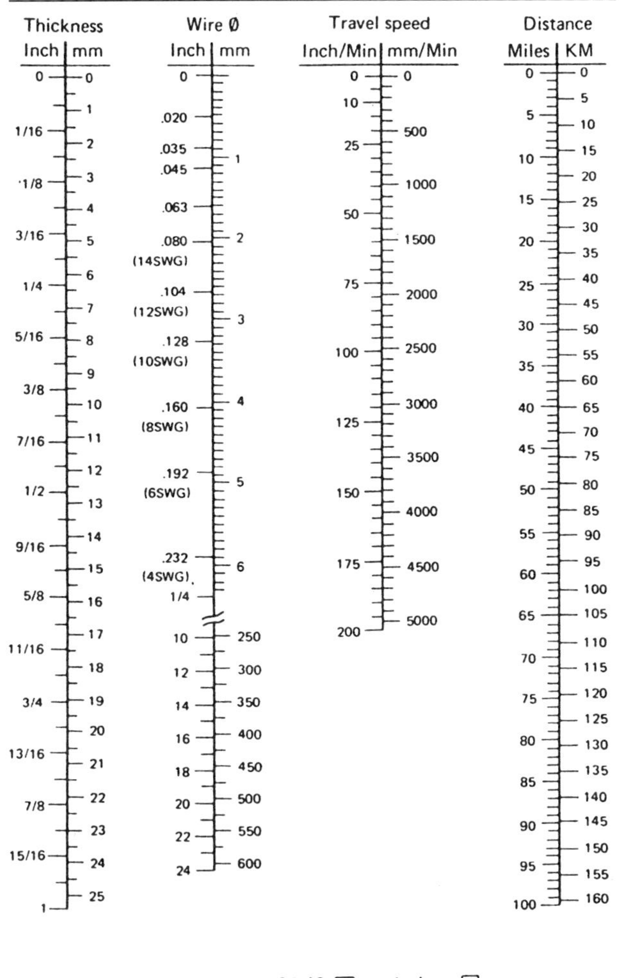

For use with electronic calculators—First enter the conversion constant number. Press the x button. Enter the known quantity of the dimension. Press the =. The desired value of the dimension will appear on the display.

25.40	×	___ inch	=	___ mm
304.8	×	___ feet	=	___ mm
.0393	×	___ mm	=	___ inch
.00328	×	___ mm	=	___ feet
.0621	×	___ km	=	___ miles
1.609	×	___ miles	=	___ Kilometers
.4233	×	___ in/min	=	___ mm/second
2.362	×	___ mm/s	=	___ in/min.

Table A.5 Flow Rate and Liquid Measure—Approximate Conversions

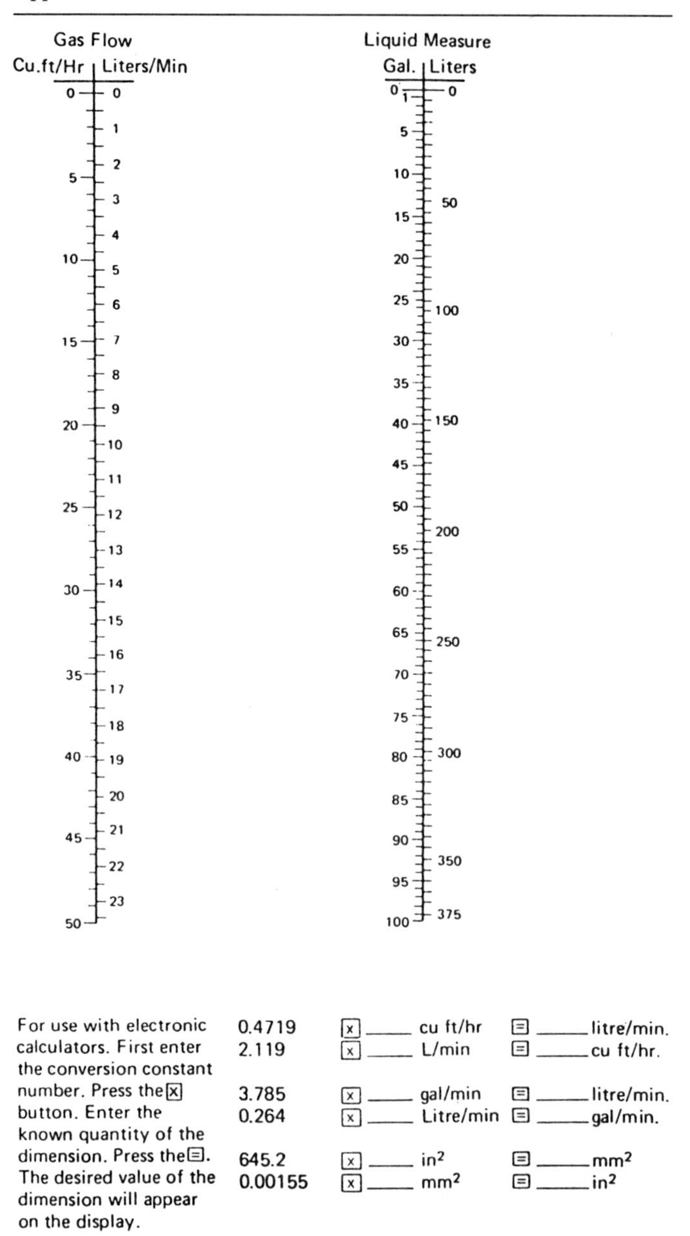

For use with electronic calculators. First enter the conversion constant number. Press the [x] button. Enter the known quantity of the dimension. Press the [=]. The desired value of the dimension will appear on the display.

Constant				
0.4719	[x] ____	cu ft/hr	[=] ____	litre/min.
2.119	[x] ____	L/min	[=] ____	cu ft/hr.
3.785	[x] ____	gal/min	[=] ____	litre/min.
0.264	[x] ____	Litre/min	[=] ____	gal/min.
645.2	[x] ____	in^2	[=] ____	mm^2
0.00155	[x] ____	mm^2	[=] ____	in^2

Table A.6 Weight, Pressure, and Load—Approximate Conversions

Weight

Lbs.	KG
1, 5, 10, 15, 20, 25, 30, 35, 40, 45, 50, 55, 60, 65, 70, 75, 80, 85, 90, 95, 100	0, 5, 10, 15, 20, 25, 30, 35, 40

Pressure

Lbs.Inch²	KG/CM²
0, 1, 5, 10, 15, 20, 25, 30, 35, 40, 45, 50, 55, 60, 65, 70, 75, 80, 85, 90, 95, 100	0, 1, 2, 3, 4, 5, 6, 7

Load

Lbs.Inch²	KG/MM²
0, 1000, 5000, 10000, 15000, 20000, 25000, 30000, 35000, 40000, 45000, 50000, 55000, 60000, 65000, 70000, 75000, 80000, 85000, 90000, 95000, 100000	0, 1, 10, 20, 30, 40, 50, 60, 70

Load

ksi	MPa
10, 20, 30, 40, 50, 60, 70, 80, 90, 100, 110, 120, 130, 140, 150, 160, 170, 180, 190, 200	100, 200, 300, 400, 500, 600, 700, 800, 900, 1000, 1100, 1200, 1300, 1400

Table A.7 Temperature and Impact Values—Approximate Conversions

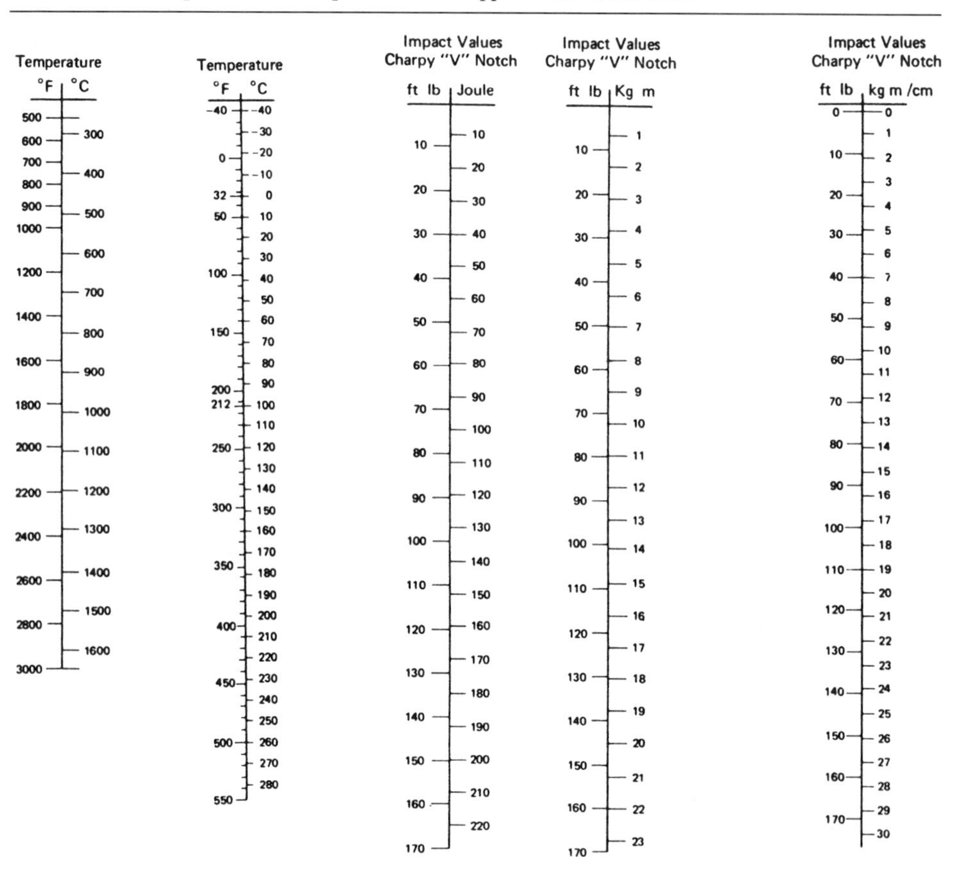

For use with electronic calculators—First enter the conversion constant number. Press the ☒ button. Enter the known quantity of the dimension. Press the ⊟. The desired value of the dimension will appear on the display.

0.1383 ☒ ___ ft-lbs ⊟ ___ kg-m
7.233 ☒ ___ kg-m ⊟ ___ ft-lbs

1.356 ☒ ___ ft-lbs ⊟ ___ Joule
.7376 ☒ ___ Joule ⊟ ___ ft-lbs

.8 ☒ ___ kg-m ⊟ ___ kg-m/cm
0.1728 ☒ ___ ft.lbs ⊟ ___ kg-m/cm²
5.787 ☒ ___ kg-m/cm² ⊟ ___ ft-lbs

Table A.8 Impact Values

Foot-pounds	Kilogram-meters	Joules	Foot-pounds	Kilogram-meters	Joules	Foot-pounds	Kilogram-meters	Joules	Foot-pounds	Kilogram-meters	Joules
1	0.14	1.36	26	3.60	35.26	51	7.05	69.16	76	10.51	103.06
2	0.28	2.71	27	3.73	36.61	52	7.19	70.51	77	10.65	104.41
3	0.42	4.07	28	3.87	37.97	53	7.33	71.87	78	10.78	105.77
4	0.55	5.42	29	4.01	39.32	54	7.47	73.22	79	10.92	107.12
5	0.69	6.78	30	4.15	40.68	55	7.60	74.58	80	11.06	108.48
6	0.83	8.14	31	4.29	42.04	56	7.74	75.94	81	11.20	109.84
7	0.97	9.49	32	4.42	43.39	57	7.88	77.29	82	11.34	111.19
8	1.11	10.85	33	4.56	44.75	58	8.02	78.65	83	11.48	112.55
9	1.24	12.20	34	4.70	46.10	59	8.16	80.00	84	11.61	113.90
10	1.38	13.56	35	4.84	47.46	60	8.30	81.36	85	11.75	115.26
11	1.52	14.92	36	4.98	48.82	61	8.43	82.72	86	11.89	116.62
12	1.66	16.27	37	5.12	50.17	62	8.57	84.07	87	12.03	117.97
13	1.80	17.63	38	5.25	51.53	63	8.71	85.43	88	12.17	119.33
14	1.94	18.98	39	5.39	52.88	64	8.85	86.78	89	12.31	120.68
15	2.07	20.34	40	5.53	54.24	65	8.99	88.14	90	12.44	122.04
16	2.21	21.70	41	5.67	55.60	66	9.13	89.50	91	12.58	123.40
17	2.35	23.05	42	5.81	56.95	67	9.26	90.85	92	12.72	124.75
18	2.49	24.41	43	5.95	58.31	68	9.40	92.21	93	12.86	126.11
19	2.63	25.76	44	6.08	59.66	69	9.54	93.56	94	13.00	127.46
20	2.77	27.12	45	6.22	61.02	70	9.68	94.92	95	13.13	128.82
21	2.90	28.48	46	6.36	62.38	71	9.82	96.28	96	13.27	130.18
22	3.04	29.83	47	6.50	63.73	72	9.95	97.63	97	13.41	131.53
23	3.18	31.19	48	6.64	65.09	73	10.09	98.99	98	13.55	132.89
24	3.32	32.54	49	6.78	66.44	74	10.23	100.34	99	13.69	134.24
25	3.46	33.90	50	6.91	67.80	75	10.37	101.70	100	13.83	135.60

Table A.9 Wire Gauge Diameter Conversion

U.S. steel wire gauge No.	in.	mm	U.S. steel wire gauge No.	in.	mm
7/0's	0.4900	12.447	20	0.0348	0.8839
6/0's	0.4615	11.7221	21	0.0317	0.8052
5/0's	0.4305	10.9347	22	0.0286	0.7264
4/0's	0.3938	10.0025	23	0.0258	0.6553
3/0's	0.3625	9.2075	24	0.0230	0.5842
2/0's	0.3310	8.4074	25	0.0204	0.5182
0	0.3065	7.7851	26	0.0181	0.4597
1	0.2830	7.1882	27	0.0173	0.4394
2	0.2625	6.6675	28	0.0162	0.4115
3	0.2437	6.1899	29	0.0150	0.381
4	0.2253	5.7226	30	0.0140	0.3556
5	0.2070	5.2578	31	0.0132	0.3353
6	0.1920	4.8768	32	0.0128	0.3251
7	0.1770	4.4958	33	0.0118	0.2997
8	0.1620	4.1148	34	0.0104	0.2642
9	0.1483	3.7668	35	0.0095	0.2413
10	0.1350	3.429	36	0.0090	0.2286
11	0.1205	3.0607	37	0.0085	0.2159
12	0.1055	2.6797	38	0.0080	0.2032
13	0.0915	2.3241	39	0.0075	0.1905
14	0.0800	2.032	40	0.0070	0.1778
15	0.0720	1.8389	41	0.0066	0.1678
16	0.0625	1.5875	42	0.0062	0.1575
17	0.0540	1.3716	43	0.0060	0.1524
18	0.0475	1.2065	44	0.0058	0.1473
19	0.0410	1.0414			

Table A.10 Inches per Pound of Electrode Wire

Decimal inches	Fraction inches	Aluminum	Bronze, 10%	Bronze, silicon	Copper (deox.)	Copper nickel	Magnesium	Nickel	Steel, mild	Steel, stainless	Flux-cored steel electrodes[a]
0.020		32400	11600	10300	9800	9950	50500	9900	11100	10950	—
0.025		22300	7960	7100	6750	6820	34700	6820	7680	7550	—
0.030		14420	5150	4600	4360	4430	22400	4400	4960	4880	—
0.035		10600	3780	3380	3200	3260	16500	3240	3650	3590	—
0.040		8120	2900	2580	2450	2490	12600	2480	2790	2750	—
0.045	3/64	6410	2290	2040	1940	1970	9990	1960	2210	2170	2375
0.062	1/16	3382	1120	1070	1020	1040	5270	1030	1160	1140	1230
0.078	5/64	2120	756	675	640	650	3300	647	730	718	996
0.093	3/32	1510	538	510	455	462	2350	460	519	510	640
0.125	1/8	825	295	263	249	253	1280	252	284	279	346
0.156	5/32	530	189	169	160	163	825	162	182	179	225
0.187	3/16	377	134	120	114	116	587	115	130	127	—
0.250	1/4	206	74	66	62	64	320	63	71	70	—

[a]Not absolute; ratio of flux to sheath varies for different types of flux-cored electrodes.

Table A.11 Wire Diameter Measures

Inch Fraction	Inch Decimal	Millimeter	Inch Fraction	Inch Decimal	Millimeter
1/64	0.0158	0.3969	33/64	0.5156	13.0968
1/32	0.0312	0.7937	17/32	0.5312	13.4937
3/64	0.0469	1.1906	35/64	0.5469	13.8906
1/16	0.0625	1.5875	9/16	0.5625	14.2875
5/64	0.0781	1.9844	37/64	0.5781	14.6844
3/32	0.0937	2.3812	19/32	0.5937	15.0812
7/64	0.1094	2.7781	39/64	0.6094	15.4781
1/8	0.125	3.175	5/8	0.625	15.875
9/64	0.1406	3.5719	41/64	0.6406	16.2719
5/32	0.1562	3.9687	21/32	0.6562	16.6687
11/64	0.1719	4.3656	43/64	0.6719	17.0656
3/16	0.1875	4.7625	11/16	0.6875	17.4625
13/64	0.2031	5.1594	45/64	0.7031	17.8594
7/32	0.2187	5.5562	23/32	0.7187	18.2562
15/64	0.2344	5.9531	47/64	0.7344	18.6532
1/4	0.25	6.35	3/4	0.750	19.050
17/64	0.2656	6.7469	49/64	0.7656	19.4469
9/32	0.2812	7.1437	25/32	0.7812	19.8433
19/64	0.2969	7.5406	51/64	0.7969	20.2402
5/16	0.3125	7.9375	13/16	0.8125	20.6375
21/4	0.3281	8.3344	53/64	0.8281	21.0344
11/32	0.3437	8.7312	27/32	0.8437	21.4312
23/64	0.3594	9.1281	55/64	0.8594	21.8281
3/8	0.375	9.525	7/8	0.875	22.2250
25/64	0.3906	9.9219	57/64	0.8906	22.6219
13/32	0.4062	10.3187	29/32	0.9062	23.0187
27/64	0.4219	10.7156	59/64	0.9219	23.4156
7/16	0.4375	11.1125	15/16	0.9375	23.8125
29/64	0.4531	11.5094	61/64	0.9531	24.2094
15/32	0.4687	11.9062	31/32	0.9687	24.6062
31/64	0.4844	12.3031	63/64	0.9844	25.0031
1/2	0.500	12.700			

Table A.12 Geometric Formulas

AREA

Parallelogram = base × altitude.

Triangle = half base × altitude.

Trapezoid = half the sum of the two parallel sides × the perpendicular distance between them.

Regular polygon = half of perimeter × the perpendicular distance from the center to any one side.

Circle = square of the diameter × 0.7854.

Sector of circle = number of degrees in arc × square of radius × 0.008727.

Segment of circle = area of sector with same arc minus area of triangle formed by radii of the arc and chord of the segment.

Octagon = square of diameter of inscribed circle × 0.828.

Hexagon = square of diameter of inscribed circle × 0.866.

Sphere = area of its great circle × 4; or square of diameter × 3.1416 (π).

VOLUME

Prism = area of base × altitude.

Wedge = length of edge plus twice length of base × one-sixth of the product of the height of the wedge and the breadth of its base.

Cylinder = area of base × altitude.

Cone = area of base × one-third of altitude.

Sphere = cube of diameter × 0.5236.

MISCELLANEOUS

Diameter of circle = circumference × 0.31831.

Circumference of circle = diameter × 3.1416(π).

Index